Nanomaterials for Security

NATO Science for Peace and Security Series

This Series presents the results of scientific meetings supported under the NATO Programme: Science for Peace and Security (SPS).

The NATO SPS Programme supports meetings in the following Key Priority areas: (1) Defence Against Terrorism; (2) Countering other Threats to Security and (3) NATO, Partner and Mediterranean Dialogue Country Priorities. The types of meetings supported are generally "Advanced Study Institutes" and "Advanced Research Workshops". The NATO SPS Series collects together the results of these meetings. The meetings are co-organized by scientists from NATO countries and scientists from NATO's "Partner" or "Mediterranean Dialogue" countries. The observations and recommendations made at the meetings, as well as the contents of the volumes in the Series, reflect those of participants and contributors only; they should not necessarily be regarded as reflecting NATO views or policy.

Advanced Study Institutes (ASI) are high-level tutorial courses to convey the latest developments in a subject to an advanced-level audience.

Advanced Research Workshops (ARW) are expert meetings where an intense but informal exchange of views at the frontiers of a subject aims at identifying directions for future action.

Following a transformation of the programme in 2006, the Series has been re-named and re-organised. Recent volumes on topics not related to security, which result from meetings supported under the programme earlier, may be found in the NATO Science Series.

The Series is published by IOS Press, Amsterdam, and Springer, Dordrecht, in conjunction with the NATO Emerging Security Challenges Division.

Sub-Series

A.	Chemistry and Biology	Springer
B.	Physics and Biophysics	Springer
C.	Environmental Security	Springer
D.	Information and Communication Security	IOS Press
E.	Human and Societal Dynamics	IOS Press

http://www.nato.int/science
http://www.springer.com
http://www.iospress.nl

Series A: Chemistry and Biology

Nanomaterials for Security

edited by

Janez Bonča
J. Stefan Institute, Ljubljana, Slovenia

and

Sergei Kruchinin
Bogolyubov Institute for Theoretical Physics, Kiev, Ukraine

Springer

Published in Cooperation with NATO Emerging Security Challenges Division

Proceedings of the NATO Advanced Research Workshop on
Nanomaterials for Security
Odessa, Ukraine
August 30–September 3, 2015

Library of Congress Control Number: 2016946179

ISBN 978-94-017-7594-6 (PB)
ISBN 978-94-017-7591-5 (HB)
ISBN 978-94-017-7593-9 (e-book)
DOI 10.1007/978-94-017-7593-9

Published by Springer,
P.O. Box 17, 3300 AA Dordrecht, The Netherlands.

www.springer.com

Printed on acid-free paper

Preface

These proceedings of the NATO-ARW "Nanomaterials for Security" held at the "Londonskaya" Hotel, Odessa, Ukraine, from August 30–September 3, 2015, emerged as a result of many presentations and discussions between workshop participants.

The Odessa workshop focused on several open problems including the current state of nanomaterials and security problems.

Recent advances in nanoscience have demonstrated that fundamentally new physical phenomena arise when systems are reduced in size down to dimensions comparable with the fundamental microscopic length scales of investigated materials. There have been many significant advances in the past two years since the last workshop, and some entirely new directions of research in these fields have been undertaken. The programme of the workshop allowed presentations and opened discussions on several emerging modern research topics, such as new nanomaterials and sensors. Theoretical advances were tested against major experimental and technological achievements in related materials. There was intensive discussion in the field of nanotechnologies and safety systems that include nanosensors, nanocomposite multifunctional coatings for safety systems, bio-nanosensors, and nanoanalyzers. In the session on nanomaterials, the physical properties of graphene, carbon nanotubes, new composite materials, and spintronics were presented. Latest developments in nanotechnology and measurement techniques facilitate the detection of explosives. The most promising new materials and experimental techniques for the detection of hazardous materials including explosives are carbon nanotubes, Josephson junctions, and NMR techniques. Participants benefitted from presentations of new methods for the detection of CBRN agents using chemical and biochemical sensors. The contemporary open problems of the physics of sensors include the determination of sizes of nanoparticles, identification of particles, and determination of concentration and mobility of nanoparticles.

We are grateful to members of the International Advisory Committee for their consistent help and suggestions, in particular to Prof. F. Peeters. We are grateful to the Nobel Laureate Prof. Klaus von Klitzing for a high level of scientific talk.

We would like to thank the NATO Science Committee for the essential financial support, without which the meeting could not have taken place. We also acknowledge the National Academy of Science of Ukraine, J. Stefan Institute, Ljubljana, and Faculty of Mathematics and Physics, University of Ljubljana, Slovenia, for their generous support.

Ljubljana, Slovenia
Kiev, Ukraine
February 2016

Janez Bonča
Sergei Kruchinin

HOTEL LONDONSKAYA

Nobel Laureate K. von Klitzing and co-directors J. Bonča and S. Kruchinin

Contents

Part I Nanomaterials

1 **Atomic Collapse in Graphene** 3
D. Moldovan and F.M. Peeters

2 **Fluorination Clusters on Graphene Resolved by Conductive AFM** 19
A. Mishchenko, A. Eckmann, I.V. Grigorieva, and K.S. Novoselov

3 **Spin Relaxation in GaAs Based Quantum Dots for Security and Quantum Information Processing Applications** 25
S. Prabhakar and R. Melnik

4 **Very Sensitive Nanocalorimetry of Small Mass Systems and Glassy Materials** 35
J.-L. Garden, A. Tavakoli, T. Nguyen-Duc, A. Frydman, M. Laarraj, J. Richard, and O. Bourgeois

5 **Phase Conversion of Y-Ba-Cu-O Thin Films by Super-Oxygenation and Cu-Enrichment** 45
H. Zhang, N. Gauquelin, C. Mcmahon, D.G. Hawthorn, G.A. Botton, and J.Y.T. Wei

6 **Strong-Coupling Diagram Technique for Strong Electron Correlations** 57
A. Sherman

7 **Spin-Dependent Transport of Carbon Nanotubes with Chromium Atoms** 67
S.P. Kruchinin, S.P. Repetsky, and I.G. Vyshyvana

8 Cell Monolayer Functioning Detection Based on Quantum Polarization Effects in Langmuir–Blodgett Multi-Walled Carbon Nanotubes Films 97
H.V. Grushevskaya, N.G. Krylova, I.V. Lipnevich, T.I. Orekhovskaja, and B.G. Shulitski

9 Magnetic Properties of Cobalt and Nitrogen Co-modified Titanium Dioxide Nanocomposites 109
N. Guskos, J. Typek, G. Zolnierkiewicz, E. Kusiak-Nejman, S. Mozia, and A.W. Morawski

10 Physical Properties of $(As_2Se_3)_{1-x}$:Sn_x and $(As_4S_3Se_3)_{1-x}$:Sn_x Glasses 127
O.V. Iaseniuc, D.V. Harea, E.E. Harea, G.F. Volodina, and M.S. Iovu

11 Thermal Memory and Thermal Induced Phase Transformation in Shape Memory Alloys 141
O. Adiguzel

12 Spectroscopic Properties of Nanoceria Allowing Visualization of Its Antioxidant Action 149
V. Seminko, A. Masalov, P. Maksimchuk, V. Klochkov, I. Bespalova, O. Viagin, and Yu. Malyukin

13 Vibration Based Microstructure Replication and Analysis 159
R. Šakalys, G. Janušas, A. Palevičius, V. Grigaliūnas, and P. Palevičius

Part II Nanosensors

14 Nanotechnology and Microfluidics Based Biosensing 175
H. van Heeren and P. Salomon

15 Resistivity Sensors of Metal Oxides with Metal Nanoparticles as Catalysts 187
G.A. Mousdis, M. Kompitsas, D. Tsamakis, M. Stamataki, G. Petropoulou, and P. Koralli

16 Iono-Electronic Interface Based on Innovative Low Temperature Zeolite Coated NMOS (Circuits) for Bio-nanosensor Manufacture 201
A.S. Fiorillo, S.A. Pullano, R. Tiriolo, and J.D. Vinko

17 SQUID Detectors for Non-destructive Evaluation in Industry 215
W. Nawrocki

18 Morphological Features of Nanostructured Sensor for X-Ray and Optical Imaging, Based on Nonideal Heterojunction 227
Ie. Brytavskyi, V. Smyntyna, and V. Borschak

19 Hetero-Carbon Nanostructures as the Effective Sensors in Security Systems 239
G. Kharlamova, O. Kharlamov, M. Bondarenko, and O. Khyzhun

20 Characterization of SnO_2 Sensors Nanomaterials by Polarization Modulation Method 259
V.S. Grinevych, L.M. Filevska, V.A. Smyntyna, M.O. Stetsenko, S.P. Rudenko, L.S. Maksimenko, and B.K. Serdega

21 Diagnostic of Resonant Properties of Au-PTFE Nanostructures for Sensor Applications 267
L.S. Maksimenko, S.P. Rudenko, M.O. Stetsenko, I.E. Matyash, O.M. Mischuk, Yu.V. Kolomzarov, and B.K. Serdega

22 Metal Oxide Based Biosensors for the Detection of Dangerous Biological Compounds 281
A.V. Tereshchenko, V.A. Smyntyna, I.P. Konup, S.A. Geveliuk, and M.F. Starodub

23 Cerium Dioxide CeO_{2-x} and Orthovanadate ($Gd_{0.9}Eu_{0.1}VO_4$) Nanoparticles for Protection of Living Body from X-Ray Induced Damage 289
G. Grygorova, V. Klochkov, Ye. Mamotyuk, and Yu. Malyukin

24 Anthology and Genesis of Nanodimensional Objects and GM Food as the Threats for Human Security 297
O. Kharlamov, M. Bondarenko, O. Khyzhun, and G. Kharlamova

25 Transmission of Three Resistance Sensor Signals Over Four Wire Line with Losses 311
A. Penin and A. Sidorenko

Author Index 327

Subject Index 329

Contributors

O. Adiguzel Department of Physics, Firat University, Elazig, Turkey

I. Bespalova Institute for Scintillation Materials, Kharkiv, Ukraine

M. Bondarenko Frantsevich Institute for Problems of Materials Science of NASU, Kyiv, Ukraine

V. Borschak Odessa I.I. Mechnikov National University, Odessa, Ukraine

G.A. Botton Canadian Centre for Electron Microscopy & Brockhouse Institute for Materials Research, McMaster University, Hamilton, ON, Canada

O. Bourgeois Institut NÉEL, CNRS, Grenoble, France

University of Grenoble Alpes, Institut NEEL, Grenoble, France

Ie. Brytavskyi Odessa I.I. Mechnikov National University, Odessa, Ukraine

A. Eckmann School of Physics and Astronomy, University of Manchester, Manchester, UK

L.M. Filevska Department of Experimental Physics I.I. Mechnikov, Odessa National University (ONU), Odessa, Ukraine

A.S. Fiorillo Department of Health Sciences, School of Biomedical Engineering, University Magna Grœcia of Catanzaro, Catanzaro, Italy

A. Frydman Institut NÉEL, CNRS, Grenoble, France

University of Grenoble Alpes, Institut NEEL, Grenoble, France

Department of Physics, Bar-Ilan University, Ramat Gan, Israel

J.-L. Garden Institut NÉEL, CNRS, Grenoble, France

University of Grenoble Alpes, Institut NEEL, Grenoble, France

N. Gauquelin Canadian Centre for Electron Microscopy & Brockhouse Institute for Materials Research, McMaster University, Hamilton, ON, Canada

S.A. Geveliuk Odessa National I.I. Mechnikov University, Odessa, Ukraine

V. Grigaliūnas Institute of Materials Science, Kaunas University of Technology, Kaunas, Lithuania

I.V. Grigorieva School of Physics and Astronomy, University of Manchester, Manchester, UK

V.S. Grinevych Department of Experimental Physics I.I. Mechnikov, Odessa National University (ONU), Odessa, Ukraine

H.V. Grushevskaya Physics Department, Belarusan State University, Minsk, Belarus

G. Grygorova Institute for Scintillation Materials of NAS of Ukraine, Kharkiv, Ukraine

N. Guskos Department of Physics, West Pomeranian, University of Technology, Szczecin, Poland

D.V. Harea Institute of Applied Physics, Academy of Sciences of Moldova, Chisinau, Republic of Moldova

E.E. Harea Institute of Applied Physics, Academy of Sciences of Moldova, Chisinau, Republic of Moldova

D.G. Hawthorn Department of Physics and Astronomy, University of Waterloo, Waterloo, ON, Canada

O.V. Iaseniuc Institute of Applied Physics, Academy of Sciences of Moldova, Chisinau, Republic of Moldova

M.S. Iovu Institute of Applied Physics, Academy of Sciences of Moldova, Chisinau, Republic of Moldova

G. Janušas Department of Mechanical Engineering, Kaunas University of Technology, Kaunas, Lithuania

O. Kharlamov Frantsevich, Institute for Problems of Materials Science of NASU, Kyiv, Ukraine

G. Kharlamova Taras Shevchenko National University of Kyiv, Kyiv, Ukraine

O. Khyzhun Frantsevich Institute for Problems of Materials Science of NASU, Kyiv, Ukraine

V. Klochkov Institute for Scintillation Materials of NAS of Ukraine, Kharkiv, Ukraine

Yu.V. Kolomzarov V.E. Lashkaryov Institute of Semiconductors Physics, National Academy of Sciences of Ukraine (NASU) Nauky Prospect, Kyiv, Ukraine

M. Kompitsas NHRF-National Hellenic Research Foundation, Theoretical and Physical Chemistry Institute-TPCI, Athens, Greece

I.P. Konup Odessa National I.I. Mechnikov University, Odessa, Ukraine

P. Koralli NHRF-National Hellenic Research Foundation, Theoretical and Physical Chemistry Institute-TPCI, Athens, Greece

S.P. Kruchinin Bogolyubov Institute for Theoretical Physics, Kiev, Ukraine

N.G. Krylova Physics Department, Belarusan State University, Minsk, Belarus

E. Kusiak-Nejman Department of Chemical and Environmental Engineering, West Pomeranian University of Technology, Szczecin, Poland

M. Laarraj Institut NÉEL, CNRS, Grenoble, France

University of Grenoble Alpes, Institut NEEL, Grenoble, France

I.V. Lipnevich Physics Department, Belarusan State University, Minsk, Belarus

P. Maksimchuk Institute for Scintillation Materials, Kharkiv, Ukraine

L.S. Maksimenko V.E. Lashkarev Institute of Semiconductors Physics, National Academy of Sciences of Ukraine (NASU) Nauky Prospect, Kyiv, Ukraine

Yu. Malyukin Institute for Scintillation Materials of NAS of Ukraine, Kharkiv, Ukraine

Ye. Mamotyuk Grigoriev Institute for Medical Radiology-National Academy of Medical Science of Ukraine, Kharkiv, Ukraine

A. Masalov Institute for Scintillation Materials, Kharkiv, Ukraine

I.E. Matyash V.E. Lashkaryov Institute of Semiconductors Physics, National Academy of Sciences of Ukraine (NASU) Nauky Prospect, Kyiv, Ukraine

C. Mcmahon Department of Physics and Astronomy, University of Waterloo, Waterloo, ON, Canada

R. Melnik MS2Discovery Interdisciplinary Research Institute, M^2NeT Laboratory, Wilfrid Laurier University, Waterloo, ON, Canada

O.M. Mischuk V.E. Lashkaryov Institute of Semiconductors Physics, National Academy of Sciences of Ukraine (NASU) Nauky Prospect, Kyiv, Ukraine

A. Mishchenko School of Physics and Astronomy, University of Manchester, Manchester, UK

D. Moldovan Departement Fysica, Universiteit Antwerpen, Antwerpen, Belgium

A.W. Morawski Department of Chemical and Environmental, Engineering, West Pomeranian, University of Technology, Szczecin, Poland

G.A. Mousdis NHRF-National Hellenic Research Foundation, Theoretical and Physical Chemistry Institute-TPCI, Athens, Greece

S. Mozia Department of Chemical and Environmental Engineering, West Pomeranian University of Technology, Szczecin, Poland

W. Nawrocki Faculty of Electronics and Telecommunications, Poznan University of Technology, Poznan, Poland

T. Nguyen-Duc Institut NÉEL, CNRS, Grenoble, France

University of Grenoble Alpes, Institut NEEL, Grenoble, France

K.S. Novoselov School of Physics and Astronomy, University of Manchester, Manchester, UK

T.I. Orekhovskaja Physics Department, Belarusan State University, Minsk, Belarus

A. Palevičius Department of Mechanical Engineering, Kaunas University of Technology, Kaunas, Lithuania

P. Palevičius Department of Mathematical Modelling, Kaunas University of Technology, Kaunas, Lithuania

F.M. Peeters Departement Fysica, Universiteit Antwerpen, Antwerpen, Belgium

A. Penin "D. Ghitu" Institute of Electronic Engineering and Nanotechnologies, Academy of Sciences of Moldova, Chisinau, Moldova

G. Petropoulou NHRF-National Hellenic Research Foundation, Theoretical and Physical Chemistry Institute-TPCI, Athens, Greece

S. Prabhakar MS2Discovery Interdisciplinary Research Institute, M^2NeT Laboratory, Wilfrid Laurier University, Waterloo, ON, Canada

S.A. Pullano Department of Health Sciences, School of Biomedical Engineering, University Magna Grœcia of Catanzaro, Catanzaro, Italy

S.P. Repetsky Taras Shevchenko Kiev National University, Kiev, Ukraine

J. Richard Institut NÉEL, CNRS, Grenoble, France

University of Grenoble Alpes, Institut NEEL, Grenoble, France

S.P. Rudenko V.E. Lashkarev Institute of Semiconductors Physics, National Academy of Sciences of Ukraine (NASU) Nauky Prospect, Kyiv, Ukraine

R. Šakalys Department of Mechanical Engineering, Kaunas University of Technology, Kaunas, Lithuania

P. Salomon enablingMNT, Berlin, Germany

V. Seminko Institute for Scintillation Materials, Kharkiv, Ukraine

B.K. Serdega V.E. Lashkarev Institute of Semiconductors Physics, National Academy of Sciences of Ukraine (NASU) Nauky Prospect, Kyiv, Ukraine

A. Sherman Institute of Physics, University of Tartu, Tartu, Estonia

B.G. Shulitski Belarusan State University of Informatics and Radioelectronica, Minsk, Belarus

A. Sidorenko "D. Ghitu" Institute of Electronic Engineering and Nanotechnologies, Academy of Sciences of Moldova, Chisinau, Moldova

V.A. Smyntyna Department of Experimental Physics I.I. Mechnikov, Odessa National University (ONU), Odessa, Ukraine

M. Stamataki Electrical and Computer Engineering Department, National Technical University of Athens, Athens, Greece

M.F. Starodub National University of Life and Environmental Sciences, Kyiv, Ukraine

M.O. Stetsenko V.E. Lashkarev Institute of Semiconductors Physics, National Academy of Sciences of Ukraine (NASU) Nauky Prospect, Kyiv, Ukraine

A. Tavakoli Institut NÉEL, CNRS, Grenoble, France

University of Grenoble Alpes, Institut NEEL, Grenoble, France

A.V. Tereshchenko Odessa National I.I. Mechnikov University, Odessa, Ukraine

R. Tiriolo Department of Health Sciences, School of Biomedical Engineering, University Magna Grœcia of Catanzaro, Catanzaro, Italy

D. Tsamakis Electrical and Computer Engineering Department, National Technical University of Athens, Athens, Greece

H. van Heeren enablingMNT The Netherlands, Dordrecht, The Netherlands

O. Viagin Institute for Scintillation Materials, Kharkiv, Ukraine

J.D. Vinko HERA Physics, Velletri, Italy

G.F. Volodina Institute of Applied Physics, Academy of Sciences of Moldova, Chisinau, Republic of Moldova

I.G. Vyshyvana Institute of High Technologies, Taras Shevchenko Kiev National University, Kiev, Ukraine

J.Y.T. Wei Department of Physics, University of Toronto, Toronto, ON, Canada

Canadian Institute for Advanced Research, Toronto, ON, Canada

H. Zhang Department of Physics, University of Toronto, Toronto, ON, Canada

G. Zolnierkiewicz Department of Physics, West Pomeranian University of Technology, Szczecin, Poland

Part I
Nanomaterials

Chapter 1
Atomic Collapse in Graphene

D. Moldovan and F.M. Peeters

Abstract When the charge Z of an atom exceeds the critical value of 170, it will undergo a process called atomic collapse which triggers the spontaneous creation of electron-positron pairs. The high charge requirements have prevented the observation of this phenomenon with real atomic nuclei. However, thanks to the relativistic nature of the carriers in graphene, the same physics is accessible at a much lower scale. The atomic collapse analogue in graphene is realized using artificial nuclei which can be created via the deposition of impurities on the surface of graphene or using charged vacancies. These supercritically charged artificial nuclei trap electrons in a sequence of quasi-bound states which can be observed experimentally as resonances in the local density of states.

1.1 Introduction

Since 2004 when graphene was mechanically exfoliated [1], it has attracted a lot of scientific interest due to its unique properties. Its two-dimensional crystal lattice hosts massless Dirac fermions [2] which move with a Fermi velocity of about 1/300 the speed of light with a linear spectrum close to the K and K' points of the Brillouin zone [3]. Its unique properties include the anomalous integer quantum Hall effect in the presence of a magnetic field [4, 5], room-temperature ballistic transport [6] and the Klein paradox, i.e. unusually high transmission through classically forbidden regions [7]. Combined with excellent mechanical strength [8], this makes graphene a very promising material for future electronic devices. However, the electrical properties of graphene are highly sensitive to charged impurities [9–11] which make Coulomb interaction a topic of significant interest.

In 1928, Paul Dirac proposed a relativistic formulation of the quantum mechanics of the electron [12], which predicted that relativistic quantum effects can tip the delicate balance of stable atomic orbitals. When the charge of the nucleus exceeds a certain critical value, atomic states are rendered unstable: the energy levels of

---

D. Moldovan • F.M. Peeters (✉)
Departement Fysica, Universiteit Antwerpen, Groenenborgerlaan 171,
B-2020 Antwerpen, Belgium
e-mail: francois.peeters@uantwerpen.be

J. Bonča, S. Kruchinin (eds.), *Nanomaterials for Security*, NATO Science for Peace and Security Series A: Chemistry and Biology, DOI 10.1007/978-94-017-7593-9_1

bound states dive into the lower continuum and the spontaneous generation of positrons is expected [13–15]. The semiclassical trajectories of these unstable states spiral into the nucleus, hence the name *atomic collapse*. This is a long sought-after phenomenon in quantum electrodynamics (QED). The critical charge requires extremely heavy nuclei with $Z > 170$, which has prevented the observation of this phenomenon notwithstanding several attempts using collisions of heavy ions [16, 17].

The unusually high carrier mobility in graphene is attributed to the fact that the Dirac quasiparticles cannot be confined by electrostatic potentials or bound by charged impurities [3, 18, 19]. However, when the impurity charge exceeds a critical value it triggers a fundamental change in the electronic properties, manifest by the emergence of quasi-bound localized states in its vicinity [20–22]. This effect is analogous to the atomic collapse as predicted almost a century ago. In graphene where the fine structure constant is much larger, the critical Coulomb coupling can be reached for relatively modest charge Z [20–23]. Still, even in graphene this regime has been elusive because screening reduces the effective coupling rendering the charge subcritical [24, 25].

1.1.1 Atomic Collapse with Heavy Nuclei

Classical physics cannot properly explain the structure of an atom. Because the energy is unbounded, an electron in orbit around a nucleus would continually lose energy, assume a spiraling orbit and eventually collide with the nucleus. Quantum mechanics solves this problem and naturally explains stable atomic orbits. Given a nucleus with the atomic number Z, an orbiting electron feels the Coulomb potential $U = -Ze^2/r$. Its kinetic energy is described by $K = p^2/2m$ where we can substitute the uncertainty principle to find $K \approx \hbar^2/mr^2 \propto 1/r^2$, while the potential part is $U \propto -1/r$. Being interested in conditions around the nucleus, we take the limit $r \to 0$ to find that $|K| \gg |U|$ and thus the total energy $E = K + U \to \infty$. An electron would need an infinite amount of energy to find itself at the position of the nucleus. By this fact a collision with the nucleus is made impossible and quantum mechanics leaves only one possible outcome: electron orbits must be stable.

However, when heavy nuclei (with large Z) are considered, quantum mechanics alone does not give an accurate description and relativity must be taken into account. The relativistic electron will find itself inside the same Coulomb potential as before, but now its kinetic energy must be described as $K = cp$, where c is the speed of light. Substituting the uncertainty principle, we find that $K \propto 1/r$ which is the same as the potential energy but with the opposite sign. Taking the limit $r \to 0$ will not yield a definitive result this time, because both the potential and kinetic parts are of the same order. This opens the possibility of a finite energy electron state at the nucleus with disruptive results: atomic collapse.

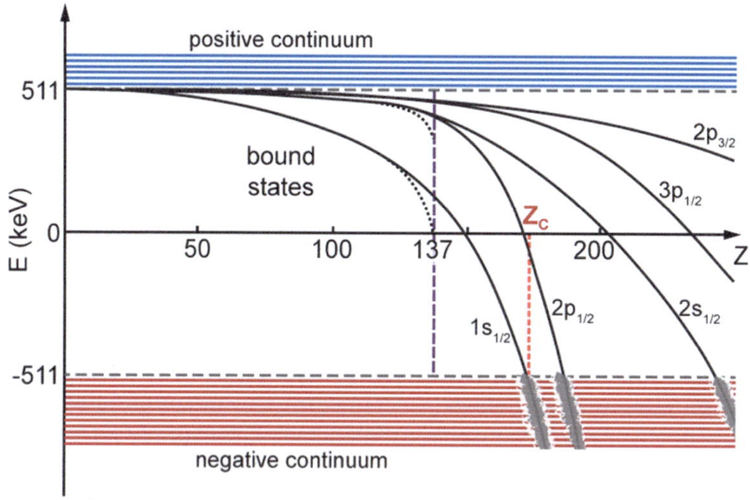

Fig. 1.1 Schematic drawing of the atomic bounding energies as function of the atomic number Z and the level diving process in the supercritical regime. The *dotted curves* indicate the solutions for a point charge while the *solid curves* take into account the finite size of the nucleus

Solving the Dirac equation for an atomic nucleus modeled as a point charge reveals the energy of the 1S atomic bound state, $E_{1S} = m_e\sqrt{1-(Z\alpha)^2}$ [15], where $m_e \approx 511$ keV is the electron rest mass and $\alpha \approx 1/137$ is the fine structure constant (see Fig. 1.1). Beyond the value of $Z_c\alpha = 1$, i.e. $Z_c \approx 137$, the energy becomes complex. This "collapses" the wave function and the bound state ceases to exist. However, this outcome is reached by assuming the nucleus to be a point charge, which is quite unrealistic especially since it is expected to house more that a hundred protons. Taking into account the finite size of the nucleus will truncate the Coulomb potential thereby removing the divergence. This correction extends the stability beyond $Z \approx 137$ [13, 15]. The 1S state continues to exist up to the new critical value of $Z_c \approx 170$. At that point, as it enters the supercritical regime, the electron state leaves the discrete spectrum and dives into the positron continuum. There, the bound state acquires a finite lifetime, transforming it into a narrow resonance [14, 15]. The electron state is coupled to the positron continuum via tunneling which gives rise to spontaneous positron emission which is the expected measurable signature of atomic collapse.

Because of the extremec requirements, a stable nucleus with supercritical charge cannot be found in nature, but it is possible to create one, at least for a short time. Artificial nuclei with charge values $Z > 170$ can be realized in high-energy collisions of very heavy ions. While such experiments with Uranium atoms did confirm several features of the near-collapse regime, direct proof of supercritical positron emission was not found [16, 17].

1.1.2 Atomic Collapse Analogue in Graphene

The relativistic nature of the charge carriers in graphene allow it to display the same atomic collapse physics of QED but at a much smaller energy scale. The speed of light is substituted with the much smaller Fermi velocity $v_F \approx c/300$, which has a profound effect on the effective fine structure constant:

$$\alpha_G = \frac{c}{v_F}\frac{1}{\kappa}\alpha \approx \frac{2.2}{\kappa} \approx 1. \tag{1.1}$$

Here $\kappa \approx 2.5$ is the dielectric constant of graphene. Because the effective fine structure constant α_G is much larger than α, the value of the critical charge Z_C is expected to be much smaller, on the order of unity. This means that modest charged impurities in graphene could play the role of supercritical nuclei.

It is important to keep in mind that the effect in QED concerns real particles while the Dirac-like band structure of graphene refers to quasiparticle excitation in an effective low-energy approximation. Graphene is two-dimensional and massless while QED is three-dimensional and massive. The switch to graphene also changes the energy scale from MeV to sub-eV. Nevertheless, the physical analogue is undeniable and graphene has the key advantage: experimental accessibility. In addition, graphene experiments allow for further study by varying other parameters such as the back-gate voltage, defects or the number of layers. The results of this exploration have an effect on more than just theoretical QED, but also the understanding of confinement and transport in graphene.

1.2 Theoretical Model

The tight-binding description for the π_z orbitals of carbon in graphene is given by the Hamiltonian

$$H_0 = t\sum_{\langle i,j\rangle}\left(a_i^\dagger b_j + \text{H.c.}\right) + t'\sum_{\langle\langle i,j\rangle\rangle}\left(a_i^\dagger b_j + b_i^\dagger b_j + \text{H.c.}\right), \tag{1.2}$$

where operators $a_i^\dagger(a_i)$ and $b_i^\dagger(b_i)$ create (annihilate) an electron on site $\mathbf{R}_i$ of sublattice A and B, respectively. Here $t \approx -2.8\,\text{eV}$ is the nearest neighbor ($\langle i,j\rangle$) hopping energy and $t' \approx 0.1\,\text{eV}$ is the next-nearest neighbor ($\langle\langle i,j\rangle\rangle$) hopping energy. It is sometimes assumed that the contribution of the next-nearest hopping term is negligible, especially in the low energy approximation. However, it plays an important role in graphene since it breaks electron-hole symmetry which is responsible for various effects which can be observed experimentally, especially in the presence of localized defects, as we shall see.

Considering only low-energy states close to the K point and nearest neighbor interactions ($t' = 0$), the tight-binding Hamiltonian Eq. (1.2) may be simplified so that the equation $H\Psi = E\Psi$ reduces to the massless Dirac–Weyl equation

$$-i\hbar v_F \boldsymbol{\sigma} \cdot \nabla \Psi(\boldsymbol{r}) = E\Psi(\boldsymbol{r}), \tag{1.3}$$

where σ_i are the Pauli matrices, $v_F = 3a_{cc}t/2 \approx 10^6$ m/s is the Fermi velocity, $a_{cc} \approx 0.142$ nm is the carbon-carbon distance. The wave functions $\Psi(\boldsymbol{r}) = [A(\boldsymbol{r}), B(\boldsymbol{r})]^T$ are two-component spinors where the localization of the electron on sublattices A and B plays the role of a pseudospin.

Solving the problem for pristine graphene yields the linear dispersion relation: $E(k) = \pm\hbar v_F|k|$. This corresponds to the dispersion relation of ultrarelativistic particles $E = c|p|$ but with the much smaller Fermi velocity v_F instead of the speed of light c. Furthermore, graphene's band structure describes electron and hole states which are an analogue of the relativistic electrons and positrons (particle and antiparticle states of QED).

1.2.1 Coulomb Impurity

In the presence of a charged impurity, a local energy term must be added to the tight-binding Hamiltonian (1.2),

$$H_{\text{imp}} = \sum_i V_i^A a_i^\dagger a_i + \sum_i V_i^B b_i^\dagger b_i, \tag{1.4}$$

where $V_i^{A,B}$ is the impurity potential at site $\mathbf{R}_i$. A Coulomb center of charge Z generates the potential $V(r) = -\beta/r$ where $\beta \equiv Ze^2/\kappa\hbar v_F$ is the dimensionless coupling constant.

As before, considering the effective low-energy approximation of the total Hamiltonian $H = H_0 + H_{\text{imp}}$, we find that the system is governed by the wave equation

$$\hbar v_F \left(-i\boldsymbol{\sigma} \cdot \nabla - \frac{\beta}{r}\right) \Psi(\boldsymbol{r}) = E\Psi(\boldsymbol{r}). \tag{1.5}$$

The equation describes the behavior of Dirac fermions in a single valley of graphene's Brillouin zone. Because of the long-range nature of the Coulomb field, intervalley processes are not relevant and the problem may be solved independently within each valley.

Due to the axial symmetry of the potential, Eq. (1.5) is separable in cylindrical coordinates. The total angular momentum is conserved which allows us to use the eigenstates of the momentum operator $J_z = L_z + \sigma_z/2$, where L_z is the orbital angular momentum [26],

$$\Psi_j(r) = \frac{1}{\sqrt{r}} \begin{pmatrix} e^{i(j-1/2)\varphi} a_j(r) \\ ie^{i(j+1/2)\varphi} b_j(r) \end{pmatrix}. \tag{1.6}$$

Here $j = \pm 1/2, \pm 3/2, \ldots$ are the eigenvalues of J_z. Equation (1.5) therefore reduces to,

$$\begin{bmatrix} \varepsilon + \beta/r & -(\partial_r + j/r) \\ (\partial_r - j/r) & \varepsilon + \beta/r \end{bmatrix} \begin{bmatrix} a_j(r) \\ b_j(r) \end{bmatrix} = 0, \tag{1.7}$$

where $\varepsilon = E/\hbar v_F$. This coupled pair of first order differential equations can be reduced to two decoupled second order equations. In the limit $r \to 0$ the solution behaves as

$$\varphi_j(r) \sim r^\gamma, \tag{1.8}$$

where

$$\gamma = \sqrt{j^2 - \beta^2}. \tag{1.9}$$

This reveals a problem for the lowest angular momentum ($j = \pm 1/2$), because γ becomes imaginary if $\beta > \beta_c = 1/2$. In this case the solution oscillates endlessly towards the center as $e^{i \log r}$. From a classical perspective (see Fig. 1.2) this can be understood as a critical angular momentum above which the orbits spiral and fall into the potential origin [22].

1.2.2 *Supercritical Regime*

The quantum-mechanical problem is ill-defined for a point charge in the supercritical regime ($\beta > 1/2$). An additional boundary condition must be introduced to cut off the potential at short distances. This is analogous to the introduction of the finite size of the nucleus in QED [14]. The modified potential reads

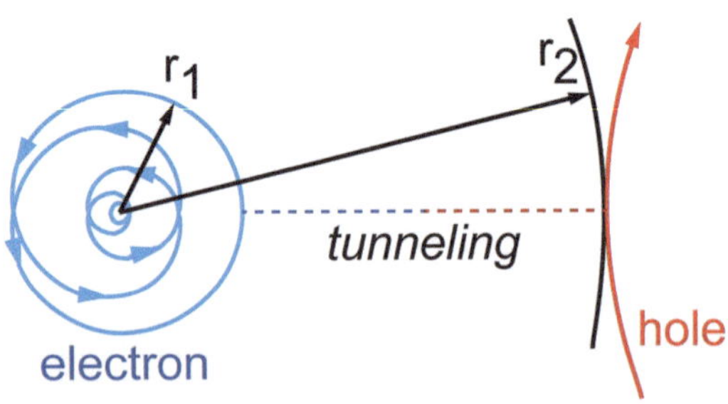

Fig. 1.2 Quasiclassical trajectories of atomic collapse in graphene. Collapsing electron coupled to escaping hole trajectory

$$V(r) = \begin{cases} -\hbar v_F \frac{\beta}{r_0}, & \text{if } r \le r_0, \\ -\hbar v_F \frac{\beta}{r}, & \text{if } r > r_0, \end{cases} \tag{1.10}$$

where r_0 is the cutoff. For graphene the natural minimal cutoff length originates from the lattice structure, i.e. $r_0 = a_{cc}$, the distance between two nearest carbon atoms. A larger cutoff may be taken (depending on the nature of the charge center) but the lattice cutoff is a natural minimum distance required to regularize the potential and permit the solution to extend into the supercritical regime.

The charge may also be located a significant distance away from the graphene plane [27] in which case the Coulomb potential may be regularized in the alternative form of $V(r) = -\hbar v_F \beta / \sqrt{r^2 + d^2}$, which corresponds to a charge β located a distance d from the graphene plane.

With the potential regularized with cutoff r_0, the spectrum of the supercritical states behaves as [22]

$$E_n \approx -\hbar v_F \frac{\beta}{r_0} e^{-\pi n / \sqrt{\beta^2 - \beta_c^2}}. \tag{1.11}$$

Note that there is a singularity at the critical value $\beta = \beta_c$ and the states are only defined for $\beta > \beta_c$. The energy scale is set by the regularization cutoff r_0. Because there is no gap in the energy spectrum, true bound states are not possible so these are quasi-bound states (resonances).

For a measurable feature of the supercritical states, we turn to the local density of states (LDOS),

$$\text{LDOS}(E, r) = \frac{4}{\pi \hbar v_F} \sum_i |\Psi_i(k_E, r)|^2, \tag{1.12}$$

where $k_E = |E| / \hbar v_F$ and the sum runs over the relevant angular momentum channels in the continuum. The supercritical quasi-bound state appears as a resonance peak in the LDOS located at negative energy below the Dirac point, as shown in Fig. 1.3. As the charge β is increased, the resonance broadens as it moves down to lower energy. On the other hand, its spatial extent is inversely proportional to $|E|$. This feature can be measured experimentally. It corresponds to the inner electron part of the collapse wave function. This is in contrast to the QED heavy ion experiments which look for the outer part of the wavefunction: the outgoing positron.

1.3 Artificial Nuclei

The resonance peak which appears in the LDOS of a supercritical charge is a clear, experimentally observable signature of atomic collapse in graphene. The LDOS can be measured with scanning tunneling microscopy (STM). An atomically sharp tip

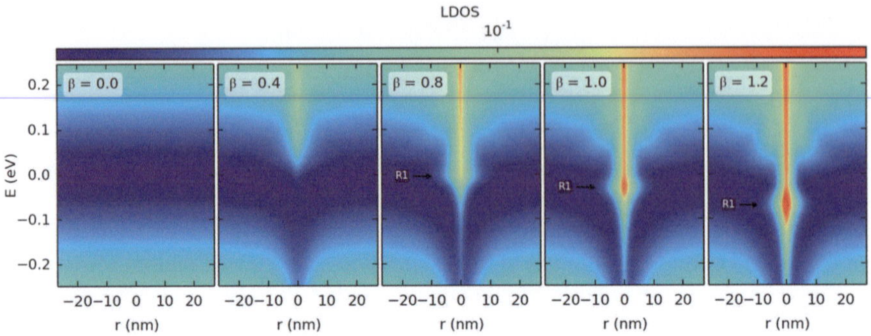

Fig. 1.3 Colormap of the LDOS as a function of position and energy. The charge is located in the center ($r = 0$). The coupling β varies from 0 to 1.2 as labeled in the subplots. The $R1$ label indicates the collapse resonance which appears in the supercritical regime $\beta > 0.5$

scans over a sample of graphene and measures the electric current I the tip draws from the surface as a consequence of the tunnelling effect. The current depends on the relative voltage V between the tip and the sample. The derivative of I with respect to V (differential conductance) is proportional to the local density of states

$$\text{LDOS}(E, r) \propto \frac{dI}{dV}. \tag{1.13}$$

Even though the critical value is relatively modest in graphene, introducing a sufficiently large impurity charge on graphene's surface is still challenging [24]. Because graphene is a good conductor it is difficult to deposit and maintain a charge on its surface. It is common to find impurities in graphene samples, but the charge of a single impurity atom turned out to be insufficient. Thus it is necessary to create more intense charge centers which can serve the role of artificial nuclei for atomic collapse in graphene.

1.3.1 Charged Impurity Cluster

In order to successfully create a supercritical charge, multiple Calcium dimers need to be deposited on the surface of a graphene sheet as was done in the experiment by Wang et al. [25] Individual dimers have subcritical charge, but in the experiment they are used like movable charge centers using an STM tip to assemble a cluster of supercritical charge. For the theoretical model, the previously introduced potential cutoff r_0 is modified in order to match the size of the cluster ($\approx$1 nm).

The results of the experiment are shown in Fig. 1.4. The insets show the positions of the Calcium dimers as seen by the STM. The curves depict the differential conductance dV/dI, i.e. LDOS, as a function of the bias voltage V for different positions of the STM tip. A change of V corresponds to changing the energy and thus

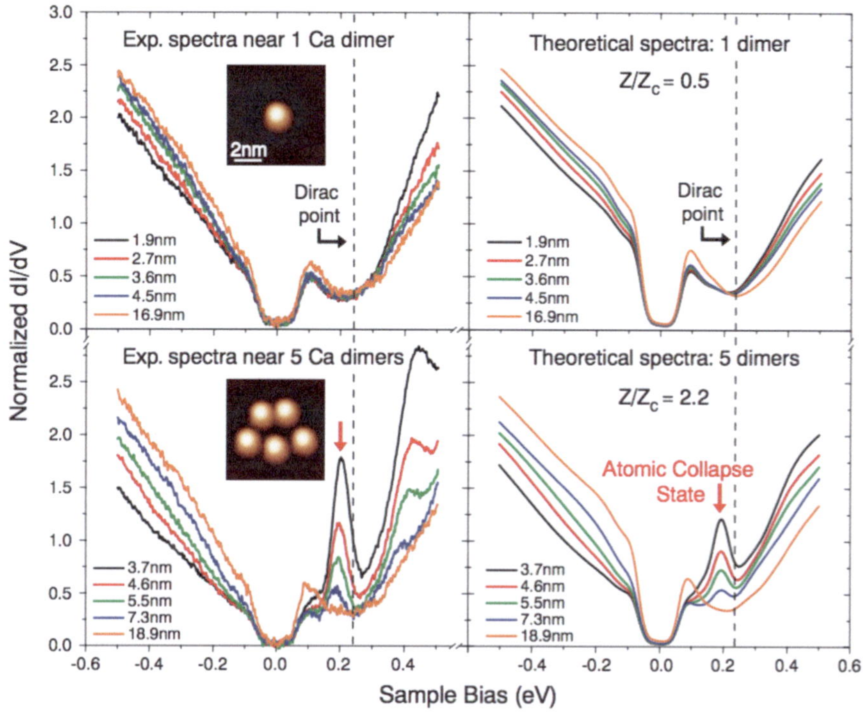

Fig. 1.4 STM measurements (*left*) and calculation (*right*) of dI/dV for graphene with 1 and 5 charge centers: *Ca* dimers as shown in the *inset*. The various curves correspond to different distances from the center of the charge (Adapted from Wang et al. [25])

allows to scan the band structure. The gap in the LDOS is caused by phonon-assisted inelastic tunneling and is not of interest. Instead, the main feature is the appearance of a clear narrow maximum in the vicinity of the charge center in the case of 5 dimers. This is the expected evidence of an atomic collapse state in graphene with a supercritical impurity charge.

1.3.2 Charged Vacancy

In order to achieve a supercritical charge a large number of charge centers must be placed in a small area which is difficult due to strong Coulomb repulsion. Alternatively, it has been shown that a vacancy in graphene can stably host a positive charge [28, 29]. The charge can be made supercritical, allowing the vacancy to take on the role of an artificial nucleus for atomic collapse in graphene. Its effectiveness is further enhanced because it is located directly in the plane of graphene instead of on top of the surface.

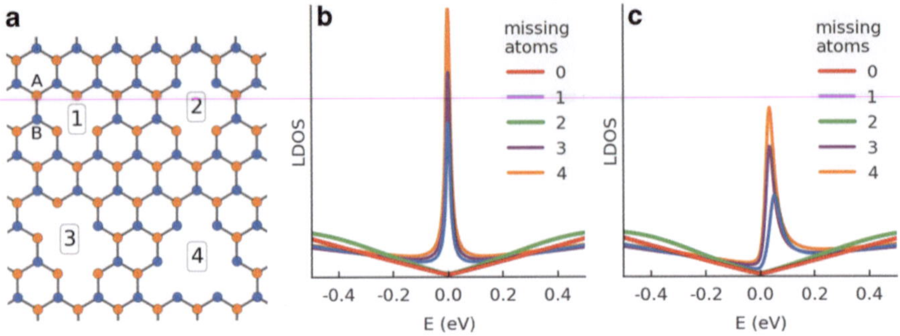

Fig. 1.5 (**a**) Illustration of vacancies with 1 to 4 missing atoms. (**b**) Simulated LDOS at 1 nm from the center of each type of vacancy including only the first nearest neighbor hopping. (**c**) The same results including the next-nearest hopping term (broken electron-hole symmetry)

A vacancy is the absence of an atom at a given site in graphene. Vacancies can be created by sputtering graphene with He^+ ions [30]. The bombardment of ions physically removes carbon atoms from the graphene sheet. This has an effect both on the conduction band electrons as well as on the structural bonds. We shall first consider the effect of removing the π_z orbital at a lattice site, i.e. a charge-neutral vacancy.

Graphene with vacancies can still be described by the tight-binding Hamiltonian (1.2) where the hoppings to vacant sites are forbidden. A vacancy may consist of $N_v = N_A + N_B$ missing carbon atoms, where N_A and N_B correspond to the number of atoms removed from sublattices A and B, respectively. In the presence of electron-hole symmetry ($t' = 0$), introducing a vacancy with $N_A \neq N_B$ will break the sublattice symmetry and create a zero energy state which is quasilocalized near the vacancy. In addition, this state exists only on a single sublattice, corresponding to the one with the lowest number of removed atoms [31].

Figure 1.5a illustrates four types of vacancies labeled 1 to 4 according to the number of missing carbon atoms. The LDOS in Fig. 1.5b is shown for each vacancy as well as for pristine graphene (0 missing atoms). Vacancies with 1, 3 and 4 missing atoms exhibit a very high intensity peak in the LDOS at the Dirac point which corresponds to the quasilocalized zero energy state. The vacancy with 2 missing atoms retains a V-shaped LDOS with only a slightly increased slope compared to pristine graphene. This is because it preserves the sublattice symmetry with one missing atom from each sublattice. The other vacancies have a sublattice difference $N_B - N_A$ of 1, 1 and -2 for vacancies 1, 3 and 4, respectively.

It is possible to write down an analytical wave function of the zero energy mode introduced by a single-atom vacancy in the continuum limit [32]

$$\Psi(x, y) \simeq \frac{e^{i\boldsymbol{K}'\boldsymbol{r}}}{x + iy} + \frac{e^{i\boldsymbol{K}\boldsymbol{r}}}{x - iy}. \tag{1.14}$$

The state is quasilocalized around the vacancy in the center and its amplitude decays with distance as $1/r$. However, this vacancy state exists in this form only in the presence of electron-hole symmetry. This is also the reason why the vacancy peaks in Fig. 1.5b are sharp and symmetric. Once electron-hole symmetry is broken, the quasilocalized vacancy state turns into a resonance and the solution can no longer be obtained analytically. Instead, a numerical solution of the tight-binding Hamiltonian is required.

Including the next-nearest neighbor hopping term ($t' \neq 0$) breaks electron-hole symmetry. The transformation from a zero energy state to a resonance means that the resulting LDOS peak will broaden and shift slightly, as seen in Fig. 1.5c. The linewidth and the displacement (from zero energy) are both proportional to the electron-hole asymmetry (i.e. the value of the t' parameter). The LDOS vacancy peaks move to positive energy in response to the higher density of holes introduced by the t' term.

Apart from the just described effect on the conduction band, the removal of a carbon atom also affects the structural bonds. One possible outcome is that the lattice will locally undergo a bond reconstruction thereby transforming the usual hexagonal lattice to pentagon/heptagon structures or other similar local reconstructions in the vicinity of the vacancy [33]. On the other had, the disrupted bonds may remain as dangling bonds. The structure around the vacancy retains the general hexagonal shape, but the carbon atoms on the edge of the vacancy are allowed to relax. Subsequently, this lattice relaxation produces a positively charged vacancy with effective charge $Z/\kappa \approx +1|\mathrm{e}|$, where $|e|$ is the elementary charge [28].

In order to create artificial nuclei, the focus is on single-atom vacancies which have not been passivated by trapped ions. These vacancies can be identified by their triangular structure in STM topography. Measuring the LDOS also identifies the low energy vacancy peak of a single-atom vacancy (see Fig. 1.5c) which distinguishes it from the two-atom vacancy that lacks this feature.

After bombarding the graphene sheet with an ion beam, the created vacancy is mostly charge neutral. The charge is deposited by applying voltage pulses with an STM tip [29]. It is well known that such pulses can functionalize atoms, tailor the local structure or change the charge state [34, 35]. Both positive and negative voltage pulses cause a gradual buildup of a positive charge at the vacancy. The sign of the STM voltage pulses does not matter because the pulses help the vacancy structure to relax, thus indirectly exposing the positive charge on the dangling bonds of the edge atoms. The charge is stable and remains unchanged as long as the experiment is kept cold. The gradual charge buildup is monitored with scanning tunneling spectroscopy (STS) and Landau level (LL) spectroscopy, and compared to numerical simulations.

The evolution of the LDOS with β is presented in Fig. 1.6, as calculated from the tight-binding model up to the next-nearest neighbor hopping. At low β the spectrum consists of a single peak which results from the uncharged vacancy. With increasing β the vacancy peak (VP) broadens and its energy becomes more negative, all the while remaining tightly localized on the vacancy site. Upon exceeding the critical value, $\beta > 0.5$, a new branch, labeled $R1$, emerges below the Dirac point. This is the counterpart of the 1S collapse state in atoms [13, 14]. The $R1$ collapse resonance

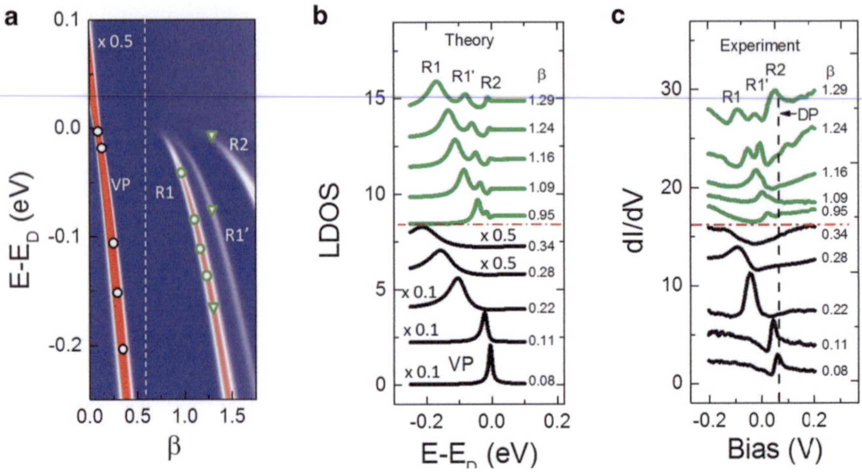

Fig. 1.6 (**a**) Simulated map of the evolution of the spectra with β. The intensity scale in the vacancy peak (VP) regime is divided by 2 to facilitate the comparison with the atomic collapse (AC) regime. The symbols represent the energies of the VP and AC peaks from panel C. (**b**) Simulated spectra for the β values in (c). Curves are vertically offset for clarity. (**c**) Evolution of STS with charge (increasing from bottom to top). Each curve is marked with its corresponding β value. *The horizontal dashed line* separates between spectra in the subcritical and supercritical regimes. *The vertical dashed line* represents the bulk DP measured far from any vacancy (From Mao et al. [29])

is clearly distinguishable from the VP by its significantly larger spatial extent. With increasing β, $R1$ develops a satellite, $R1'$, which tracks its evolution with β. While $R1$ is a universal feature of AC states, $R1'$ is due to the locally broken sublattice symmetry and is peculiar to the supercritically charged single-atom vacancy. As β further increases more branches emerge, starting with $R2$ which is the equivalent of the 2S AC state.

The form of the truncated potential, Eq. (1.10), takes the finite size of the charge into account via the parameter r_0. For the charged vacancy, based on experimental data, this parameter is set to 0.5 nm. It is smaller than the size of the impurity cluster. The smaller charge area enhances the effect making the $R2$ collapse resonance experimentally observable.

The effect of the size of the charge is further investigated in Fig. 1.7a. Reducing the size of r_0 increases the strength of the potential and in the limit $r_0 \rightarrow 0$ it corresponds to a point charge. As we move closer to this limit, the energy levels $R1$ and $R1'$ diverge: $R1$ dives to very low energy while increasing its intensity, but $R1'$ stays close to the Dirac point while losing intensity. $R1$ and $R1'$ are mostly made up of opposite sublattice states. As shown in Fig. 1.7b, sublattice A is closer to the center with a radius of a_{cc}, while the minimum radius of sublattice B is farther away at radius a. This defines the minimum radii of the $R1$ and $R1'$ resonances. Figure 1.7c shows a zoomed in view and the vertical lines correspond to the radii

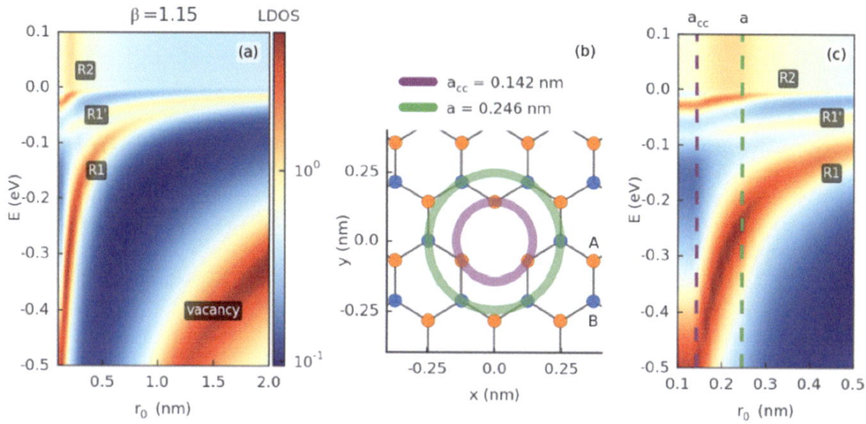

Fig. 1.7 (**a**) LDOS as function of the r_0 parameter for $\beta = 1.15$. (**b**) Illustration of the model system around the vacancy. The *circles* indicate minimum radii of sublattices A and B. (**c**) Zoomed in view of panel (**a**) close to the center. *The vertical lines* correspond to the radii from panel (**b**)

from Fig. 1.7b. Resonance $R1'$ corresponds to a state with a larger radius. It loses most of its intensity in the region where $a_{cc} < r_0 < a$. Resonance $R1$ significantly increases its intensity in the region $r_0 < a$ and saturates at a_{cc}.

1.4 Summary

Atomic collapse is a long sought-after phenomenon in quantum electrodynamics which is not observable with real atomic nuclei because of the extremely large charge requirements. Graphene offers the opportunity to observe the same physics at a much smaller charge and energy scale. As the charge is increased, the interaction undergoes a transition into a supercritical regime where electrons are trapped in a sequence of quasi-bound states which resemble an artificial atom. In order to reach the supercritical regime an artificial nucleus with sufficient charge must be constructed. This can be done by assembling a cluster of charged impurities. Alternatively a single-atom vacancy can stably host a local charge that is built up by applying voltage pulses with the tip of a scanning tunneling microscope. The quasi-bound electron states are detected as a strong enhancement of the local density of states within a disc centered on the charge.

Acknowledgements This work was supported by the European Science Foundation (ESF) under the EUROCORES Program EuroGRAPHENE within the project CONGRAN, the Flemish Science Foundation (FWO-Vl) and the Methusalem Funding of the Flemish Government. We acknowledge the very fruitful collaboration with the Eva Andrei group at Rutgers University.

References

1. Novoselov KS, Geim AK, Morozov SV, Jiang D, Zhang Y, Dubonos SV, Grigorieva IV, Firsov AA (2004) Electric field effect in atomically thin carbon films. Science 306(5696):666
2. Novoselov KS, Geim AK, Morozov SV, Jiang D, Katsnelson MI, Grigorieva IV, Dubonos SV, Firsov AA (2005) Two-dimensional gas of massless Dirac fermions in graphene. Nature 438(7065):197
3. Castro Neto AH, Peres NMR, Novoselov KS, Geim AK (2009) The electronic properties of graphene. Rev Mod Phys 81(1):109
4. Zhang Y, Tan YW, Stormer HL, Kim P (2005) Experimental observation of the quantum Hall effect and Berry's phase in graphene. Nature 438(7065):201
5. Novoselov KS, Jiang Z, Zhang Y, Morozov SV, Stormer HL, Zeitler U, Maan JC, Boebinger GS, Kim P, Geim AK (2007) Room-temperature quantum Hall effect in graphene. Science 315(5817):1379
6. Gunlycke D, Lawler H, White C (2007) Room-temperature ballistic transport in narrow graphene strips. Phys Rev B 75(8):085418
7. Katsnelson MI, Novoselov KS, Geim AK (2006) Chiral tunnelling and the Klein paradox in graphene. Nat Phys 2(9):620
8. Lee C, Wei X, Kysar JW, Hone J (2008) Measurement of the elastic properties and intrinsic strength of monolayer graphene. Science 321(5887):385
9. Peres NMR, Guinea F, Castro Neto AH (2006) Electronic properties of disordered two-dimensional carbon. Phys Rev B 73(12):125411
10. Ando T (2006) Screening effect and impurity scattering in monolayer graphene. J Phys Soc Jpn 75(7):074716
11. Hwang E, Adam S, Sarma S (2007) Carrier transport in two-dimensional graphene layers. Phys Rev Lett 98(18):186806
12. Dirac PAM (1928) The quantum theory of the electron. Proc R Soc A Math Phys Eng Sci 117(778):610
13. Pomeranchuk I, Smorodinsky Y (1945) About the energy levels of systems with $Z > 137$. J Phys 9:97
14. Zeldovich Y, Popov VS (1972) Electronic structure of superheavy atoms. Sov Phys Usp 14:673
15. Greiner W, Müller B, Rafelski J (1985) Quantum electrodynamics of strong fields. Springer, Berlin/Heidelberg
16. Schweppe J, Gruppe A, Bethge K, Bokemeyer H, Cowan T, Folger H, Greenberg J, Grein H, Ito S, Schule R, Schwalm D, Stiebing K, Trautmann N, Vincent P, Waldschmidt M (1983) Observation of a peak structure in positron spectra from U+Cm collisions. Phys Rev Lett 51(25):2261
17. Cowan T, Backe H, Begemann M, Bethge K, Bokemeyer H, Folger H, Greenberg J, Grein H, Gruppe A, Kido Y, Klüver M, Schwalm D, Schweppe J, Stiebing K, Trautmann N, Vincent P (1985) Anomalous positron peaks from supercritical collision systems. Phys Rev Lett 54(16):1761
18. Kotov VN, Uchoa B, Pereira VM, Guinea F, Castro Neto AH (2012) Electron–electron interactions in graphene: current status and perspectives. Rev Mod Phys 84(3):1067
19. Das Sarma S, Adam S, Hwang EH, Rossi E (2011) Electronic transport in two-dimensional graphene. Rev Mod Phys 83(2):407
20. Khalilov VR, Ho CL (1998) Dirac electron in a Coulomb field in 2+1 dimensions. Mod Phys Lett A 13(08):615
21. Pereira VM, Nilsson J, Castro Neto AH (2007) Coulomb impurity problem in graphene. Phys Rev Lett 99(16):166802
22. Shytov AV, Katsnelson MI, Levitov LS (2007a) Atomic collapse and quasi–Rydberg states in graphene. Phys Rev Lett 99(24):246802
23. Shytov AV, Katsnelson MI, Levitov LS (2007b) Vacuum polarization and screening of supercritical impurities in graphene. Phys Rev Lett 99(23):236801

24. Wang Y, Brar VW, Shytov AV, Wu Q, Regan W, Tsai HZ, Zettl A, Levitov LS, Crommie MF (2012) Mapping Dirac quasiparticles near a single Coulomb impurity on graphene. Nat Phys 8(9):653
25. Wang Y, Wong D, Shytov AV, Brar VW, Choi S, Wu Q, Tsai HZ, Regan W, Zettl A, Kawakami RK, Louie SG, Levitov LS, Crommie MF (2013) Observing atomic collapse resonances in artificial nuclei on graphene. Science 340(6133):734
26. DiVincenzo DP, Mele EJ (1984) Self-consistent effective-mass theory for intralayer screening in graphite intercalation compounds. Phys Rev B 29(4):1685
27. Luican-Mayer A, Kharitonov M, Li G, Lu C, Skachko I, Gonçalves AMB, Watanabe K, Taniguchi T, Andrei EY (2014) Screening charged impurities and lifting the orbital degeneracy in graphene by populating Landau levels. Phys Rev Lett 112(3):036804
28. Liu Y, Weinert M, Li L (2015) Determining charge state of graphene vacancy by noncontact atomic force microscopy and first-principles calculations. Nanotechnology 26(3):035702
29. Mao J, Jiang Y, Moldovan D, Li G, Watanabe K, Taniguchi T, Masir MR, Peeters FM, Andrei EY (2015) Realization of a tunable artificial atom at a charged vacancy in graphene. arXiv:1508.07667
30. Lehtinen O, Kotakoski J, Krasheninnikov AV, Tolvanen A, Nordlund K, Keinonen J (2010) Effects of ion bombardment on a two-dimensional target: atomistic simulations of graphene irradiation. Phys Rev B 81(15):153401
31. Pereira VM, Lopes dos Santos JMB, Castro Neto AH (2008) Modeling disorder in graphene. Phys Rev B 77(11):115109
32. Pereira VM, Guinea F, Lopes dos Santos JMB, Peres NMR, Castro Neto AH (2006) Disorder induced localized states in graphene. Phys Rev Lett 96(3):036801
33. Ding F (2005) Theoretical study of the stability of defects in single-walled carbon nanotubes as a function of their distance from the nanotube end. Phys Rev B – Condens Matter Mater Phys 72(24):1
34. Repp J (2004) Controlling the charge state of individual gold adatoms. Science 305(5683):493
35. Zhao AD, Li QX, Chen L, Xiang HJ, Wang WH, Pan S, Wang B, Xiao XD, Yang JL, Hou JG, Zhu QS (2005) Controlling the Kondo effect of an adsorbed magnetic ion through its chemical bonding. Science 309:1542

Chapter 2
Fluorination Clusters on Graphene Resolved by Conductive AFM

A. Mishchenko, A. Eckmann, I.V. Grigorieva, and K.S. Novoselov

Abstract Partially fluorinated graphene samples were studied by conductive AFM. The conductivity is found to be strongly non-uniform with low conductivity patches of about 10 nm. We interpret this observation as formation of low conducting clusters of fluorinated graphene separated by areas of pristine material. This is the first direct observation of clustering of covalently bonded impurities on graphene surface.

2.1 Introduction

Graphene has opened a floodgate for other 2D materials to be discovered and studied [1, 2]. One of the possible ways to create new two-dimensional (2D) crystals is to use graphene as a molecular scaffolding and attach other species to it. Several chemical modifications of graphene has been realised up to date, including graphane (with hydrogen atoms attached to each carbon) [3] or fluorographene [4]. The latter is of particular interest for applications as fluorographene is stable, has a large band gap, mechanically strong and optically transparent.

Partially fluorinated graphene can also be of significant interest. Thus, it was recently shown that partial fluorination induces paramagnetism in graphene [5]. Fitting of the magnetisation data with Brillouin function led to the conclusion that number of paramagnetic centres is three orders of magnitude less than the measured number of fluorine adatoms in the samples. To explain this surprising result authors suggested that fluorine atoms on graphene form clusters due to low migration barriers for fluorine adatoms [6–8] as well as enhanced chemical activity of ripples in graphene [6, 7]. However, no direct observations of such clusters were reported so far.

In this work we employed conductive atomic force microscopy (CAFM) to visualise fluorine clusters on partially fluorinated graphene. This technique allows one to measure the local conductivity of a sample. Since large gap in the electronic

A. Mishchenko (✉) • A. Eckmann • I.V. Grigorieva • K.S. Novoselov
School of Physics and Astronomy, University of Manchester, Manchester M13 9PL, UK
e-mail: artem.mishchenko@gmail.com

J. Bonča, S. Kruchinin (eds.), *Nanomaterials for Security*, NATO Science for Peace and Security Series A: Chemistry and Biology, DOI 10.1007/978-94-017-7593-9_2

spectrum has been observed in the fluorinated graphene [4], the fluorinated patches are expected to have low conductivity, thus, showing themselves in the conductivity maps.

2.2 Experimental

Graphene flakes were exfoliated [9] on oxygen plasma cleaned quartz substrate by micromechanical cleavage and fluorinated by exposure to XeF_2 crystals at room temperature under argon for a few hours [4]. At room temperatures such relatively short exposure results only in partial fluorination. Quality of pristine graphene flakes and extent of fluorination was monitored by Raman spectroscopy. After fluorination Cr/Au (4 nm/50 nm) contacts were deposited through shadow mask (to avoid contamination of the surface by the resist) by electron-beam evaporation.

Topography and current images were obtained with AFM Nanoscope Dimension V (Bruker) in contact mode with conductive Pt/Ir coated cantilevers PPP-CONTPt (Nanosensors). Current was measured at fixed bias of 0.1 V applied to the tip via 2400 Source/Meter (Keithey). Fluorinated graphene (and control samples of pristine graphene) was grounded through 1 MΩ limiting resistor and voltage drop across this resistor measured by 2184 A Nanovoltmeter (Keithey) was fed back to AFM. Images were obtained in ambient conditions with scan rate 0.2 Hz and force applied $\sim$5–10 nN (ca. 60 nN together with capillary force).

2.3 Results

The scheme of experimental setup is shown on Fig. 2.1a. Topographic and current images for partially fluorinated graphene are presented on the Fig. 2.1b, d, respectively. For comparison, current map of pristine graphene is shown on Fig. 2.1c. Topographic images of partially fluorinated graphene show height variation on the order of 1 nm (Fig. 2.1b) in good agreement with STM imaging results obtained for pristine graphene on SiO_2 substrate [10]. Lateral dimensions of observed corrugations (as estimated from height-height correlation analysis) were of the order of 50 nm, in good agreement with STM investigation of graphene on SiO_2 substrate [11]. These corrugations are larger and higher than expected for graphene ripples, and most probably determined by the underlying quartz substrate [10, 11].

Conductivity maps of pristine graphene do not show any sharp features, Fig. 2.1c. Typical resistance between the tip and graphene is of the order of 100 kΩ and the local resistance variations are of the order of 1–2 %. As the observed resistance is much larger than the typical sheet resistance of graphene (6 kΩ) we conclude that it is limited by the tunnelling resistance between the tip and graphene. At the same time, no significant dependence on the force between the cantilever and the surface has been observed – in line with the previous investigations [12, 13].

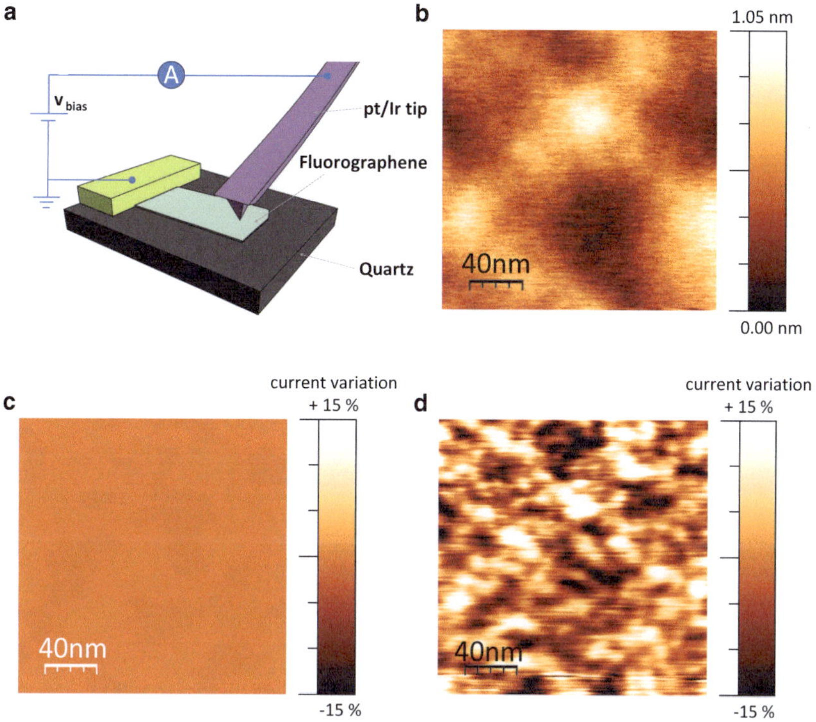

Fig. 2.1 Scheme of conductive AFM setup (**a**), partially fluorinated graphene, topography image (**b**), pristine graphene, current map (**c**), partially fluorinated graphene, current map (**d**)

Typical resistance of the fluorinated samples was much higher – in the region of several MΩ (depending on the level of fluorination). Current images also show fine structure (thus, for the sample presented on Fig. 2.1d – resistance variation from 4 to 11 MΩ). Degree of fluorination in this case was low, since sample was still conductive (low resistance features were 3–5 MΩ), which corresponds well to results obtained for lateral transport in partially fluorinated samples [4]. The lateral scale of the conductivity variations is much smaller than the typical undulations in topography, Fig. 2.1d (for clarity, current images were flattened and equalized).

Because in our experiment graphene was supported on the substrate, fluorination was most probably one-sided. In the case of one-side functionalization of graphene the disordered fluorination should be favoured by the presence of ripples and substrate-induced corrugations in graphene [6]. We interpret the low-conductivity parts to be fluorinated patches and the high-conductivity regions to be the areas of pristine graphene. The fact that even the patches of pristine graphene are still of relatively low conductivity is explained by the fact that we are measuring not only the local conductivity, but also the conductivity of the whole fluorinated flake between the tip and the metal contact. The latter is limited by the resistance of the percolating network of pristine graphene among the fluorinated patches. Also, the

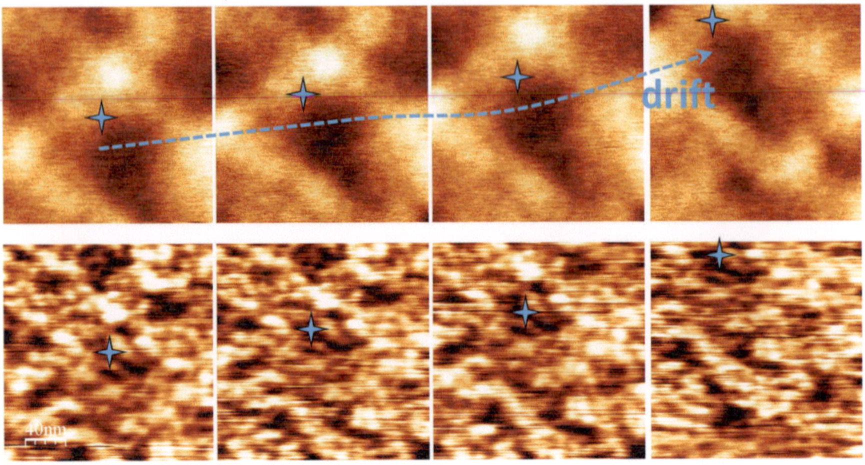

Fig. 2.2 Several images of partially fluorinated graphene, obtained sequentially: topography (*top row*) and current maps (*bottom row*)

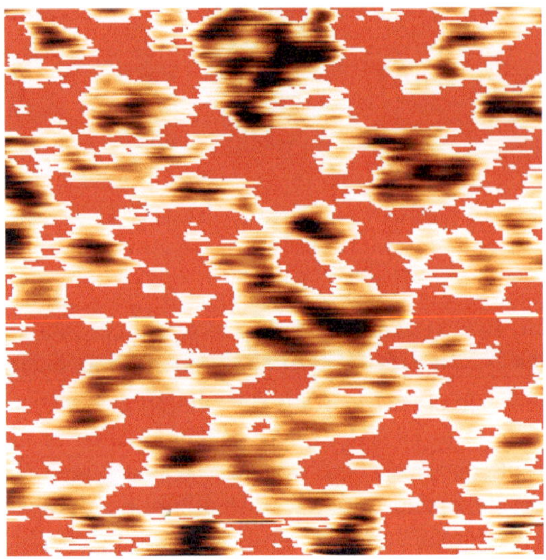

Fig. 2.3 Representative current map after performing "flooding" analysis in WSXM package [15]

resistance of the narrow areas of pristine graphene is likely to be limited by the size quantisation – similarly as for the case of graphene nanoribbons [14].

Although there is a weak correlation between topographic image and contrast in current map at large scale (30–60 nm), Fig. 2.1b, d, size and shape of high resistance clusters on the current map are completely different from topography image. Several images, obtained sequentially, show the same features, Fig. 2.2, thus excluding contributions of electrical or mechanical noise to this fine structure. From flooding analysis of several images (example on Fig. 2.3) first-neighbour distance between low-conductance clusters was estimated to be 9.1 ± 0.6 nm. This value corresponds well to the size of ripples in graphene [16] and to the size of fluorographene clusters calculated from magnetization measurements [5].

2.4 Conclusion

In conclusion: we confirmed experimentally clustering of adatoms on graphene, induced by local curvature of graphene. This opens up possibility to perform spatially-controlled functionalization of graphene, which, together with strain engineering could lead to variety of nanoelectronics applications [17, 18].

Acknowledgements This work was supported by the European Research Council, European Commision FP7, Engineering and Physical Research Council (UK), the Royal Society, U.S. Office of Naval Research, U.S. Air Force Office of Scientific Research, and the Körber Foundation. A.M. aknowledges support from the EPSRC Early Career Fellowship.

References

1. Novoselov KS, Jiang D, Schedin F, Booth TJ, Khotkevich VV, Morozov SV, Geim AK (2005) Two-dimensional atomic crystals. Proc Natl Acad Sci U S A 102:10451–10453
2. Novoselov KS (2011) Nobel Lecture: graphene: materials in the flatland. Rev Mod Phys 83:837–849
3. Elias DC, Nair RR, Mohiuddin TMG, Morozov SV, Blake P, Halsall MP, Ferrari AC, Boukhvalov DW, Katsnelson MI, Geim AK, Novoselov KS (2009) Control of graphene's properties by reversible hydrogenation: evidence for graphane. Science 323:610–613
4. Nair RR, Ren W, Jalil R, Riaz I, Kravets VG, Britnell L, Blake P, Schedin F, Mayorov AS, Yuan S, Katsnelson MI, Cheng H-M, Strupinski W, Bulusheva LG, Okotrub AV, Grigorieva IV, Grigorenko AN, Novoselov KS, Geim AK (2010) Fluorographene: a two-dimensional counterpart of Teflon. Small 6:2877–2884
5. Nair RR, Sepioni M, Tsai IL, Lehtinen O, Keinonen J, Krasheninnikov AV, Thomson T, Geim AK, Grigorieva IV (2012) Spin-half paramagnetism in graphene induced by point defects. Nat Phys 8:199–202. doi:10.1038/nphys2183
6. Boukhvalov DW, Katsnelson MI (2009) Enhancement of chemical activity in corrugated graphene. J Phys Chem C 113:14176–14178
7. Sl O, Torrent-Sucarrat M, Solà M, Geerlings P, Ewels CP, Lier GV (2010) Reaction mechanisms for graphene and carbon nanotube fluorination. J Phys Chem C 114:3340–3345
8. Şahin H, Topsakal M, Ciraci S (2011) Structures of fluorinated graphene and their signatures. Phys Rev B 83:115432
9. Novoselov KS, Geim AK, Morozov SV, Jiang D, Zhang Y, Dubonos SV, Grigorieva IV, Firsov AA (2004) Electric field effect in atomically thin carbon films. Science 306:666–669
10. Stolyarova E, Rim KT, Ryu S, Maultzsch J, Kim P, Brus LE, Heinz TF, Hybertsen MS, Flynn GW (2007) High-resolution scanning tunneling microscopy imaging of mesoscopic graphene sheets on an insulating surface. Proc Natl Acad Sci 104:9209–9212
11. Ishigami M, Chen JH, Cullen WG, Fuhrer MS, Williams ED (2007) Atomic structure of graphene on SiO2. Nano Lett 7:1643–1648
12. Lee GH, Yu YJ, Lee C, Dean C, Shepard KL, Kim P, Hone J (2011) Electron tunneling through atomically flat and ultrathin hexagonal boron nitride. Appl Phys Lett 99:243114

13. Britnell L, Gorbachev RV, Jalil R, Belle BD, Schedin F, Katsnelson MI, Eaves L, Morozov SV, Mayorov AS, Peres NMR, Castro Neto AH, Leist J, Geim AK, Ponomarenko LA, Novoselov KS (2012) Electron tunneling through ultrathin boron nitride crystalline barriers. Nano Lett 12:1707
14. Han MY, Ozyilmaz B, Zhang YB, Kim P (2007) Energy band-gap engineering of graphene nanoribbons. Phys Rev Lett 98:206805
15. Horcas I, Fernandez R, Gomez-Rodriguez JM, Colchero J, Gomez-Herrero J, Baro AM (2007) WSXM: a software for scanning probe microscopy and a tool for nanotechnology. Rev Sci Instrum 78:013705
16. Meyer JC, Geim AK, Katsnelson MI, Novoselov KS, Booth TJ, Roth S (2007) The structure of suspended graphene sheets. Nature 446:60–63
17. Guinea F, Katsnelson MI, Geim AK (2010) Energy gaps and a zero-field quantum Hall effect in graphene by strain engineering. Nat Phys 6:30–33
18. Severin N, Dorn M, Kalachev A, Rabe JP (2011) Replication of single macromolecules with graphene. Nano Lett 11:2436–2439

Chapter 3
Spin Relaxation in GaAs Based Quantum Dots for Security and Quantum Information Processing Applications

S. Prabhakar and R. Melnik

Abstract We report new three-dimensional modeling results of the band structure calculation of GaAs/$Al_{0.3}Ga_{0.7}As$ quantum dots (QDs) in presence of externally applied magnetic and electric fields along z-direction. We explore the influence of spin-orbit coupling in the effective g-factor of electrons in such QDs for possible application in security devices, encrypted data and quantum information processing. We estimate the relaxation rate in QDs caused by piezo-phonons.

3.1 Introduction

Several research proposals for the design of robust semiconductor spintronic devices for quantum logic gates in quantum information processing and security applications is based on the accurate manipulation of the effective g-factor of electrons with electric fields and estimation of the spin relaxation rate [14–16, 20]. One may expect a larger spin relaxation or decohenrence time than the gate operation time for the possible implementation of quantum dots devices in quantum information processing [1, 2]. Authors in Refs. [6, 11] have measured long spin relaxation times of 0.85 ms in GaAs QDs by pulsed relaxation rate measurements and 20 ms in InGaAs QDs by optical orientation measurements. These experimental studies in QDs confirm that the manipulation of spin-flip rates by spin-orbit coupling is important for the design of spintronics logic and other devices [7, 9, 16].

The spin-orbit coupling consists of the Rashba [3] and the linear Dresselhaus [5] terms that arise from structural inversion asymmetry along the growth direction and the bulk inversion asymmetry of the crystal lattice [1, 2, 24]. The electric and magnetic fields tunability of the electron g-factor in gated III-V semiconductor QDs with the Rashba and Dresselhaus spin-orbit couplings were explored in

---

S. Prabhakar (✉) • R. Melnik
MS2Discovery Interdisciplinary Research Institute, M ^{2}NeT Laboratory, Wilfrid Laurier University, Waterloo, ON, N2L 3C5 Canada
e-mail: sprabhakar@wlu.ca; rmelnik@wlu.ca

J. Bonča, S. Kruchinin (eds.), *Nanomaterials for Security*, NATO Science for Peace and Security Series A: Chemistry and Biology, DOI 10.1007/978-94-017-7593-9_3

Refs. [12, 15, 22–24]. It has also been noted that in-plane anisotropy due to gate potential also influences the effective Landé g-factor of electrons in QDs [15, 17, 27]. In this paper, we consider three-dimensional $GaAs/Al_{0.3}Ga_{0.7}As$ conical QDs that are epitaxially formed at the heterojunction. We study the variation in the g-factor of electrons with electric and magnetic fields. We also estimate the phonon mediated spin-flip rate of electron spin states.

The paper is organized as follows. In Sect. 3.2, we provide a description of the theoretical model of three-Dimensional semiconductor QDs in presence of externally applied electric and magnetic fields. In Sect. 3.3, we give details of the diagonalization technique used for finding the eigenvalues and eigenstates for electrons in QDs. In Sect. 3.4, we discuss the dependency of the effective g-factor and spin relaxation rate with magnetic fields. Finally, in Sect. 3.5, we summarize our results.

3.2 Theoretical Model

We consider a $GaAs/Al_{0.3}Ga_{0.7}As$ QD epitaxially formed at the heterojunction. In presence of magnetic field applied along z-direction, we write the total Hamiltonian of this quantum dot as: [1, 9, 19, 21, 24]

$$H = H_0 + H_R + H_D, \tag{3.1}$$

where H_R and H_D are Hamiltonians associated with the Rashba and Dresselhaus spin-orbit couplings and

$$H_0 = \frac{\mathbf{P}^2}{2m} + V(x, y, z) + eE_z z + \frac{1}{2} g_o \mu_B \sigma_z B, \tag{3.2}$$

where $\mathbf{P} = \mathbf{p} + e\mathbf{A}$ is the kinetic momentum operator, $\mathbf{p} = -i\hbar(\partial_x, \partial_y, \partial_z)$ is the canonical momentum operator, $\mathbf{A} = B(-y, x, 0)$ is the vector potential, m is the effective mass of the electron in the conduction band, μ_B is the Bohr magneton and $\boldsymbol{\sigma} = (\sigma_x, \sigma_y, \sigma_z)$ are the Pauli spin matrices. Also, $V(x, y, z)$ is the confinement potential which is zero in the GaAs region and 0.3 eV in the barrier material, and E_z is the applied external field along z-direction. The Hamiltonians associated with the Rashba and Dresselhaus spin-orbit couplings can be written as [3, 5, 24]

$$H_R = \frac{\alpha_R}{\hbar}\left(\sigma_x P_y - \sigma_y P_x\right), \tag{3.3}$$

$$H_D = \frac{\alpha_D}{\hbar}\left(-\sigma_x P_x + \sigma_y P_y\right), \tag{3.4}$$

where the strengths of the Rashba and Dresselhaus spin-orbit couplings are characterized by the parameters α_{R} and α_{D}. They are given by

$$\alpha_{\mathrm{R}} = \gamma_{\mathrm{R}} eE, \quad \alpha_{\mathrm{D}} = 0.78\gamma_{\mathrm{D}} \left(\frac{2me}{\hbar^2}\right)^{2/3} E^{2/3}, \tag{3.5}$$

where γ_{R} and γ_{D} are the Rashba and Dresselhaus coefficients. Now we write $H|\Psi\rangle = \varepsilon|\Psi\rangle$ in terms of a coupled eigenvalue problem, consisting of two equations, in the basis states $|\psi_1\rangle$ and $|\psi_2\rangle$ as

$$\begin{aligned}&\Big[-\frac{\hbar^2}{2m}\left(\partial_x^2+\partial_y^2+\partial_z^2\right)+\frac{1}{8}m\omega_c^2\left(x^2+y^2\right)-\frac{i\hbar\omega_c}{2}\left(-y\partial_x+x\partial_y\right)+V\left(x,y,z\right)\\&+eE_z z+\frac{1}{2}g_0\mu_{\mathrm{B}}B\sigma_z\Big]|\psi_1\rangle+\Big[\alpha_{\mathrm{R}}eE_z\{\partial_x-i\partial_y+\frac{m\omega_c}{2\hbar}\left(x-iy\right)\},\\&+\alpha_{\mathrm{D}}\Big\{i\partial_x-\partial_y+\frac{m\omega_c}{2\hbar}\left(y-ix\right)\Big\}\Big]|\psi_2\rangle=\varepsilon|\psi_1\rangle,\end{aligned} \tag{3.6}$$

$$\begin{aligned}&\Big[-\frac{\hbar^2}{2m}\left(\partial_x^2+\partial_y^2+\partial_z^2\right)+\frac{1}{8}m\omega_c^2\left(x^2+y^2\right)-\frac{i\hbar\omega_c}{2}\left(-y\partial_x+x\partial_y\right)+V\left(x,y,z\right)\\&+eE_z z+\frac{1}{2}g_0\mu_{\mathrm{B}}B\sigma_z\Big]|\psi_2\rangle+\Big[\alpha_{\mathrm{R}}eE_z\{-\partial_x+i\partial_y+\frac{m\omega_c}{2\hbar}\left(x+iy\right)\},\\&+\alpha_{\mathrm{D}}\Big\{-i\partial_x+\partial_y+\frac{m\omega_c}{2\hbar}\left(y+ix\right)\Big\}\Big]|\psi_1\rangle=\varepsilon|\psi_2\rangle,\end{aligned} \tag{3.7}$$

where $\omega_c = eB/m$.

We now turn to the calculation of the phonon induced spin relaxation rate at absolute zero temperature. Following Ref. [18], the interaction between electrons and piezo-phonons can be written as [9, 10, 13, 29]

$$u_{ph}^{\mathbf{q}\alpha}\left(\mathbf{r},t\right) = \sqrt{\frac{\hbar}{2\rho V\omega_{\mathbf{q}\alpha}}}e^{i\left(\mathbf{q}\cdot\mathbf{r}-\omega_{q\alpha}t\right)}eA_{\mathbf{q}\alpha}b_{\mathbf{q}\alpha}^{\dagger} + \mathrm{H.c.} \tag{3.8}$$

Here, ρ is the crystal mass density, V is the volume of the QDs, $b_{\mathbf{q}\alpha}^{\dagger}$ creates an acoustic phonon with wave vector $\mathbf{q}$ and polarization $\hat{e}_\alpha$, where $\alpha = l, t_1, t_2$ are chosen as one longitudinal and two transverse modes of the induced phonon in the dots. Also, $A_{\mathbf{q}\alpha} = \hat{q}_i\hat{q}_k e\beta_{ijk}e_{\mathbf{q}\alpha}^j$ is the amplitude of the electric field created by phonon strain, where $\hat{\mathbf{q}} = \mathbf{q}/q$ and $e\beta_{ijk} = eh_{14}$ for $i \neq k, i \neq j, j \neq k$. The polarization directions of the induced phonon are $\hat{e}_l = (\sin\theta\cos\phi, \sin\theta\sin\phi, \cos\theta)$,

$\hat{e}_{t_1} = (\cos\theta\cos\phi, \cos\theta\sin\phi, -\sin\theta)$ and $\hat{e}_{t_2} = (-\sin\phi, \cos\phi, 0)$. Based on the Fermi Golden Rule, the phonon induced spin transition rate in the QDs is given by [10, 24]

$$\frac{1}{T_1} = \frac{2\pi}{\hbar} \int \frac{d^3\mathbf{q}}{(2\pi)^3} \sum_{\alpha=l,t} |M(\mathbf{q}\alpha)|^2 \delta\left(\hbar s_\alpha \mathbf{q} - \varepsilon_f + \varepsilon_i\right), \tag{3.9}$$

where s_l,s_t are the longitudinal and transverse acoustic phonon velocities in QDs. Under the dipole approximation, we write (3.9) as [10, 25]

$$\frac{1}{T_1} = c\left(|M_x|^2 + |M_y|^2 + |M_z|^2\right), \tag{3.10}$$

where

$$c = \frac{2\,(eh_{14})^2\left(\varepsilon_f - \varepsilon_i\right)^3}{35\pi\hbar^4\rho}\left(\frac{1}{s_l^5} + \frac{4}{3}\frac{1}{s_t^5}\right), \tag{3.11}$$

$$M_x = \langle\psi_i|x|\psi_f\rangle, \tag{3.12}$$

$$M_y = \langle\psi_i|y|\psi_f\rangle, \tag{3.13}$$

$$M_z = \langle\psi_i|z|\psi_f\rangle. \tag{3.14}$$

In the above expression, we use $c = c_l I_{xl} + 2c_t I_{xt}$, where $c_\alpha = \frac{q^2e^2}{(2\pi)^2\hbar^2 s_\alpha}|\varepsilon_{q\alpha}|^2$, $|\varepsilon_{q\alpha}|^2 = \frac{q^2\hbar}{2\rho\omega_{q\alpha}}$. For longitudinal phonon modes similar to Refs. [7, 10], we have $|A_{q,l}|^2 = 36h_{14}^2\cos^2\theta\sin^4\theta\sin^2\phi\cos^2\phi$. Likewise, for transverse phonon modes, we have $|A_{q,t}|^2 = 2h_{14}^2[\cos^2\theta\sin^2\theta + \sin^4\theta\left(1 - 9\cos^2\theta\right)\sin^2\phi\cos^2\phi]$.

3.3 Computational Method

The geometry of GaAs conical QDs with wetting layer is shown in Fig. 3.1. The QDs are surrounded by host barrier material $Al_{0.3}Ga_{0.7}As$. We diagonalize the total Hamiltonian H using the finite-element method [4]. In typical examples reported here, the geometry contains on the order of 56729 elements. We impose Dirichlet boundary conditions at the outside boundaries to let the wavefunction to vanish and Neumann boundary conditions at the internal boundaries to let the wavefunction to follow continuity equation. Then we find eigenvalues and eigenfunctions by solving two coupled equations (3.6) and (3.7).

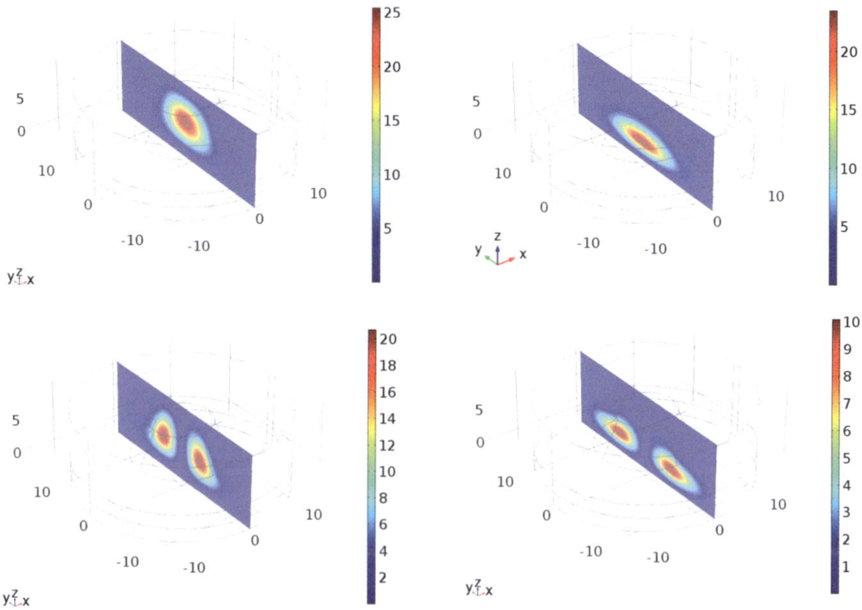

Fig. 3.1 Illustration of ground state wavefunctions (*upper panel*) and first excited state wavefunctions (*lower panel*) of electrons in GaAs/AlGaAs QDs. *Left column* QD wavefunctions correspond to the externally applied electric field, $E_z = 0.001$ V/nm, while *right column* QD wavefunctions correspond to the externally applied electric field, $E_z = 0.1$ V/nm. Evidently, penetration of wavefunctions from the QD region to the barrier materials leads to the variation in the Lande g-factor of electrons that can be utilized to design spintronic devices for security and quantum information processing

3.4 Results and Discussions

In Fig. 3.1, we have plotted probability distribution, $|\psi_1|^2 + |\psi_2|^2$ (arbitrary unit) of ground (upper panel) and first excited state (lower panel) wavefunctions of electron at $E_z = 0.001$ V/nm (left panel) and $E_z = 0.1$ V/nm (right panel). As we can see, the penetration of electron wavefunctions into the barrier material is enhanced for larger values of the applied electric field along z-direction. This leads to the variation in the effective g-factor (see also Fig. 3.3) of electrons in $GaAs/Al_{0.3}Ga_{0.7}As$ QDs that can be utilized to design quantum logic gates for applications in security and quantum information processing. In Fig. 3.2, we have plotted a cross section along z-direction of the edge potential and probability distribution ($|\psi_1|^2 + |\psi_2|^2$) of ground state

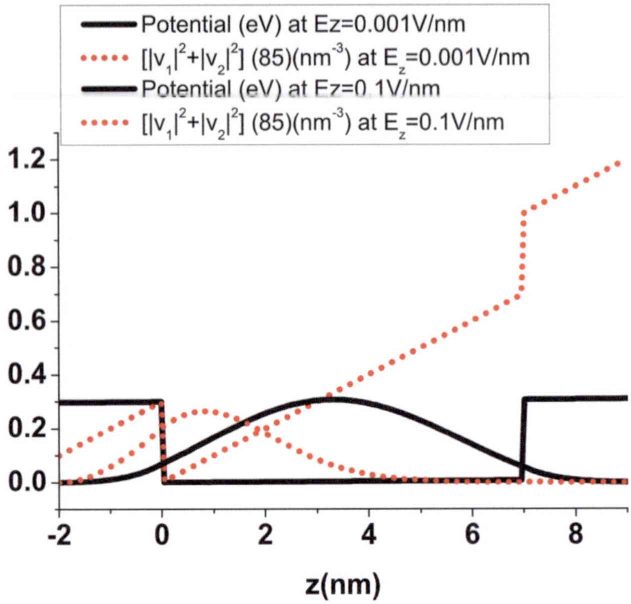

Fig. 3.2 Conduction band edge and its corresponding normalized wavefunctions vs distance from $GaAs/Al_{0.3}Ga_{0.7}As$. Again, penetration of wavefunctions in the barrier material at large electric fields can be seen (also see Ref. [8])

wavefunctions of electrons at $E_z = 0.001\,V/nm$ (left panel) and $E_z = 0.1\,V/nm$ (right panel). Here we again see that the penetration of electron wavefunctions into the barrier material is enhanced for larger values of the applied electric field along z-direction (for experiment results, see Ref. [8]). In Fig. 3.3 (upper panel), we have plotted the effective g-factor, $g = (\varepsilon_1 - \varepsilon_2)/\mu_B B$ of electron vs magnetic fields. Here, the variation of effective g-factor of electrons in quantum dots with electric field can be seen and can be applied to make quantum logic gates for application in spintronic devices. At large magnetic fields, we find the level crossing due to admixture of spin and orbital states (see lower panel of Fig. 3.3). In Fig. 3.4, we have plotted piezo-phonon mediated spin and orbital relaxation vs magnetic fields in $GaAs/Al_{0.3}Ga_{0.7}As$ QDs at $E_z = 0.1\,V/nm$. We see that the relaxation rate is enhanced with magnetic fields and can approach to the orbital relaxation rate at large magnetic fields where the decoherence time is greatly reduced due to the level crossing of orbital and spin states. Such an ideal location should be avoided during the design of spintronic devices for application in security and quantum information processing (Table 3.1).

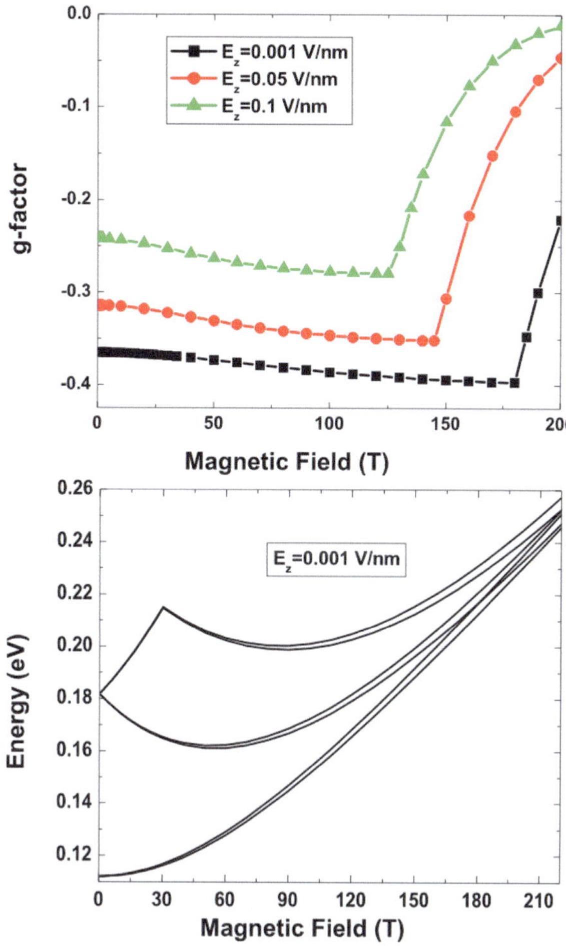

Fig. 3.3 (*upper*) Electron effective g-factor vs magnetic field in GaAs/$Al_{0.3}Ga_{0.7}As$ QDs. Level crossing can be observed at large magnetic fields which vary with the applied external electric fields along z-direction. (*lower*) Band diagram of electron in GaAs/$Al_{0.3}Ga_{0.7}As$ QDs vs magnetic field. The band crossing is seen at large magnetic field, $B \approx 185\,\text{T}$

3.5 Conclusion

Based on finite element implementation, we have provided three-dimensional modeling results for the tuning of the effective g-factor of electrons with spin orbit coupling in GaAs/$Al_{0.3}Ga_{0.7}As$ QDs and shown that the level crossing can be observed at large magnetic fields due to the admixture of spin and orbit states. We also estimated the relaxation rate caused by piezo-phonon and shown that the spin-hot spot can be observed at the level crossing point. Decoherence time is greatly reduced at the level crossing point. Thus, We suggest to avoid such an ideal location in the design of GaAs based QD devices for possible applications in spintronics, security, encrypted data and quantum information processing.

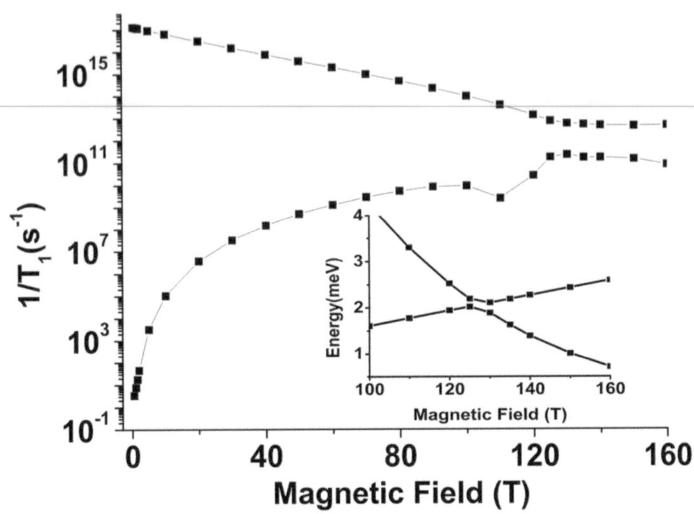

Fig. 3.4 Piezo-phonon mediated spin relaxation vs magnetic fields in GaAs/$Al_{0.3}Ga_{0.7}As$ QDs at $E_z = 0.1$ V/nm

Table 3.1 The material constants used in our calculations are taken from Refs. [19, 26, 28]

Parameters	GaAs	$Al_{0.3}Ga_{0.7}As$
g_0	−0.44	0.4
m	$0.067m_0$	$0.088m_0$
α_R [nm^2]	0.044	0.022
α_D [eVnm^3]	0.026	0.0076
eh_{14} [10^{-5} erg/cm]	2.34	0.54
s_l [10^5 cm/s]	5.14	
s_t [10^5 cm/s]	3.03	
ρ [g/cm^3]	5.3176	

References

1. Bulaev DV, Loss D (2005) Spin relaxation and decoherence of holes in quantum dots. Phys Rev Lett 95:076805
2. Bulaev DV, Loss D (2005) Spin relaxation and anticrossing in quantum dots: Rashba versus dresselhaus spin-orbit coupling. Phys Rev B 71:205324
3. Bychkov YA, Rashba EI (1984) Oscillatory effects and the magnetic susceptibility of carriers in inversion layers. J Phys C: Solid State Phys 17:6039
4. Comsol Multiphysics version 5.1. www.comsol.com
5. Dresselhaus G (1955) Spin-orbit coupling effects in zinc blende structures. Phys Rev 100:580
6. Elzerman JM, Hanson R, Willems van Beveren LH, Witkamp B, Vandersypen LMK, Kouwenhoven LP (2004) Single-shot read-out of an individual electron spin in a quantum dot. Nature 430:431
7. Golovach VN, Khaetskii A, Loss D (2004) Phonon-induced decay of the electron spin in quantum dots. Phys Rev Lett 93:016601

8. Jiang HW, Yablonovitch E (2001) Gate-controlled electron spin resonance in GaAs/Al_xGa_{1-x}As heterostructures. Phys Rev B 64:041307
9. Khaetskii AV, Nazarov YV (2000) Spin relaxation in semiconductor quantum dots. Phys Rev B 61:12639
10. Khaetskii AV, Nazarov YV (2001) Spin-flip transitions between zeeman sublevels in semiconductor quantum dots. Phys Rev B 64:125316
11. Kroutvar M, Ducommun Y, Heiss D, Bichler M, Schuh D, Abstreiter G, Finley, JJ (2004) Optically programmable electron spin memory using semiconductor quantum dots. Nature 432:81
12. Nowak MP, Szafran B, Peeters FM, Partoens B, Pasek WJ (2011) Tuning of the spin-orbit interaction in a quantum dot by an in-plane magnetic field. Phys Rev B 83:245324
13. Olendski O, Shahbazyan TV (2007) Theory of anisotropic spin relaxation in quantum dots. Phys Rev B 75:041306
14. Prabhakar S, Melnik R (2015) Electric field control of spin splitting in III–V semiconductor quantum dots without magnetic field. Eur Phys J B 88:273
15. Prabhakar S, Raynolds JE (2009) Gate control of a quantum dot single-electron spin in realistic confining potentials: anisotropy effects. Phys Rev B 79:195307
16. Prabhakar S, Raynolds J, Inomata A, Melnik R (2010) Manipulation of single electron spin in a gaas quantum dot through the application of geometric phases: the Feynman disentangling technique. Phys Rev B 82:195306
17. Prabhakar S, Raynolds JE, Melnik R (2011) Manipulation of the landé g factor in inas quantum dots through the application of anisotropic gate potentials: exact diagonalization, numerical, and perturbation methods. Phys Rev B 84:155208
18. Prabhakar S, Melnik R, Bonilla LL (2012) The influence of anisotropic gate potentials on the phonon induced spin-flip rate in GaAs quantum dots. Appl Phys Lett 100:023108
19. Prabhakar S, Melnik R, Bonilla LL (2013) Electrical control of phonon-mediated spin relaxation rate in semiconductor quantum dots: Rashba versus dresselhaus spin-orbit coupling. Phys Rev B 87:235202
20. Prabhakar S, Melnik R, Bonilla L (2014) Gate control of Berry phase in III–V semiconductor quantum dots. Phys Rev B 89:245310
21. Prabhakar S, Melnik R, Inomata A (2014) Geometric spin manipulation in semiconductor quantum dots. Appl Phys Lett 104:142411
22. Pryor CE, Flatté ME (2006) Landé g factors and orbital momentum quenching in semiconductor quantum dots. Phys Rev Lett 96:026804
23. Pryor CE, Flatté ME (2007) Erratum: Landé g factors and orbital momentum quenching in semiconductor quantum dots. Phys Rev Lett 99:179901
24. Sousa R, Sarma S (2003) Gate control of spin dynamics in III-V semiconductor quantum dots. Phys Rev B 68:155330
25. Stano P, Fabian J (2006) Orbital and spin relaxation in single and coupled quantum dots. Phys Rev B 74:045320
26. Stern F, Sarma S (1984) Electron energy levels in GaAs-$Ga_{1-x}Al_x$As heterojunctions. Phys Rev B 30:840
27. Takahashi S, Deacon RS, Yoshida K, Oiwa A, Shibata K, Hirakawa K, Tokura Y, Tarucha S (2010) Large anisotropy of the spin-orbit interaction in a single InAs self-assembled quantum dot. Phys Rev Lett 104:246801
28. von Allmen P (1992) Conduction subbands in a GaAs/Al_xGa_{1-x}As quantum well: comparing different k·p models. Phys Rev B 46:15382
29. Woods LM, Reinecke TL, Lyanda-Geller Y (2002) Spin relaxation in quantum dots. Phys Rev B 66:161318

Chapter 4
Very Sensitive Nanocalorimetry of Small Mass Systems and Glassy Materials

J.-L. Garden, A. Tavakoli, T. Nguyen-Duc, A. Frydman, M. Laarraj, J. Richard, and O. Bourgeois

Abstract Nanocalorimetry is a technique that deals with any thermal measurement methods in which either the samples to be studied have a size in the range of the nanometer scale or the measured energies involved are of the order of the nanojoule or below Garden et al (Acta 492:16–28, 2009). In this paper, we show the results of two nanocalorimetric experiments. The first one is related to the measurement of specific heat on ultra-thin small systems (thin films) at low temperature. It is shown that such measurement can be sensitive to less than one monolayer of materials. The second one illustrates the efficiency of calorimetric studies sensitive at the nanoJoule on complex system (polymeric glass) at room temperature. We then discuss the potentiality of these experimental methods in the field of security: the measurement of either a very small mass or very small quantity of energy for the detection of tiny thermal events.

4.1 Introduction

The calorimetric techniques are the experimental methods that allows the measurement of heat exchanges during a phase transition or any change in matter (temperature, pressure, volume etc.). It is a versatile method used for materials characterization that spans a wide range of domains going from low temperature physics to room temperature biology [1]. Part of the calorimetric techniques are dedicated to the measurement of the specific heat, one of the most important thermal properties of materials along with the thermal conductivity. The specific heat of a

J.-L. Garden (✉) • A. Tavakoli • T. Nguyen-Duc • M. Laarraj • J. Richard • O. Bourgeois
Institut NÉEL, CNRS, 25 avenue des Martyrs, F-38042 Grenoble, France

University of Grenoble Alpes, Institut NEEL, F-38042 Grenoble, France
e-mail: jean-luc.garden@neel.cnrs.fr; olivier.bourgeois@neel.cnrs.fr

A. Frydman
Institut NÉEL, CNRS, 25 avenue des Martyrs, F-38042 Grenoble, France

University of Grenoble Alpes, Inst NEEL, F-38042 Grenoble, France

Department of Physics, Bar-Ilan University, Ramat Gan 52900, Israel

J. Bonča, S. Kruchinin (eds.), *Nanomaterials for Security*, NATO Science for Peace and Security Series A: Chemistry and Biology, DOI 10.1007/978-94-017-7593-9_4

system reveals its intrinsic property by quantifying the amount of heat necessary to increase the temperature of the system by 1 K. Hence, nanocalorimetry is the part of calorimetry allowing the measurement of very small systems (nanometer scale) or very small quantities of heat (nanoJoule range). As an illustration, we will present in this article two sensitive nanocalorimetric measurements of the specific heat of very different systems: the first concerns the measurement of the specific heat of very thin superconducting layers (few mono-atomic layers). The second is about the measurement of very fine specific heat variation during the relaxation of a glassy polymeric system. These two examples of nanocalorimetric measurement will enlighten its general interest in the thermal detection of small systems or small thermal events.

4.2 Nanocalorimetric Measurement of Nanometric Thick Layers

The purpose of this first part is to illustrate how sensitive specific heat measurements can help detecting a very small mass change. By using ac calorimetry, we can indeed detect the heat capacity of materials in Joule per Kelvin, linearly dependent on the mass of the sample. In this experiment, based on a silicon membrane sensor [2], we demonstrate that the measurement of mono-atomic layer of materials is possible. Necessarily, measuring the specific heat of very thin film is a major experimental challenge. Indeed, it involves the measurement of very small mass samples requiring a very high sensitivity, here at low temperature. This will be made possible thanks to recent experimental developments of highly sensitive specific heat measurement equipment [3–5].

4.2.1 Experimental Set-Up

A new apparatus has been made to measure the heat capacity of quench condensed thin films. Quench condensed films are obtained by thermally evaporating materials in-situ directly on the thermal sensor [5–8] at low temperature. This is allowing the deposition of a fraction of monolayer of materials in order to test the actual sensitivity of the calorimetric cell; This has been done here for superconducting materials but has been also performed for magnetic one [9]. The system has been tested by evaporating pure Pb (99.99 %) on the membrane that is regulated at 8 K. The fabrication of the calorimetric sensor has been described elsewhere, we will only highlight its very peculiarities [2, 3, 5]. The sensitive part is composed of a thin silicon membrane (5 μm thick) on which a NbN thermometer [10] and a Cu heater have been lithographied to allow thermal measurements (see Fig. 4.1). The total heat capacity measurement that we will show below represents the heat capacity of the membrane itself along with the thermometer and heater on which the heat

Fig. 4.1 (**a**). Picture of the bottom part of the evaporation chamber. The baskets used for the quench condensation of Pb materials can be seen. (**b**). Picture of the sample holder part were the silicon membrane is located on the back of the regulation stage and the quartz crystal used for the measurement of mass sample. (**c**). A photograph of a silicon membrane after an evaporation located between the heater and the thermometer of the membrane. The electrodes used for the electrical characterization of the evaporated samples are seen on the two sides of the membrane

capacity of the sample is added. We keep the absolute values measured for the sake of simplicity.

4.2.2 Results and Discussions

In this experiment, we have done sequential evaporations and measure the heat capacity for every of them (see Fig. 4.2). The size of the sample is 3 mm by 2 mm, and hence since a monolayer (ML) of Pb represents 0.18 nm, 1 ML corresponds approximately to 10 ng. Each thermal evaporation of metal can be easily distinguished from the previous one from the C_p data. As expected, the heat capacity increases as the thickness (and the number of ML) of the thin films increases. It is important to stress here that only that experiment allows the measurement of ultra-thin layer thanks to the cold evaporation of materials. Any evaporation at room temperature would severely damage the quality of the sample rendering any data interpretation very difficult.

The table gathers the different experimental values that can be extracted from the measurement at 5.55 K for the different evaporation of Pb (see Table 4.1). The successive materials depositions (between the 5th and the 8th) show clearly the quality of the measurement. Having a sensitivity of the order of $\pm$ 0.01 nJ/K, it is

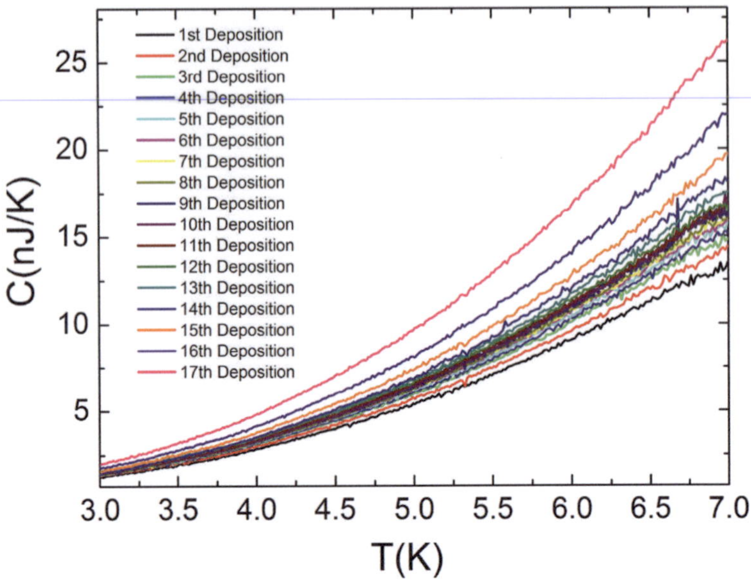

Fig. 4.2 Heat capacity measurement versus temperature for various successive evaporation of Pb on the silicon membrane. The heat capacity is continuously increasing as materials is added. Each evaporation from the 3rd to next correspond approximately to less than 1 ML

Table 4.1 This table gathers the different experimental values that can be extracted from the measurement at 5.55 K. The first column gives the number of the evaporation, the second the overall heat capacity (including the contribution of the sensor), the third column the thickness of the film, the fourth the number of ML, the fifth the mass (in microgram) and the last column shows the sensitivity in J/K

Evap. ♯	C_p (nJ/K)	t (nm)	ML	m (μg)	ΔC_p
5th	8.37	10.3	57.2	0.48	±0.01
6th	8.50	10.46	58.1	0.487	±0.01
7th	8.54	10.51	58.4	0.49	±0.01
8th	8.62	10.61	58.9	0.494	±0.01

clear that a film thickness smaller than 1 ML can be detected. To illustrate more clearly that point, we have calculated in the table the corresponding number of monolayers measured as a function of the sequential depositions. It appears that this experimental set-up is able to detect a variation of mass of less than 1 ML; basically on the order of 0.5 ML. This number is definitely beyond the state of the art in terms of highly sensitive specific heat measurement at low temperature.

Similar developments have been done using silicon nitride membrane as the mechanical support for samples instead of silicon. We have also recently demonstrated very sensitive measurement using this new sensor since mass of less than 1 ng can be detected.

4.2.3 *Conclusions*

We have developed an innovative equipment that allows the measurement of heat capacity of cold evaporated thin films that is sensitive to sub-monolayer change of materials thickness. It can be used for the study of Bi, Sn, In, Pb or any other materials that can be quench-condensed at low temperature. This experiment show the possibility of performing the C_p measurements successively by increasing the thickness of the films in-situ. Very or ultra-thin films can be measured using this experiment especially for the study of heat capacity signature of phase transition in very low dimensional systems. It is also a clear demonstration that ultra-small mass variation can be detected from a calorimetric tool that has been optimized; this is paving the way for new detection schemes based on highly sensitive thermal measurements.

4.3 Nanocalorimetric Measurement of Glassy Polymer

The glass transition is the transformation of a liquid that has been super-cooled (by high cooling rates for example) into a disordered solid called a glass. From a fundamental point of view, this transformation is still opened to questions particularly because glassy systems are outside thermodynamic equilibrium. In this case, some thermodynamic state variables of the glass depend on time. For example, the enthalpy or entropy of a glass decreases along time whether the system is maintained at constant pressure and temperature. Calorimetry is a usual method to study the glass transition. In particular, during the glass transition (vitrification) the heat capacity of the system undergoes a jump. During successive heating, the jump is recovered along with an enthalpy peak which reflects the decrease in energy that has occurred during the previous cooling. This quantity is analogous to the energy furnished to the system in order to recover the initial state (liquid state). Classical calorimetry fails to give new insights on this phenomenon partly because its lack of resolution. Here we present the results of nanocalorimetric experiments on a classical polymeric glass-former, the poly(vinyl acetate) PVAc. The main result reside in the recording of very fine decreases of the heat capacity during aging at a constant temperature below the glass transition temperature, Tg.

4.3.1 *The Nanocalorimeter*

The nanocalorimeter has been described in the reference [11]. It is composed of three different parts:

- the first one is constituted by a microsensor made by means of microfabrication technologies (see Fig. 4.3). It is composed of two half-cells, one containing a thermometer and the other containing a heater. The sample is taken as a sandwich

Fig. 4.3 Half-cell of the nanocalorimeter with patterned thin film heater in copper-nickel on a 25 μm thick polyimide membrane stuck on a copper piece

between these two half-cells ensuring good thermal contacts between sample, thermometer and heater. An oscillating thermal power is generated into the sample by the heater, and the resulting temperature oscillation is recorded with the thermometer at the same frequency;

- the second one is an home-made low noise and highly stable electronic detection chain allowing oscillating power generation along with oscillating temperature detection at the same frequency with very high sensitivity;
- the third one is an home-made thermal surroundings dedicated to the sensor and the electronic chain. It is composed of different thermally regulated shields under vacuum inside an external stage in stainless steel.

The nanocalorimetric method of detection is the so-called ac-calorimetry [12] which consisting in the recording of a temperature oscillation (amplitude and phase) with the thermometer after supplying an ac-power in the cell by means of a heater (see Fig. 4.3). The temperature oscillation is recorded versus time and numerical fast Fourier transform (FFT) provides the amplitude and phase. The measurement is carried out following different experimental protocols (cooling, heating, aging, ...) having different thermal histories. In particular, the amplitude of the sample temperature oscillations gives rise to the modulus of a complex heat capacity by means of the equation:

$$\left|C_p^*\right| = \frac{P_0}{\omega \left|T_{ac}\right|}. \tag{4.1}$$

4.3.2 Results

In Fig. 4.4, we present the modulus of the complex heat capacity and the phase of the temperature oscillation recorded with the nanocalorimeter during a heating ramp ($dT/dt = +0.2$ K/min) of the glassy polymeric sample (PVAc) undergoing

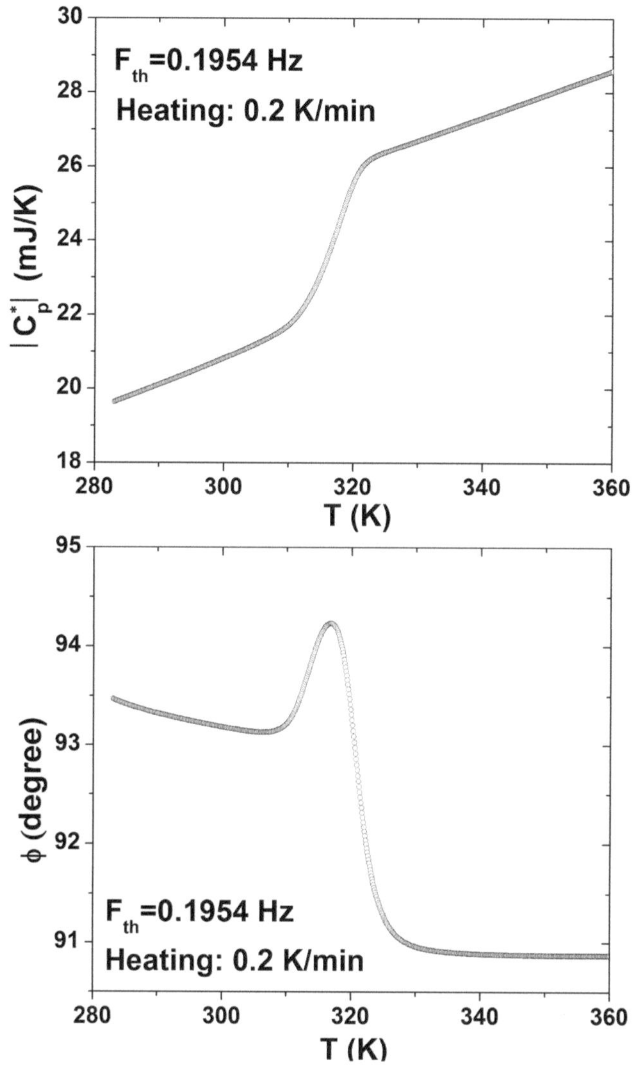

Fig. 4.4 The modulus of the complex heat capacity and the phase of the temperature oscillation are represented as a function of temperature during a heating ramp of +0.2 K/min between 283 and 360 K. The frequency of the temperature oscillation is of 0.1954 Hz with an amplitude of the order of 450 mK. The sample has been previously cooled between these two temperatures at a rate of −1.2 K/min

a glass transition. The modulus of the complex heat capacity shows a typical jump during heating due to the unfreezing of the molecular movements when the glass is transforming into a liquid. The phase exhibits a anomaly with a peak superimposed to the jump. The high resolution of the measurement is illustrated by the quality of the baselines either in the glassy or in the liquid state for the modulus and the phase.

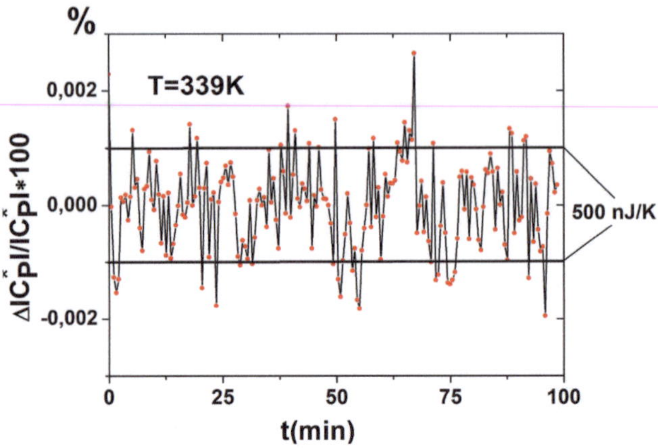

Fig. 4.5 Resolution in percent on the modulus of the complex heat capacity when the temperature oscillations are recorded during 100 min at a constant temperature of 339 K

In Fig. 4.5, the resolution in percent is directly shown on the modulus of the complex heat capacity when the measurement is carried out during 100 min at a constant temperature of 339 K. The resolution is the noise (peak to peak) divided by the absolute value of the modulus of the complex heat capacity for each experimental point $\Delta \left|C_P^*\right| / \left|C_P^*\right|$. The graph is directly given in percent showing that the nanocalorimeter is about one thousand times better than classical temperature oscillating calorimeters having a resolution of about 1 %. The minimum heat capacity variation during a thermal event that can be detected with the calorimeter is above 500 nJ/K.

In Fig. 4.6, we present the slow decrease of the modulus of the complex heat capacity along time when the PVAc sample is maintained at a constant temperature of 297 K during more than 3 h. The graph is presented using a logarithmic scale showing that the decrease slows down progressively but does not stop. This decrease is a consequence of the aging process (decrease of the glass energy) when the temperature is maintained constant below the glass transition temperature. This is typical of a glassy system. The point that we want to stress here is the significant variation of C_p during such a time interval. The relative variation is of the order of 1.2 % on the aging time. That is to say that this decrease cannot be measured with classical calorimeters having a resolution of few percents. It is important to mention that, over the same time window, no temperature drift of the sample has been recorder. As a conclusion, these two last examples highlight the capability of the nanocalorimeter of detecting very fine heat capacity variations during the evolution in time of slowly relaxing complex systems.

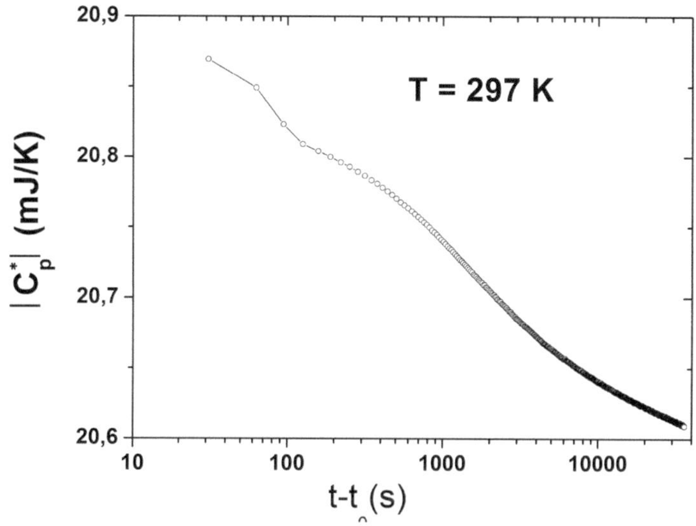

Fig. 4.6 Modulus of the complex heat capacity along time in logarithm scale during a aging experiment at the constant temperature of 297 K

4.4 Discussion

In this paper, we have presented two different experiments in the fields of nanocalorimetry. The first one deals with measurement of very thin layers at low temperature while the second one deals with measurement of complex systems at room temperature. These typical examples show that it is possible to reach the nanojoule per kelvin range with high sensitive calorimetry. By definition, calorimetry measure any kind of energy variation with temperature of any materials or any phase transformation of materials (specific heat, enthalpy, entropy changes etc.). We thus can imagine that such highly sensitive method of detection can find applications in the fields of security by its capability of measuring down-scaled nanosystems. We can however list different topics where nanocalorimetry can help in the detection process:

- detection of fine molecular interactions in liquids (pharmacy, drug detection, contamination of water, doping substances detection, ...);
- detection of small amount of toxic gas (gas sensor, ...);
- detection of small quantities (particles, gas) of explosives.

Acknowledgements We acknowledge technical supports from Nanofab, the Cryogenic and the Electronic facilities and the Pole Capteur Thermométrique et Calorimétrie of Institut Néel for these experiments. Funding for this project was provided by LANEF, Laboratoire d'Excellence, for the senior grant of Aviad Frydman, and by the CNRS (Mission pour l'interdisciplinarité) by means of the call DEFI Instrumentation aux Limites.

References

1. Garden J-L, Guillou H, Lopeandia AF, Richard J, Heron J-S, Souche GM, Ong FR, Vianay B, Bourgeois O (2009) Thermodynamics of small systems by nanocalorimetry: from physical to biological nano-objects. Thermochim Acta 492:16–28
2. Fominaya F, Fournier T, Gandit P, Chaussy J (1997) Nanocalorimeter for high resolution measurements of low temperature heat capacities of thin films and single crystals. Rev Sci Instrum 68:4191
3. Bourgeois O, Skipetrov SE, Ong F, Chaussy J (2005) Attojoule calorimetry of mesoscopic superconducting loops. Phys Rev Lett 94:057007
4. Ong FR, Bourgeois O (2007) Topology effects on the heat capacity of mesoscopic superconducting disks. Europhys Lett 79:67003
5. Poran S, Molina-Ruiz M, Gerardin A, Frydman A, Bourgeois O (2014) Specific heat measurement set-up for quench condensed thin superconducting films. Rev Sci Instrum 85:053903
6. Garno JP (1978) Simple high-vacuum evaporation system with low-temperature substrate. Rev Sci Instrum 49:1218
7. Bourgeois O, Dynes RC (2002) Strong coupled superconductor in proximity with a quench-condensed ferromagnetic Ni film: a search for oscillating T-c. Phys Rev B 65:144503
8. Bourgeois O, Frydman A, Dynes RC (2002) Inverse proximity effect in a strongly correlated electron system. Phys Rev Lett 88:186403
9. Molina Ruiz M, Lopeandia AF, Pi F, Givord D, Bourgeois O, Rodriguez-Viejo J (2011) Evidence of finite-size effect on the Neel temperature in ultrathin layers of CoO nanograins. Phys Rev B 83:140407(R)
10. Bourgeois O, André E, Macovei C, Chaussy J (2006) Liquid nitrogen to room-temperature thermometry using niobium nitride thin films. Rev Sci Instrum 77:126108
11. Laarraj M, Adhiri R, Ouaskit S, Moussetad M, Guttin C, Richard J, Garden J-L (2015) Highly sensitive pseudo-differential ac-nanocalorimeter for the study of the glass transition. Rev Sci Instrum 86:115110
12. Sullivan P, Seidel G (1968) Steady-state AC-temperature calorimetry. Phys Rev 173:679

Chapter 5
Phase Conversion of Y-Ba-Cu-O Thin Films by Super-Oxygenation and Cu-Enrichment

H. Zhang, N. Gauquelin, C. Mcmahon, D.G. Hawthorn, G.A. Botton, and J.Y.T. Wei

Abstract The superconducting critical temperature (T_c) of hole-doped cuprates tends to increase with their lattice complexity, which is generally correlated with higher states of oxidation. For $YBa_2Cu_3O_{7-\delta}$ (YBCO-123), it is known that solid-state reaction in ultrahigh-pressure oxygen can induce the formation of more complex and oxidized phases such as $Y_2Ba_4Cu_7O_{15-\delta}$ (YBCO-247) and $Y_2Ba_4Cu_8O_{16}$ (YBCO-248). In this study, we apply this super-oxygenation concept of oxide materials synthesis to thin films which, owing to their large surface-to-volume ratio, are more thermodynamically reactive than bulk samples. Epitaxial thin films of YBCO-123 were grown by pulsed laser-ablated deposition on $(LaAlO_3)_{0.3}(Sr_2TaAlO_6)_{0.7}$ substrates, and then annealed in 500 atm of oxygen at 800 °C. The high-pressure annealing was done in conjunction with Cu-enrichment by solid-state diffusion, as an additional driving force for phase conversion. Resistivity was measured to determine the T_c and to assess the amount of disorder in the films. Transmission electron microscopy and x-ray absorption spectroscopy were used to probe the local lattice structure and oxygen stoichiometry. Data taken on the super-oxygenated films show clear formation of YBCO-247 and YBCO-248, as well as distinct intergrowths of $YBa_2Cu_5O_9$, a novel phase of YBCO that has three CuO chains per unit cell.

H. Zhang
Department of Physics, University of Toronto, 60 St. George St., Toronto, M5S1A7, ON, Canada

N. Gauquelin • G.A. Botton
Canadian Centre for Electron Microscopy & Brockhouse Institute for Materials Research, McMaster University, 1280 Main Street West, Hamilton, L8S4M1, ON, Canada

C. Mcmahon • D.G. Hawthorn
Department of Physics and Astronomy, University of Waterloo, Waterloo, N2L3G1, ON, Canada

J.Y.T. Wei (✉)
Department of Physics, University of Toronto, 60 St. George St., Toronto, M5S1A7, ON, Canada

Canadian Institute for Advanced Research, Toronto, M5G1Z8, ON, Canada
e-mail: lucian@covaci.org

J. Bonča, S. Kruchinin (eds.), *Nanomaterials for Security*, NATO Science for Peace and Security Series A: Chemistry and Biology, DOI 10.1007/978-94-017-7593-9_5

5.1 Introduction

The Y-Ba-Cu-O (YBCO) family of compounds is exceptional among the cuprate superconductors by virtue of the one-dimensional CuO chains in its lattice structure. These CuO chains give YBCO its unique orthorhombicity, [1] in-plane anisotropy, [2–4] oxygen variability, [5] and rich phase diagram. Variation in the number of CuO chains per unit cell gives rise to different phases of YBCO. The three most common phases are $YBa_2Cu_3O_{7-\delta}$ (YBCO-123), $Y_2Ba_4Cu_7O_{15-\delta}$ (YBCO-247) and $Y_2Ba_4Cu_8O_{16}$ (YBCO-248). As shown in Fig. 5.1, YBCO-123 and YBCO-248 have single and double CuO chains, respectively, while YBCO-247 consists of alternating single and double chains. These phases are stable in different regimes of temperature and oxygen pressure, and have different optimal T_c [6–10]. In particular, it is known that YBCO-247 and YBCO-248 can be synthesized via solid-state reaction in ultrahigh-pressure oxygen [9, 10]. More exotic phases of YBCO with other arrangements of the CuO chains are also possible, and have been observed as atomic-scale defect structures in nominally YBCO-123 samples [11].

Recent studies have shown that the CuO chains have rich physics in their own right, instead of merely serving as charge reservoirs for Cooper pairing in the CuO_2 planes. It is believed that the chains can host charge density wave order, [12–16] break charge confinement within the CuO_2 planes, [17, 18] as well as proximity-couple with the CuO_2 planes to produce $d + s$ pairing symmetry [19–23]. Most interestingly, it was reported that double-CuO chains can sustain superconductivity on their own when the CuO_2 planes are rendered insulating by Pr-substitution in YBCO-247 [24–27]. This chain-based superconductivity appears to involve pairing of electrons instead of holes [28], and to be connected with Luttinger liquid physics [25]. To study the CuO chains in more detail, it is desirable to have thin-film samples, as they have well-defined crystalline axes. Thin films are also more thermodynamically active than bulk samples, by virtue of the large surface-to-volume ratio and epitaxial strain from the substrate [29, 30], and thus more likely to convert between the different YBCO phases.

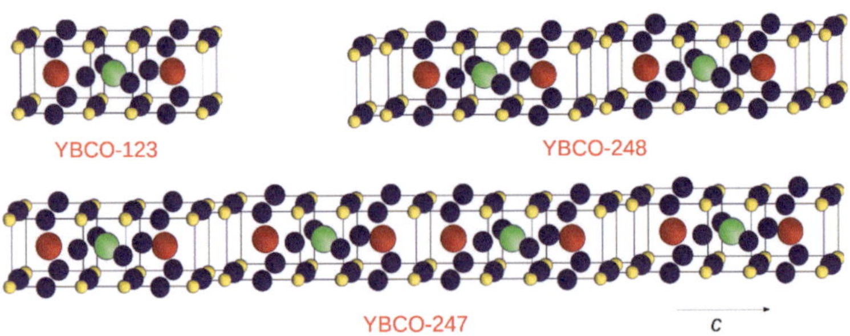

Fig. 5.1 The lattice structures of YBCO-123, YBCO-248 and YBCO-247, with the Cu, Y, and Ba atoms color-labeled as *yellow, green, and red*, respectively

In this work, we apply this super-oxygenation concept of oxide materials synthesis to YBCO thin films, in order to better stabilize and more closely study the higher-oxidation phases of YBCO. As-grown YBCO-123 thin films were annealed in high-pressure (HP) oxygen together with Cu-enrichment by solid-state diffusion. Scanning transmission electron microscopy (STEM) images revealed the formation of YBCO-247 and YBCO-248 phases in the HP-annealed films. X-ray diffraction (XRD), x-ray absorption spectroscopy (XAS) and resistivity measurements showed that higher Cu-enrichment resulted in more complete conversion to the chain-rich phases of YBCO. The XAS data also indicated that the CuO_2 planes in the HP-annealed films are underdoped, which helps to explain the generally lower T_c reported in bulk YBCO-247. In some of our super-oxygenated films, we also observed distinct intergrowths of $YBa_2Cu_5O_9$, a novel phase of YBCO that has three CuO chains per unit cell.

5.2 Experimental

The YBCO-123 thin films used in our study were epitaxially grown on (001)-oriented $(LaAlO_3)_{0.3}(Sr_2TaAlO_6)_{0.7}$ (LSAT) substrates using pulsed laser-ablated deposition (PLD). All the films used in this study were 50 nm thick. YBCO-123 targets of >99.9 % chemical purity and >90 % material density were used. Our PLD system is equipped with a 248 nm KrF excimer laser with a repetition rate of 2 Hz and fluence of ~2 J/cm^2, and the deposition was done at 800 °C in 200 mTorr of O_2. Following deposition, each film was annealed in situ by slow cooling from 800 °C to 300 °C at 12 °C/min in 760 Torr of O_2 to fully oxygenate the YBCO-123 layer. The post-annealing was done with a commercial high-pressure furnace at 800 °C in 500 atm of O_2. The annealing time was kept under 12 h to minimize cation inter-diffusion between film and substrate. During annealing, each film was buried in a powder mixture of YBCO-123 and CuO, to prevent material sublimation and to encourage phase conversion via Cu-enrichment. In this study we focus on two films that were HP-annealed using two different Cu-enrichment recipes, one with 2:1 and the other with 4:1 molar ratio of CuO to YBCO-123, respectively denoted as Sample A and Sample B.

Cross-sectional STEM images of the thin films were taken using a FEI Titan 80–300 microscope fitted with a high-brightness field emission gun and CEOS aberration correctors for both condenser and objective lens aberrations. The microscope operated at 200 keV in scanning mode using the high-angle annular dark-field (HAADF) imaging method, which is sensitive to the atomic-number contrast. The elemental mapping by the HAADF imaging was corroborated by atomic-scale electron energy loss spectroscopy (EELS) [31]. XRD was done using the $\Theta - 2\Theta$ method with a Bruker D8 DISCOVER X-ray diffractometer. Finally, electrical resistance (R) was measured as a function of temperature (T) using the ac lock-in technique with four parallel Ag-strips sputtered onto the film. The R vs. T data was taken between 300 and 4.2 K using a liquid-helium dipper probe, with the temperature monitored by a Cernox sensor to 0.1 K accuracy.

5.3 Results and Discussions

Figure 5.2 shows the STEM, XRD and R vs. T data taken on a typical as-grown film. Panel (a) is a high-resolution HAADF-STEM image, showing epitaxial interface and pure YBCO-123, i.e. containing only single CuO chains with no visible intergrowth. Double CuO-chain intergrowths were occasionally seen in some of the as-grown films, but only in trace amounts. The STEM data shown in panel (a) is well corroborated by the XRD data shown in panel (b). All the expected XRD peaks associated with the c-axis of either YBCO-123 or LSAT are present, and there are no impurity peaks within the resolution of our instrument. By relating the YBCO-123 (005)- and (007)-peaks with $2\Theta = 38.58°$ and $2\Theta = 55.08°$, respectively, we find the c-axis lattice parameter of our YBCO film to be 11.67 Å, consistent with fully oxygenated YBCO-123 [32]. Finally, as shown by the R vs. T plot in panel (c), the film is superconducting below $T_c \sim 90$ K, consistent with fully oxygenated YBCO-123. The extrapolated zero-temperature residual resistivity $\rho_0 \sim 55\ \mu\Omega$-cm. The residual-resistance ratio (RRR) between 290 and 100 K is ~2.7, which is within the range of values reported in the literature [33].

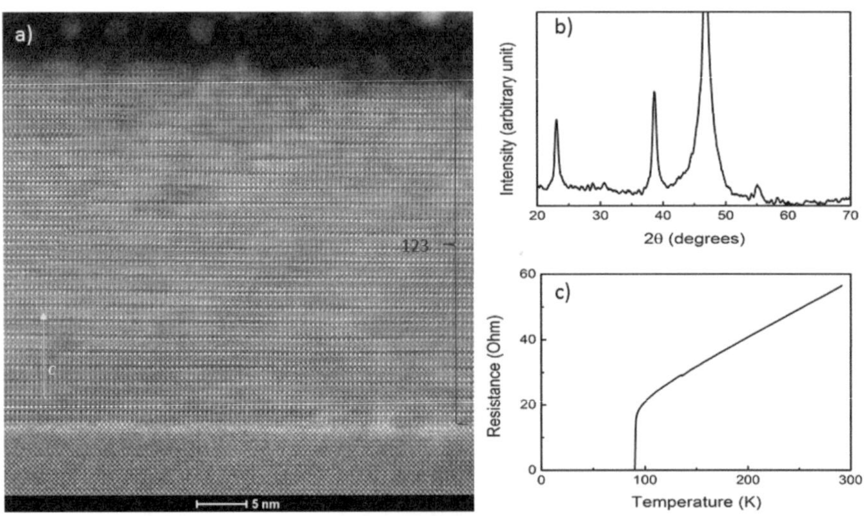

Fig. 5.2 Data taken on a typical as-grown YBCO film. Panel (**a**) shows a high-resolution HAADF-STEM image, Panel (**b**) shows the XRD pattern and panel (**c**) shows the R vs. T plot. No CuO intergrowths are evident in the STEM and XRD data

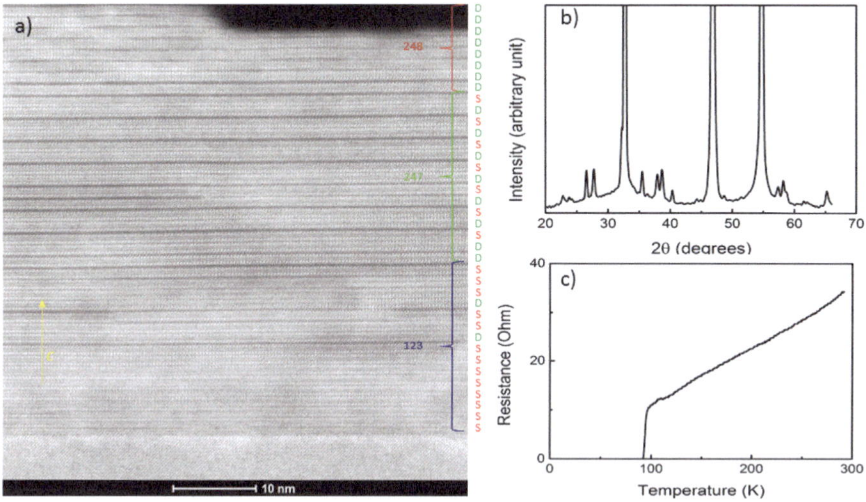

Fig. 5.3 Data taken on Sample A, which was HP-annealed together with 2:1 Cu-enrichment. Panel (**a**) shows a high-resolution HAADF-STEM image, panel (**b**) shows the XRD pattern and panel (**c**) shows the *R* vs. *T* plot. The STEM and XRD data show clear formation of YBCO-247 and YBCO-248 phases as a result of the post-annealing

After the as-grown YBCO-123 films were annealed in HP oxygen with Cu-enrichment, they showed partial conversion to YBCO-247 and YBCO-248. The YBCO phase conversion can be represented by the following equations:

$$2YBa_2Cu_3O_7 + 2CuO \xrightarrow{\text{high-pressure } O_2} Y_2Ba_4Cu_8O_{16},$$

$$2YBa_2Cu_3O_7 + CuO \xrightarrow{\text{high-pressure } O_2} Y_2Ba_4Cu_7O_{15}.$$

Figure 5.3a is an STEM image taken on Sample A, showing a highly epitaxial film-substrate interface with no noticeable cation inter-diffusion. Atomic-scale EELS mapping of the post-annealed films (data not shown) also showed no evidence of cation inter-diffusion [31]. In Fig. 5.3a, we see a gradation of YBCO phases, with YBCO-248 occurring near the surface, followed by YBCO-247 below that and then YBCO-123 near the film-substrate interface. This gradation of phases is understandable, as it is easier for both oxygen and Cu ions to diffuse into the top part of the film than to the bottom part of the film.

Figure 5.3b shows the XRD pattern taken on Sample A. Peaks associated with YBCO-123, YBCO-247 and YBCO-248 are all present, indicating a mixture of phases, as expected from the STEM image. There are two unidentified peaks, at $2\Theta = 27.7°$ and $2\Theta = 57.35°$. All the YBCO-247 peaks are quite weak, even though YBCO-247 constitutes the majority of the film volume according to the STEM image. The weakness of these peaks can be understood in terms of the complex unit cell of YBCO-247 (see Fig. 5.1), which have two sets of single-CuO chains and two sets of double-CuO chains. The *c*-axis lattice parameter of YBCO-247 (50.68 Å) is almost double that of YBCO-248 (27.24 Å) and more than four time that of YBCO-123 (11.67 Å), making YBCO-247 inherently more difficult to detect by XRD. Specifically, Fig. 5.3a shows only three complete unit cells YBCO-247 that can contribute to Bragg diffraction. Therefore, although YBCO-247 is the majority phase by volume here, YBCO-247 is the minority phase by number of unit cells.

Figure 5.3c shows the R vs. T plot of Sample A. This HP-annealed film shows a similarly sharp but higher superconducting transition at $T_c \sim 92$ K, while its $\rho_0 \sim 0\,\mu\Omega$-cm and RRR $\sim$ 3.1, in contrast to $\rho_0 \sim 55\,\mu\Omega$-cm and RRR $\sim$ 2.7 for the as-grown film. These observations indicate that HP-oxygen annealing together with Cu-enrichment tend to improve the superconductivity by reducing disorder in the film. We note that the large amounts of YBCO-247 and YBCO-248 seen in the STEM and XRD data do not lower the resistively-measured T_c. This is because electrical transport can probe the T_c of only the least resistive part of the film. Also it is conceivable that the YBCO-123 regions under the YBCO-248 and YBCO-247 regions become less disordered after the post-annealing, thus providing a higher-T_c current path.

The phase conversion induced by HP-oxygen annealing in YBCO films is enhanced upon further Cu-enrichment. Figure 5.4 compares the XRD data for an as-grown film versus the data plots for Sample A and Sample B. For Sample

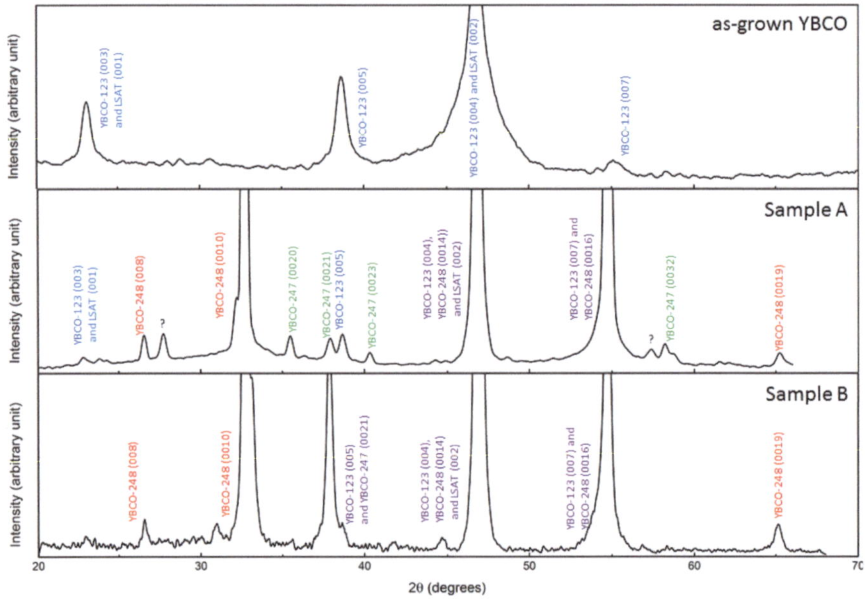

Fig. 5.4 Comparison of the XRD data of an as-grown film versus *Sample A* and *Sample B*, which were HP annealed with 2:1 and 4:1 Cu enrichment of CuO:YBCO-123 respectively

B, almost all the YBCO-247 peaks have disappeared, as well as the previously unidentified peak seen in Sample A. Both the (005)- and (007)-peaks for YBCO-123 that were seen in the as-grown film have shifted to lower values by 0.4° and 0.7° respectively. This decrease in the peak angles indicates an elongation of the c-axis lattice parameter, which is known to co-occur with a decrease in the YBCO orthorhombicity [34]. In our experiment, this decrease may be caused by the heteroepitaxial strain between YBCO-123 and YBCO-248 [30], the latter being less orthorhombic. These observations, along with the appearance of strong YBCO-248 peaks, indicate that most of the YBCO-123 in the film has converted to YBCO-248. As further evidence for this bulk conversion, the T_c of Sample B was measured to be ~80 K, which is consistent with the predominance of YBCO-248, i.e. without enough YBCO-123 left to provide a higher-T_c current path.

Figure 5.5 shows the corrected and normalized XAS spectra taken on our YBCO films. The XPS spectra with **E** // c are shifted up for clarity. Panel (a) shows the Cu L_3 absorption edge spectra, which arises from the 2p → 3d transition [35]. In the **E** // c data, the peak near 931 eV is mostly associated with the Cu(1) atoms in the CuO chains [36]. It is clear that Sample B has the strongest peak, followed by Sample A, and the as-grown film has the weakest peak. This observation is consistent with our expectation that stronger Cu-enrichment during the HP annealing process introduces more CuO chains into the YBCO structure. On the other hand, in the **E** // ab spectra, the main peak at 931 eV is mostly associated with the Cu(2) atoms in the planes, and do not show much variation between different samples. Slightly above the main peak

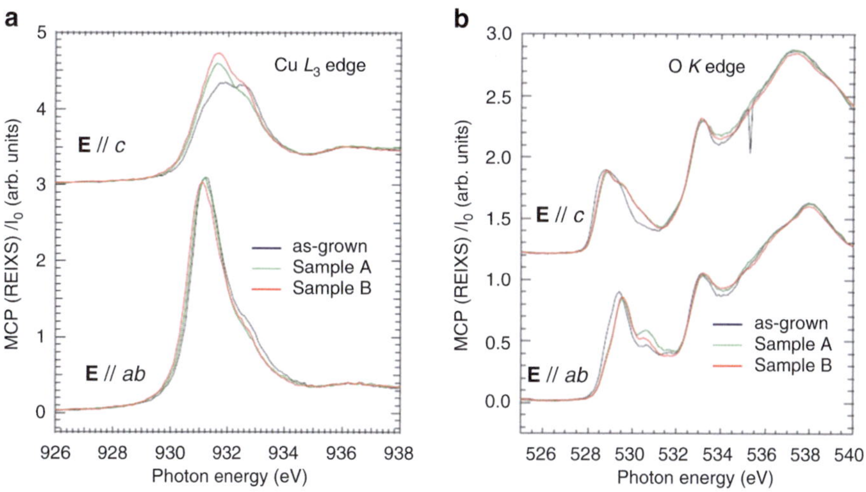

Fig. 5.5 Corrected and normalized XAS spectra taken on both as-grown and HP-oxygen annealed YBCO films. Panel (**a**) shows the spectra for the Cu L_3 edge. Panel (**b**) shows the spectra for the O K edge. The XAS spectra shows that there are more chain states in the annealed films. There is also evidence that the CuO_2 planes are underdoped in the annealed films, with the sample containing YBCO-247 being the most underdoped

in the **E**// *ab* spectra, there is a shoulder at 932 eV that is caused by the presence of doped itinerant holes in the planes [37–40]. Surprisingly, Our data shows that both HP-annealed YBCO films have fewer holes in the planes than does the as-grown film, because the shoulder is weakened after annealing. Sample A, which contains both YBCO-247 and YBCO-248 in additional to YBCO-123, appears to be more underdoped in the planes than Sample B, which lacks YBCO-247. Since YBCO-248 is known to be inherently underdoped with little oxygen variablity [41], it is reasonable that the as-grown film is less underdoped than any of the HP-annealed film, all of which contain YBCO-248. More interestingly, our data seems to suggest that the film containing YBCO-247 is more underdoped than the films without YBCO-247, because the shoulder at 932 eV is the weakest in Sample A. This observation could explain the generally lower T_c reported in pure YBCO-247 samples [42, 43].

The O *K* edge absorption edge spectra is shown in Fig. 5.5b. The peak at ~529 eV in the **E**// *c* data represents the O $2p_z$ states from apical O(4) sites [36], and this peak in both HP-annealed films are broader than in the as-grown film, implying that HP annealing increases the number of holes in the apical sites. In the **E**// *ab* spectra, the peaks at ~529 eV and ~530 eV represent the Zhang–Rice singlet band (ZRSB) and the upper Hubbard band (UHB) respectively, and a shifting of spectral weight from the ZRSB to the UHB indicates a reduction of doped holes in the planes [36]. The as-grown sample has the weakest UHB spectral weight relative to the ZRSB, while Sample A, which contains large regions of YBCO-247, has the strongest UHB. This observation is again consistent with the XPS data for the Cu L_3 absorption edge, which shows that the sample containing pure YBCO-123 is the least underdoped, and that the sample containing a mixture of YBCO-123, YBCO-248 and YBCO-247 is the most underdoped.

In some of our HP annealed YBCO films, we observed intergrowths of triple-CuO chains, as shown by the high-resolution STEM image in Fig. 5.6a. These regions of nanoscale intergrowths constitute a novel YBCO phase that contains three CuO chains per unit cell, as described by the formula $YBa_2Cu_5O_{9-\delta}$ (YBCO-125) whose lattice structure is shown in Fig. 5.5b. The phase conversion from YBCO-123 to YBCO-125 can be represented by the following equation:

$$YBa_2Cu_3O_7 + 2CuO \xrightarrow{\text{high-pressure } O_2} YBa_2Cu_5O_9.$$

Since each additional layer of CuO chain increases the YBCO *c*-axis lattice parameter by ~1.9 Å [44], the *c*-axis lattice parameter of YBCO-125 is expected to be ~15.5 Å. Figure 5.6c shows the XRD plot of an YBCO film containing the YBCO-125 phase. Apart from the peaks at 23.1° and 47.3° that are associated with the LSAT substrate, the location of all other peaks do not correspond to either YBCO-123, YBCO-247 or YBCO-248. However, we cannot conclusively attribute any of the peaks to the YBCO-125 phase, since their locations yield a lattice parameter close to, but not exactly equal to, the expected value for YBCO-125. More specifically, two of the strongest peaks in Fig. 5.6c are at 28.1° and 39.9°

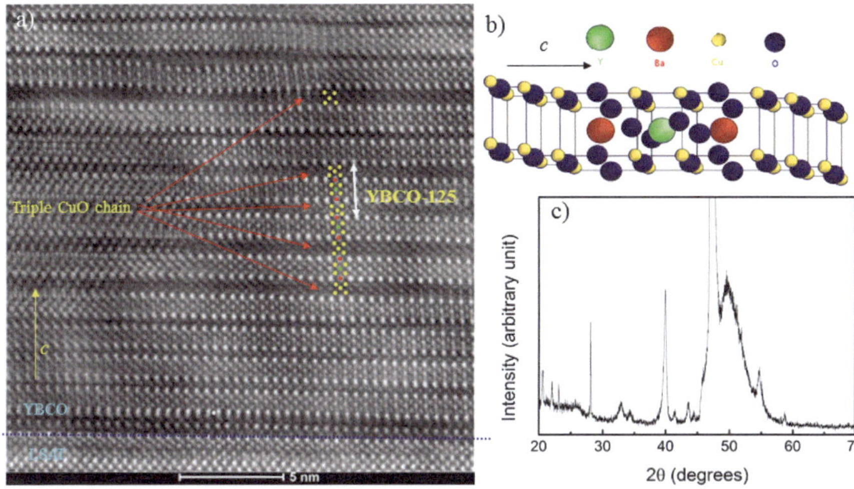

Fig. 5.6 Evidence of YBCO-125 formation, with the lattice structure shown in panel (**b**). Panel (**a**) shows an STEM image containing regions with three CuO chains per unit cell. Panel (**c**) shows the XRD pattern of the sample containing YBCO-125. The locations of the XRD peaks are different from those of either YBCO-123, YBCO-247 or YBCO-248

which, although far away from any of the YBCO-123 peaks, deviate appreciably from the expected (005)- and (007)-peaks of YBCO-125, i.e. at 28.8° and 40.7° respectively.

Regarding YBCO-125, it is worth noting that this triple-CuO chain phase of YBCO may give rise to higher T_c, since the T_c in cuprates is generally correlated with the distance between CuO_2 planes [45–47]. Therefore, successful isolation of the YBCO-125 phase will provide a unique opportunity to explore novel roles that multiple-CuO chains can play in high T_c superconductivity, by manipulating the coupling between CuO_2 planes as well the interplay between CuO chains and CuO_2 planes. We are now working to refine our super-oxygenation and Cu-enrichment procedure, in order to further isolate the novel YBCO-125 phase and to better study its physical properties.

5.4 Conclusion

In summary, we have studied phase conversion of YBCO thin films under super-oxygenation and Cu-enrichment. The YBCO-123 films were epitaxially grown by PLD on LSAT substrate, and then annealed at 800 C in 500 atm of O_2 buried in a powder mixture of YBCO-123 and CuO. STEM images taken on the post-annealed films showed clear formation of YBCO-247 and YBCO-248 phases. Resistivity, XRD and XAS measurements showed that higher Cu-enrichment resulted in greater conversion to YBCO phases that are richer in CuO chains. The XAS data also showed evidence that the CuO_2 planes in both YBCO-247 and YBCO-248 are underdoped, which helps to explain the generally lower T_c reported for YBCO-247. Finally, we observed regions of YBCO-125 formed by intergrowths of triple-CuO chains in some of our high-pressure annealed films. These results demonstrate that the technique of super-oxygenation can be applied in conjunction with Cu-enrichment on YBCO thin films, to explore novel roles that the CuO chains can play in YBCO and to discover more complex YBCO phases with potentially higher T_c.

Acknowledgements This work is supported by NSERC, CFI-OIT and CIFAR. The electron microscopy work was carried out at the Canadian Centre for Electron Microscopy, a National Facility supported by McMaster University and NSERC. The XAS work was performed at the Canadian Light Source, which is supported by NSERC, NRC, CIHR, and the University of Saskatchewan.

References

1. Jorgensen JD, Veal BW, Paulikas AP, Nowicki LJ, Crabtree GW, Claus H, Kwok WK (1990) Structural properties of oxygen-deficient $YBa_2Cu_3O_{7-\delta}$. Phys Rev B 41:1863
2. Takenaka K, Mizuhashi K, Takagi H, Uchida S (1994) Interplane charge transport in $YBa_2Cu_3O_{7-y}$: spin-gap effect on in-plane and out-of-plane resistivity. Phys Rev B 50:6534(R)
3. Basov DN, Liang R, Bonn DA, Hardy WN, Dabrowski B, Quijada M, Tanner DB, Rice JP, Ginsberg DM, Timusk T, Basov DN, Liang R, Bonn DA, Hardy WN, Dabrowski B, Quijada M, Tanner DB, Rice JP, Ginsberg DM, Timusk T (1995) In-plane anisotropy of the penetration depth in $YBa_2Cu_3O_{7-x}$ and $YBa_2Cu_4O_8$ superconductors. Phys Rev Lett 74:598
4. Lu DH, Feng DL, Armitage NP, Shen KM, Damascelli A, Kim C, Ronning F, Shen Z-X, Bonn DA, Liang R, Hardy WN, Rykov AI, Tajima S (2001) Superconducting gap and strong in-plane anisotropy in untwinned $YBa_2Cu_3O_{7-d}$. Phys Rev Lett 86:4370
5. Jorgensen JD, Beno MA, Hinks DG, Soderholm L, Volin KJ, Hitterman RL, Grace JD, Schuller IK (1987) Oxygen ordering and the orthorhombic-to-tetragonal phase transition in $YBa_2Cu_3O_{7-x}$. Phys Rev B 36:3608
6. Karpinski J, Rusiecki S, Kaldis E, Bucher B, Jilek E (1989) Phase diagrams of $YBa_2Cu_4O_8$ and YBa_2Cu3.5O7.5 in the pressure range 1 bar < Po_2 < 3000 bar. Physica C 160:449
7. Cava RJ, Batlogg B, van Dover RB, Murphy DW, Sunshine S, Siegrist T, Remeika JP, Rietman EA, Zahurak S, Espinosa GP (1987) Bulk superconductivity at 91 K in single-phase oxygen-deficient perovskite $Ba_2YCu_3O_{9-\delta}$. Phys Rev Lett 58:1676
8. Bordet P, Chaillout C, Chenavas J, Hodeau JL, Marezio M, Karpinski J, Kaldis E (1988) Structure determination of the new high-temperature superconductor $Y_2Ba_4Cu_7O_{14+x}$. Nature 334:596
9. Morris DE, Nickel JH, Wei JYT, Asmar NG, Scott JS, Scheven UM, Hultgren CT, Markelz AG, Post JE, Heaney PJ, Veblen DR, Hazen RM (1989a) Eight new high-temperature superconductors with the 1:2:4 structure. Phys Rev B 39:7347
10. Morris DE, Asmar NG, Wei JYT, Nickel JH, Sid RL, Scott JS, Post JE (1989b) Synthesis and properties of the 2:4:7 superconductors $R_2Ba_4Cu_7O_{15-x}$ (R-Y, Eu, Gd, Dy, Ho, Er). Phys Rev B 40:11406
11. Voronin GF, Degterov SA (1991) Thermodynamics of superconducting phases in the Y–Ba–Cu–O system. Physica C 176:387
12. Edwards HL, Barr AL, Markert JT, de Lozanne AL (1994) Modulations in the CuO chain layer of $YBa_2Cu_3O_{7-\delta}$: charge density waves? Phys Rev Lett 73:1154

13. Grevin B, Berthier Y, Collin G (2000) In-plane charge modulation below Tc and charge-density-wave correlations in the chain layer in $YBa_2Cu_3O_7$. Phys Rev Lett 85:1310
14. Derro DJ, Hudson EW, Lang KM, Pan SH, Davis JC, Markert JT, de Lozanne AL (2002) Nanoscale one-dimensional scattering resonances in the CuO chains of $YBa_2Cu_3O_{6+x}$. Phys Rev Lett 88:097002
15. Maki M, Nishizaki T, Shibata K, Kobayashi N, Maki M, Nishizaki T, Shibata K, Kobayashi N (2002) Electronic structure of the CuO-chain layer in $YBa_2Cu_3O_7$-studied by scanning tunneling microscopy. Phys Rev B 65:140511(R)
16. Achkar AJ, Sutarto R, Mao X, He F, Frano A, Blanco-Canosa S, Le Tacon M, Ghiringhelli G, Braicovich L, Minola M, Moretti Sala M, Mazzoli C, Liang R, Bonn DA, Hardy WN, Keimer B, Sawatzky GA, Hawthorn DG (2012) Distinct charge orders in the planes and chains of ortho-III-ordered $YBa_2Cu_3O_{6+\delta}$ superconductors identified by resonant elastic X-ray scattering. Phys Rev Lett 109:167001
17. Hussey NE, Kibune M, Nakagawa H, Miura N, Iye Y, Takagi H, Adachi S, Tanabe K (1998) Magnetic field induced dimensional crossover in the normal state of $YBa_2Cu_4O_8$. Phys Rev Lett 80:2909
18. Hussey NE, Takagi H, Iye Y, Tajima S, Rykov AI, Yoshida K (2000) Charge confinement on the CuO_2 planes in slightly overdoped $YBa_2Cu_3O_{7-\delta}$ and the role of metallic chains. Phys Rev B 61:R6475(R)
19. Breit V, Schweiss P, Hauff R, Wühl H, Claus H, Rietschel H, Erb A, Müller-Vogt G, Breit V, Schweiss P, Hauff R, Wühl H, Claus H, Rietschel H, Erb A, Müller-Vogt G (1995) Evidence for chain superconductivity in near-stoichiometric $YBa_2Cu_3O_x$ single crystals. Phys Rev B 52:R15727(R)
20. Gagnon R, Pu S, Ellman B, Taillefer L (1976) Anisotropy of heat conduction in $YBa_2Cu_3O_{6.9}$: a probe of chain superconductivity. Phys Rev Lett 78:1976
21. Srikanth H, Willemsen BA, Jacobs T, Sridhar S, Erb A, Walker E, Flükiger R, Srikanth H, Willemsen BA, Jacobs T, Sridhar S, Erb A, Walker E, Flükiger R (1997) Microwave response of $YBa_2Cu_3O_{6.95}$ crystals: evidence for a multicomponent order parameter. Phys Rev B 55:R14733(R)
22. Raffa F, Mali M, Suter A, Zavidonov AY, Roos J, Brinkmann D, Conder K (1999) Spin and charge dynamics in the Cu–O chains of $YBa_2Cu_4O_8$. Phys Rev B 60:3636
23. Ngai JH, Atkinson WA, Wei JYT (2007) Tunneling spectroscopy of *c*-Axis $Y_{1-x}Ca_xBa_2Cu_3O_{7-\delta}$ thin-film superconductors. Phys Rev Lett 98:177003
24. Matsukawa M, Yamada Y, Chiba M, Ogasawara H, Shibata T, Matsushita A, Takano Y (2004) Superconductivity in $Pr_2Ba_4Cu_7O_{15-\delta}$ with metallic double chains. Physica C 411:101
25. Berg E, Geballe TH, Kivelson SA (2007) Superconductivity in zigzag CuO chains. Phys Rev B 76:214505
26. Habaguchi T, Ōno Y, Ying Du Gh H, Sano K, Yamada Y (2011) Electronic states and superconducting transition temperature based on the tomonaga-luttinger liquid in $Pr_2Ba_4Cu_7O_{15-\delta}$. J Phys Soc Jpn 80:024708
27. Chiba T, Matsukawa M, Tada J, Kobayashi S, Hagiwara M, Miyazaki T, Sano K, Ōno Y, Sasaki T, Echigoya J (2013) Effect of magnetic field on the superconducting phase in the electron-doped metallic double-chain compound $Pr_2Ba_4Cu_7O_{15-\delta}$. J Phys Soc Jpn 82:074706
28. Matsushita A, Fukuda K, Yamada Y, Ishikawa F, Sekiya S, Hedo M, Naka T (2007) Physical properties of a new cuprate superconductor $Pr_2Ba_4Cu_7O_{15-\delta}$. Sci Technol Adv Mater 8:477
29. Su H, Welch DO, Wong-Ng W (2004) Strain effects on point defects and chain-oxygen order-disorder transition in 123 cuprate compounds. Phys Rev B 70:054517
30. Zhang H, Gauquelin N, Botton GA, Wei JYT (2013) Attenuation of superconductivity in manganite/cuprate heterostructures by epitaxially-induced CuO intergrowths. Appl Phys Lett 103:052606
31. Granstrom CR, Fridman I, Lei H-C, Petrovic C, Wei JYT (2016) Point-contact Andreev reflection spectroscopy on Bi_2Se_3 single crystals. Int J Mod Phys B 30(n13):1642002

32. Capponi JJ, Chaillout C, Hewat AW, Lejay P, Marezio M, Nguyen N, Raveau B, Soubeyroux JL, Tholence JL, Tournier R (1987) Structure of the 100 K superconductor $Ba_2YCu_3O_7$ between (5 ÷ 300) K by neutron powder diffraction. Europhys Lett 3:1301
33. Karkut MG, Perrin A, Guilloux-Viry M, Peña O, Padiou J, Sergent M (1990) In-plane orientation and superconductivity of in-situ dc sputtered $YBa_2Cu_3O_7$ films. Physica B 165–166:1473
34. Cava RJ, Hewat AW, Hewat EA, Batlogg B, Marezio M, Rabe KM, Krajewski JJ Jr, Peck WF, Rupp LW (1990) Structural anomalies, oxygen ordering and superconductivity in oxygen deficient $Ba_2YCu_3O_x$. Physica 165:419–433
35. Fink J, Nücker N, Pellegrin E, Romberg H, Alexander M, Knupfer M (1994) Electron-energy loss and X-Ray absorption spectroscopy of cuprate superconductors and related compounds. J El Spectr Rel Phen 66:395
36. Nücker N, Pellegrin E, Schweiss P, Fink J, Molodtsov SL, Simmons CT, Kaindl G, Frentrup W, Erb A, Müller-Vogt G (1995) Site-specific and doping-dependent electronic structure of $YBa_2Cu_3O_x$ probed by O 1s and Cu 2p x-ray-absorption spectroscopy. Phys Rev B 51:8529
37. Chen CT, Tjeng LH, Kwo J, Kao HL, Rudolf P, Sette F, Fleming RM (1992) Out-of-plane orbital characters of intrinsic and doped holes in $La_{2-x}Sr_xCuO$ phys. Rev Lett 68:2543
38. Pellegrin E, Nücker N, Fink J, Molodtsov SL, Gutiérrez A, Navas E, Strebel O, Hu Z, Domke M, Kaindl G, Uchida S, Nakamura Y, Markl J, Klauda M, Saemann-Ischenko G, Krol A, Peng JL, Li ZY, Greene RL (1993) Orbital character of states at the Fermi level in $La_{2-x}Sr_xCuO_4$ and $R_{2-x}Ce_xCuO_4$ (R=Nd, Sm). Phys Rev B 47:3354
39. Garg KB, Dalela S, Saini NL, Singhal RK, Jain DC, Chen CT (2003) Polarised XAS study of anomalous temperature dependence of aggregation of itinerant holes and pair formation in a $YBa_2Cu_3O_{7-d}$ single crystal. Physica C 399:98
40. Hawthorn DG, Shen KM, Geck J, Peets DC, Wadati H, Okamoto J, Huang S-W, Huang DJ, Lin H-J, Denlinger JD, Liang R, Bonn DA, Hardy WN, Sawatzky GA (2011) Resonant elastic soft x-ray scattering in oxygen-ordered $YBa_2Cu_3O_{6+\delta}$. Phys Rev B 84:075125
41. Buckley RG, Pooke DM, Tallon JL, Presland MR, Flower NE, Staines MP, Johnson HL, Meylan M, Williams GVM, Bowden M (1991) Ca- and La-substitution in $YBa_2Cu_3O_{7-\xi}$, $Y_2Ba_4Cu_7O_{15-\xi}$ and $YBa_2Cu_4O_8$. Physica C 174:383
42. Kato M, Nakanishi M, Miyano T, Shimizu T, Kakihana M, Yoshimura K, Kosuge K (1998) Microscopic studies on $Y_2Ba_4Cu_7O_{15-\delta}$ by use of TEM and NQR techniques. J Solid State Chem 139:266
43. Irizawa A, Ohmura T, Shibata T, Kato M, Yoshimura K, Kosuge K, Ito Y, Michor H, Hilscher G (2002) Collapse of superconductivity in the Y123/Y124 superlattice system $Y_2Ba_4Cu_7O_{15-\delta}$. J Phys Soc Jpn 71:574
44. Poole CK, Farach HA, Creswick RJ (1999) Handbook of superconductivity. Academic, San Diego, p 338
45. Locquet J-P, Perret J, Fompeyrine J, Mächler E, Seo JW, Van Tendeloo G (1998) Doubling the critical temperature of $La_{1.9}Sr_{0.1}CuO_4$ using epitaxial strain. Nature 394:453
46. Sato H, Tsukada A, Naito M, Matsuda A (2000) $La_{2-x}Sr_xCuO_y$ epitaxial thin films ($x = 0$ to 2): structure, strain, and superconductivity. Phys Rev B 61:12447
47. Pavarini E, Dasgupta I, Saha-Dasgupta T, Jepsen O, Andersen OK (2001) Band-structure trend in hole-doped cuprates and correlation with $T_{c\max}$. Phys Rev Lett 87:047003

Chapter 6
Strong-Coupling Diagram Technique for Strong Electron Correlations

A. Sherman

Abstract Using the strong coupling diagram technique a closed set of equations was derived for the electron Green's function of the Hubbard model. Spectral functions calculated self-consistently in the two-dimensional *t*-*U* model are in qualitative and in some cases quantitative agreement with results of Monte Carlo simulations in a wide range of electron concentrations. For three different initial bands – the semi-elliptical density of states, *t*-*U* and *t*-*t*′-*U* models – the Mott transition occurs at very close values of the Hubbard repulsion $U_c \approx \sqrt{3}\Delta/2$, where Δ is the initial bandwidth. The behavior of the Mott gap with doping and its width at half-filling depend strongly on the value of the next-nearest hopping constant t'.

6.1 Introduction

In recent years, in connection with the discovery of iron-based superconductors and investigations of the influence of electron correlations on topological insulators, the interest has increased in the multi-band generalization of the fermionic Hubbard model. For these model Hamiltonians it is desirable to find an approach which, being comparatively simple computationally, yields at least qualitatively correct results. As a possible candidate for such an approach the strong coupling diagram technique [1–3] is considered in this work. It was shown [4] that this theory is able to describe the Mott transition, and for the initial semi-elliptical density of states the critical value of the Hubbard repulsion is the same as that found in the Hubbard-III approximation [5]. To check the validity of this approach its results obtained [6, 7] in the two-dimensional *t*-*U* Hubbard model were compared with data of Monte Carlo simulations. This comparison shows not only qualitative, but in some cases quantitative agreement in positions of maxima and continua in spectral functions. The momentum distribution at half-filling, general spectral shapes, their evolution with wave vector and electron concentration in the wide range $0.7 \lesssim n \leq 1$ are also

A. Sherman (✉)
Institute of Physics, University of Tartu, Ravila 14c, 50411 Tartu, Estonia
e-mail: alekseis@ut.ee

J. Bonča, S. Kruchinin (eds.), *Nanomaterials for Security*, NATO Science for Peace and Security Series A: Chemistry and Biology, DOI 10.1007/978-94-017-7593-9_6

close. These facts, relative simplicity of calculations and ease of generalization to multi-band cases give promise that the considered approach will be useful also for more complex problems. In this article some recent results obtained in this diagram technique are briefly discussed.

6.2 Main Formulas

In the considered approach, the series expansion in powers of the kinetic term of the Hubbard Hamiltonian is used for calculating Green's functions. The elements of the arising diagram technique are on-site cumulants of electron creation and annihilation operators and hopping lines connecting cumulants on different sites. As in the weak coupling diagram technique, in the considered approach the linked-cluster theorem allows one to discard disconnected diagrams and to carry out partial summations in connected diagrams. The diagram is said to be irreducible if it cannot be divided into two disconnected parts by cutting some hopping line. The sum of all irreducible diagrams is termed the irreducible part $K(\mathbf{k}\omega)$, where $\mathbf{k}$ and ω are the wave vector and frequency, respectively. In terms of this quantity the equation for the one-electron Green's function reads

$$G(\mathbf{k}\omega) = \frac{K(\mathbf{k}\omega)}{1 - t_{\mathbf{k}} K(\mathbf{k}\omega)}, \tag{6.1}$$

where $t_{\mathbf{k}}$ is the initial dispersion. Diagrams of the first four orders in $K(\mathbf{k}\omega)$ are shown in Fig. 6.1. Here circles depict cumulants with the order determined by the number of incoming or outgoing lines. Except for external ends, these lines represent hopping constants connecting respective sites.

Notice that for the Hubbard repulsion $U \to 0$ all cumulants of orders $\nu > 1$ vanish, and with them all diagrams in $K(\mathbf{k}\omega)$ become equal to zero, except for the diagram (a) in Fig. 6.1. In the considered limit it is equal to $(\omega - \mu)^{-1}$, where μ is the chemical potential. Substituting this value into (6.1) we obtain the correct expression for Green's function of uncorrelated electrons, despite the fact that the approach is intended for the strong coupling case.

In this work $K(\mathbf{k}\omega)$ is approximated by the sum of two lowest order terms (a) and (b) in Fig. 6.1. Notice that in this approximation K does not depend on momentum. The second term includes a hopping-line loop. Using the possibility of the partial

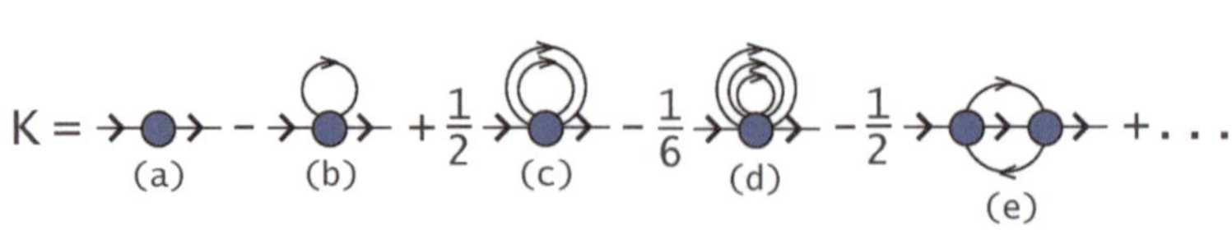

Fig. 6.1 Diagrams of the first four orders in $K(\mathbf{k}\omega)$

summation of diagrams this bare hopping line is transformed into the dressed one by including all possible combinations of diagrams (a) and (b) in it,

$$\theta(\mathbf{k}\omega) = \frac{t_{\mathbf{k}}}{1 - t_{\mathbf{k}} K(\mathbf{k}\omega)} = t_{\mathbf{k}} + t_{\mathbf{k}}^2 G(\mathbf{k}\omega). \tag{6.2}$$

Equations (6.1) and (6.2) form the closed set for calculating Green's function for given initial dispersion $t_{\mathbf{k}}$ and μ. Expressions for the respective cumulants can be found in [3].

6.3 Semi-elliptical Density of States

In the case of the semi-elliptical initial density of states

$$\rho_0(\omega) = \frac{4}{\pi\Delta}\sqrt{1 - \left(\frac{2\omega}{\Delta}\right)^2}, \quad |\omega| \le \frac{\Delta}{2} \tag{6.3}$$

the above relations can be reduced to an algebraic equation for half-filling [4, 8], and the critical repulsion U_c corresponding to the appearance of the Mott gap at the Fermi level is calculated analytically,

$$U_c = \sqrt{3}\Delta/2. \tag{6.4}$$

This value coincides exactly with the result of Hubbard-III approximation [5].

Figure 6.2 demonstrates the dimensionless quantity $\mathrm{Im}\Gamma$ connected with the density of states (DOS) $\rho(\omega) = -(\pi N)^{-1}\mathrm{Im}\sum_{\mathbf{k}} G(\mathbf{k}\omega)$ by the relation $\mathrm{Im}\Gamma =$

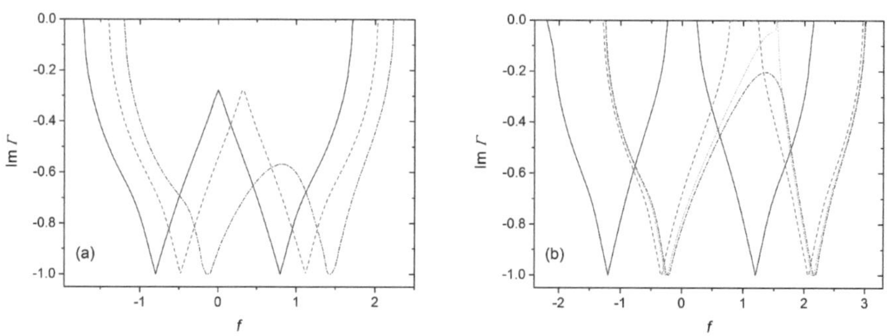

Fig. 6.2 $\mathrm{Im}\Gamma(f)$ for (**a**) $U = 0.8\Delta < U_c$, $\mu = 0.5U$ (half-filling, the electron concentration $n = 1$, *navy solid line*), $\mu = 0.3U$ ($n = 0.84$, *pink dashed line*), $\mu = 0.1U$ ($n = 0.68$, *violet dash-dotted line*) and for (**b**) $U = 1.2\Delta > U_c$, $\mu = 0.5U$ ($n = 1$, *navy solid line*), $\mu = 0.14U$ ($n = 0.75$, *pink dashed line*), $\mu = 0.11U$ ($n = 0.69$, *green dash-dot-dotted line*), $\mu = 0.1U$ ($n = 0.65$, *violet dash-dotted line*). The temperature $T = 0.001U$

$-(\pi\Delta/4)\rho$. The dimensionless frequency $f = (2/\Delta)\omega$, N is the number of sites. Due to the particle-hole symmetry of the problem only dependencies for $\mu \leq U/2$ are shown. The results are self-consistent solutions of Eqs. (6.1) and (6.2). The electron concentration $n = 2\int_{-\infty}^{\infty}[\exp(\omega/T)+1]^{-1}\rho(\omega)d\omega$, corresponding to each curve in Fig. 6.2, is indicated in the figure caption.

In the case $U < U_c$ shown in Fig. 6.2a electron correlations manifest themselves in a dip, which is located at the Fermi level at half-filling and shifts from it with doping. In this latter case the DOS is redistributed in favor of the Hubbard subband, in which the Fermi level is located – the result known from Monte Carlo simulations [9]. As U reaches U_c the DOS becomes zero at the Fermi level at half-filling. With further growth of U there appears a Mott gap around $\omega = 0$, see Fig. 6.2b. With a deviation from half-filling the gap shifts from the Fermi level and finally disappears.

For $0 < U < U_c$ the obtained solution is not a Fermi liquid, since the imaginary part of the self-energy $\Sigma(\mathbf{k}\omega) = \omega - K^{-1}(\mathbf{k}\omega)$ is nonzero at the Fermi level, while it is small for $U \ll U_c$. As U approaches the critical value $\mathrm{Im}\Sigma(\mathbf{k}\omega)$ diverges as $(U_c - U)^{-0.5}$. This result is similar to that obtained by the equation-of-motion approach [10].

6.4 Tight-Binding Initial Bands

When the hopping constants are nonzero only between nearest neighbor sites the initial band reads $t_{\mathbf{k}} = -2t[\cos(k_x a) + \cos(k_y a)]$, where a is the intersite distance. Below this quantity and the hopping constant t are set as units of length and energy, respectively. The case with this dispersion is termed the t-U model. For this model, at half-filling we face difficulties, since $\mathrm{Im}K(\omega)$ diverges near frequencies $\omega = \pm U/2$ as $[\omega^2 - (U/2)^2]^{-1}$. $K(\omega)$ coincides with Green's function for momenta meeting the condition $t_{\mathbf{k}} = 0$, and the divergence means that the function does not satisfy neither the normalization condition nor the Kramers-Kronig relations. The difficulty can be overcome by introducing an artificial broadening η, which imitates a contribution of higher-order terms and influences only widths of spectral maxima, leaving their locations, positions and widths of spectral continua intact. The obtained $\mathrm{Im}K(\omega)$ can be normalized and the real part of $K(\omega)$ can be calculated from its imaginary part using the Kramers–Kronig relation, which ensures the correct analytic properties of Green's function.

At half-filling, for the considered band the Mott transition occurs at $U_c \approx 7\Delta/8$, where $\Delta = 8$. This critical repulsion is very close to that obtained for the semi-elliptical DOS, Eq. (6.4). The frequency dependence of the DOS and its variation with doping are close to those shown in Fig. 6.2, except for the van Hove singularity, which starts to form at $\omega = 0$ when $U \to 0$ in the half-filled tight-binding band. The obtained solution is not a Fermi liquid also, since $\mathrm{Im}\Sigma(\omega = 0) \neq 0$. This quantity also diverges as $(U_c - U)^{-\nu}$ when $U \to U_c$. However, for the considered band the exponent $\nu = 0.75$ is larger than for the semi-elliptical DOS.

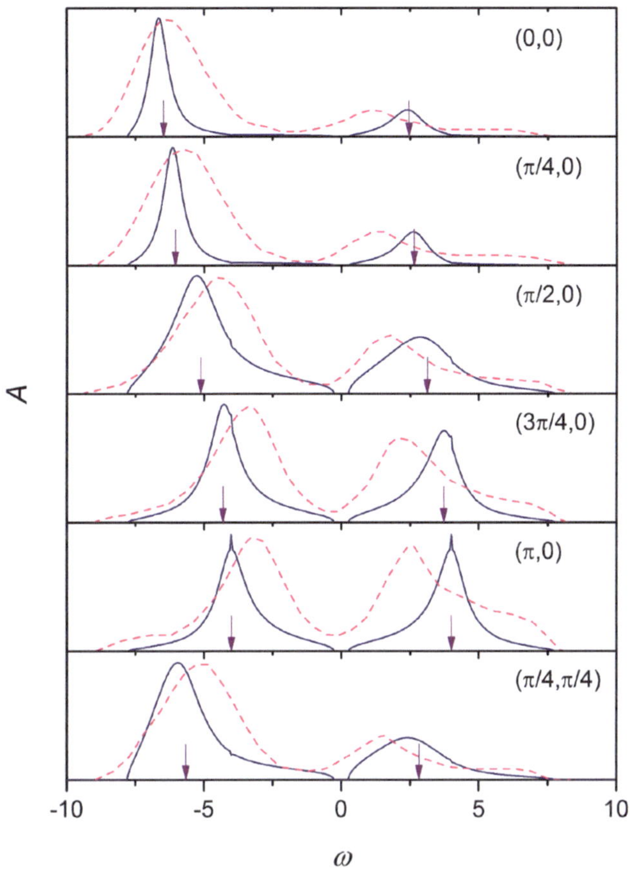

Fig. 6.3 The spectral function $A(\mathbf{k}\omega)$ calculated in a half-filled 8×8 lattice for wave vectors shown in the panels, $U = 8$ and $\eta = 0.5$ (*navy solid lines*). *Arrows* indicate positions of δ-function peaks in the Hubbard-I approximation. *Pink dashed lines* show results of Monte Carlo simulations performed in [9] for the same lattice and U, and for $T = 1$

Calculated [6] spectral functions $A(\mathbf{k}\omega) = -\pi^{-1}\text{Im}G(\mathbf{k}\omega)$ for the case of half-filling are shown in Fig. 6.3. A small lattice was chosen for comparison with Monte Carlo results of Ref. [9], performed for the same U. These results are also shown in Fig. 6.3. As seen from the figure, the agreement is wholly satisfactory in the considered case $U > U_c$ and high temperatures when magnetic ordering and spin correlations are suppressed. The used approach reproduces satisfactorily locations of maxima, widths and positions of spectral continua. The change of the artificial broadening η influences only widths of maxima. For smaller repulsions and temperatures the agreement between our and Monte Carlo results deteriorates due to the increased influence of magnetic ordering, spin and charge fluctuations, which were not taken into account by the considered diagrams. These interactions are described by diagrams containing ladders, which correspond to spin and charge susceptibilities [11]. However, even in the case of smaller U and T the theory describes qualitatively correctly positions of spectral continua, general spectral shape and changes in the locations of maxima with wave vector and U.

In Fig. 6.3 arrows indicate positions of δ-function peaks of the Hubbard-I approximation. For chemical potentials satisfying the conditions $T \ll \mu$, $T \ll U - \mu$ these frequencies can be written as $\varepsilon_{\mathbf{k},\pm} = (U + t_{\mathbf{k}} \pm \sqrt{t_{\mathbf{k}}^2 + U^2})/2 - \mu$ [12]. In the case of half-filling $\mu = U/2$. In the strong coupling diagram technique the Hubbard-I approximation is obtained if only the lowest order diagram (a) in Fig. 6.1 is retained in the irreducible part. As seen from the figure, frequencies of maxima are close to these arrows. Thus, the Hubbard-I approximation describes satisfactorily locations of maxima in the case $U \gtrsim U_c$ and $T \gtrsim t$ [9]. As is also obvious from Fig. 6.3, the Mott gap is opened in the spectral continuum.

In Fig. 6.4, the momentum distribution

$$n_{\mathbf{k}} = \int_{-\infty}^{\infty} \frac{A(\mathbf{k}\omega)}{\exp(\omega/T) + 1} d\omega$$

obtained in our calculations is compared with results of Monte Carlo simulations performed in [13]. As seen from the figure, the theory reproduces correctly variations of $n_{\mathbf{k}}$ with momentum and repulsion. The dependencies demonstrate gradual blurring of the Fermi surface with increasing U – the growing repulsion leads to an increasing occupancy in the region of the Brillouin zone, which was empty at $U = 0$.

As mentioned above, in the case of half-filling and a DOS symmetric with respect to $\omega = U/2$ there is a difficulty connected with the divergence of $\mathrm{Im}K(\omega)$ at $\omega = \pm U/2$. The difficulty can be remedied by introducing an artificial broadening. From Eqs. (6.1) and (6.2) it can be seen that if an initial DOS is not symmetric (as in the case of a nonzero next-nearest neighbor hopping constant t') and/or $n \neq 1$ the divergencies disappear. However, instead we face with another difficulty: there appear narrow gaps with nearly vertical walls in $\mathrm{Im}K(\omega)$ around $\omega = -\mu$ and $U - \mu$. Thanks to their narrowness and shape, these unphysical gaps can be easily eliminated by a linear interpolation between tops of the walls. Subsequent calculations are similar to those performed for the case of half-filling with the symmetric initial DOS.

Figure 6.5 shows our results [7] obtained in the t-U model for $n \neq 1$. We did not introduce an artificial broadening to demonstrate real spectral shapes appearing in the present approach. For comparison Monte Carlo results [9, 14] are also shown in this figure. It can be concluded that the used approach gives spectral functions in qualitative and in some cases quantitative agreement with the Monte Carlo data in the wide range of electron concentrations $0.7 \lesssim n \leq 1$. Locations of maxima, general shapes of spectra and their variation with $\mathbf{k}$ and n are as a rule close. Substantial differences in widths of maxima, observed in some spectra, may be related to the sign problem and to the numerical analytic continuation of Monte Carlo data to real frequencies.

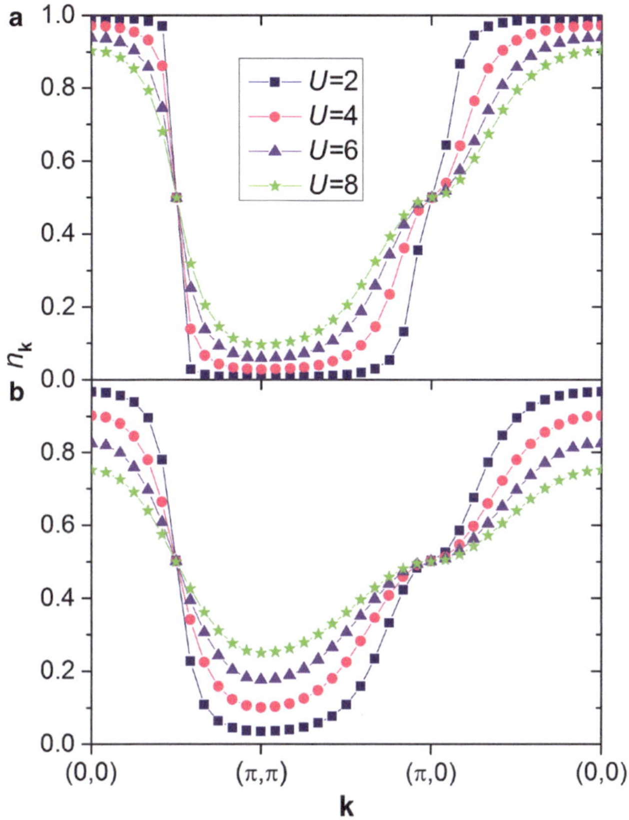

Fig. 6.4 The momentum distribution for momenta on symmetry lines and for several values of repulsion. (**a**) Monte Carlo results for T ranging from 1/32 to 1/20 [13]. (**b**) our results for $\eta = 1$. In both calculations a 24×24 lattice at half-filling was used

Arrows in Fig. 6.5 indicate frequencies of δ-function peaks in the Hubbard-I approximation. As for the case of half-filling, locations of maxima are close to these frequencies.

Figure 6.6 demonstrates the doping variation of the DOS in the model with the initial dispersion formed by nearest and next-nearest neighbor hopping integrals, $t_{\mathbf{k}} = -2[\cos(k_x) + \cos(k_y)] + 4t' \cos(k_x)\cos(k_y)$. This model is usually termed the t-t'-U model. Similar dispersions are frequently used for the description of hole-doped cuprates with the ratio t'/t close to 0.3. At half-filling for this ratio the Mott transition is observed again at U_c, which is close to the value (6.4). In contrast to the case of the t-U model, in which the Mott gap is maximal at half-filling, in this model the gap reaches its maximum value at $n \approx 1.04$ for $U = 8$ and $t' = 0.3$. For these parameters of the t-t'-U model the width of the gap is larger than in the t-U model with the same t and U (and hence with the same initial bandwidth): 0.64 and 0.54, respectively. The dependence of the gap width on $n - 1.04$ is strongly

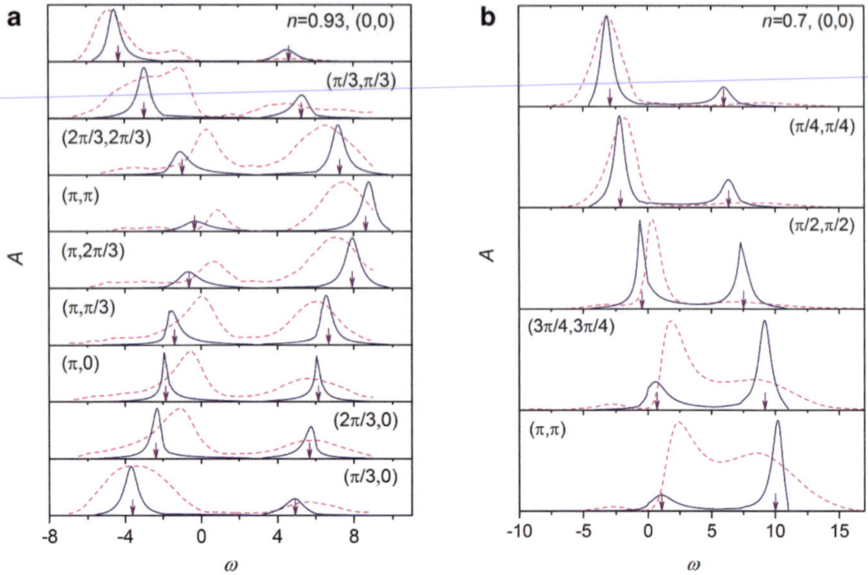

Fig. 6.5 Spectral functions for $U = 8$ and momenta shown in the panels. (**a**) $n = 0.93$, a 6×6 lattice. *Navy solid lines* are our results, *pink dashed lines* are Monte Carlo simulations for $T = 0.5$ [9]. (**b**) Same as in part (**a**), but for $n = 0.7$ and a 8×8 lattice (The Monte Carlo data are taken from [14])

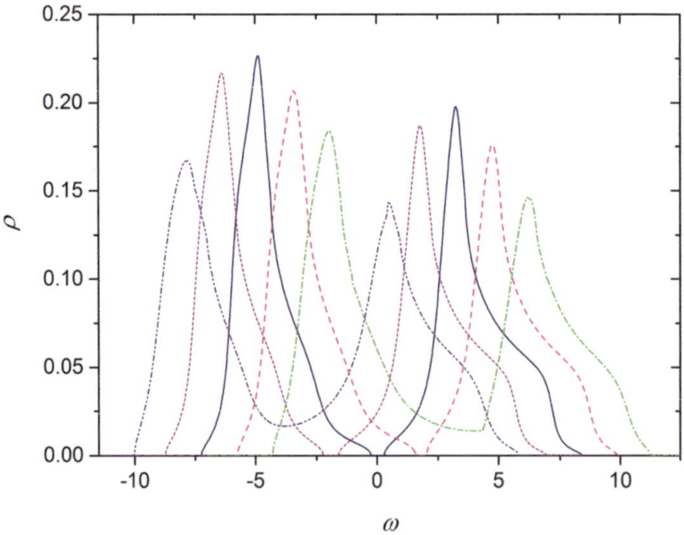

Fig. 6.6 Densities of states in the t-t'-J model for $U = 8$, $t' = 0.3$, $n = 1.26$ (*violet dash-dot-dotted line*), 1.04 (*magenta short-dashed line*), 1 (*navy solid line*), 0.99 (*pink dashed line*) and 0.87 (*green dash-dotted line*). A 240×240 lattice

asymmetric in the t-t'-J model. As seen from Fig. 6.6, the gap changes into a dip for large deviations from half-filling. Positions of maxima in spectral functions of this model are also close to those in the Hubbard-I approximation.

6.5 Concluding Remarks

In this work, equations for the electron Green's function of the repulsive Hubbard model, derived using the strong coupling diagram technique, were self-consistently solved for different electron concentrations n. Three different initial bands were considered: the semi-elliptical density of states and two tight-binding dispersions produced by the nearest neighbor hopping integral t as well as this integral and the next-nearest neighbor hopping integral t'. Terms of the first two orders in the expansion in powers of the electron kinetic energy were taken into account in the irreducible part of the equation for Green's function, and the bare internal hopping line in one of these terms was substituted with the dressed one. Comparison of spectral functions, calculated in the t-U model for the Hubbard repulsion $U = 8t$, with Monte Carlo data shows not only qualitative but in some cases quantitative agreement in positions of maxima and spectral continua. General spectral shapes, their evolution with the wave vector and electron concentration in the wide range $0.7 \lesssim n \leq 1$ are also similar. It was found that in the model with the initial semi-elliptical density of states the Mott transition occurs at $U_c = \sqrt{3}\Delta/2$, where Δ is the initial bandwidth. This value coincides exactly with the critical repulsion obtained in the Hubbard-III approximation. It was found also that in the t-U model and the t-t'-U model with $t' = 0.3t$ the Mott transition occurs at very close to this U_c values of the Hubbard repulsion. The behaviour of the Mott gap is different in the cases $t' = 0$ and $t' = 0.3t$: the maximal gap width is larger in the t-t'-U model and is reached at $n = 1.04$ rather than at half-filling, as in the case $t' = 0$. In cases of the initial tight-binding bands positions of maxima in the spectral function are close to locations of δ-function peaks of the Hubbard-I approximation.

The conducted comparison with Monte Carlo results and comparative simplicity of calculations give promise that the strong coupling diagram technique may be a useful tool in consideration of different generalizations of the Hubbard model.

Acknowledgements This work was supported by the research project IUT2-27, the European Regional Development Fund TK114 and the Estonian Scientific Foundation (grant ETF-9371).

References

1. Vladimir MI, Moskalenko VA (1990) Diagram technique for the Hubbard model. Theor Math Phys 82:301
2. Metzner W (1991) Linked-cluster expansion around the atomic limit of the Hubbard model. Phys Rev B 43:8549

3. Sherman A (2006) One-loop approximation for the Hubbard model. Phys Rev B 73:155105; Sherman A (2006) Quasiparticle states of the Hubbard model near the Fermi level. Phys Rev B 74:035104
4. Sherman A (2015) The Mott transition in the strong coupling perturbation theory. Physica B 456:35
5. Hubbard J (1964) Electron correlations in narrow energy bands. III. An improved solution. Proc R Soc Lond A 281:401
6. Sherman A (2015) Properties of the half-filled Hubbard model investigated by the strong coupling diagram technique. Int J Mod Phys B 29:1550088
7. Sherman A (2015) The Hubbard model in the strong coupling theory at arbitrary filling. Physica Status Solidi B 252:2006
8. Izyumov YA, Skryabin YN (1988) Statistical mechanics of magnetically ordered systems. Consultants Bureau, New York
9. Gröber C, Eder R, Hanke W (2000) Anomalous low-doping phase of the Hubbard model. Phys Rev B 62:4336
10. Gros C (1994) Equation-of-motion approach to the Hubbard model in infinite dimensions. Phys Rev B 50:7295
11. Sherman A, Schreiber M (2007) Magnetic susceptibility of the two-dimensional Hubbard model using a power series for the hopping constant. Phys Rev B 76:245112; Sherman A, Schreiber M (2008) Fluctuating charge-density waves in the Hubbard mode. Phys Rev B 77:155117
12. Hubbard J (1963) Electron correlations in narrow energy bands. Proc R Soc Lond A 276:238; Hubbard J (1964) Electron correlations in narrow energy bands. II. The degenerate band case. Proc R Soc Lond A 277:237
13. Varney CN et al (2009) Quantum Monte Carlo study of the two-dimensional fermion Hubbard model. Phys Rev B 80:075116
14. Haas S, Moreo A, Dagotto E (1995) Antiferromagnetically induced photoemission band in the cuprates. Phys Rev Lett 74:4281

Chapter 7
Spin-Dependent Transport of Carbon Nanotubes with Chromium Atoms

S.P. Kruchinin, S.P. Repetsky, and I.G. Vyshyvana

Abstract This paper presents a new method of describing electronic correlations in disordered magnetic crystals based on the Hamiltonian of multi-electron system and diagram method for Green's functions finding. Electronic states of a system were approximately described by self-consistent multi-band tight-binding model. The Hamiltonian of a system is defined on the basis of Kohn–Sham orbitals. Potentials of neutral atoms are defined by the meta-generalized gradient approximation (MGGA). Electrons scattering on the oscillations of the crystal lattice are taken into account. The proposed method includes long-range Coulomb interaction of electrons at different sites of the lattice. Precise expressions for Green's functions, thermodynamic potential and conductivity tensor are derived using diagram method. Cluster expansion is obtained for density of states, free energy, and electrical conductivity of disordered systems. We show that contribution of the electron scattering processes to cluster expansion is decreasing along with increasing number of sites in the cluster, which depends on small parameter. The computation accuracy is determined by renormalization precision of the vertex parts of the mass operators of electron-electron and electron-phonon interactions. This accuracy also can be determined by small parameter of cluster expansion for Green's functions of electrons and phonons. It was found the nature of spin-dependent electron transport in carbon nanotubes with chromium atoms, which are adsorbed on the surface. We show that the phenomenon of spin-dependent electron transport in a carbon nanotube was the result of strong electron correlations, caused by the presence of chromium atoms. The value of the spin polarization of electron transport is determined by the difference of the partial densities of electron states with opposite spin projection at the Fermi level. It is also determined by the difference between the relaxation times arising from different occupation numbers of single-electron states of carbon

S.P. Kruchinin
Bogolyubov Institute for Theoretical Physics, Kiev, 03680, Ukraine

S.P. Repetsky (✉)
Taras Shevchenko Kiev National University, Vladimirskaya st., 64, Kiev, 01033, Ukraine
e-mail: srepetsky@univ.kiev.ua

I.G. Vyshyvana
Institute of High Technologies, Taras Shevchenko Kiev National University, Akademika Glushkova avenue, 4g, Kiev, 02033, Ukraine

J. Bonča, S. Kruchinin (eds.), *Nanomaterials for Security*, NATO Science for Peace and Security Series A: Chemistry and Biology, DOI 10.1007/978-94-017-7593-9_7

and chromium atoms. The value of the electric current spin polarization increases along with Cr atoms concentration and magnitude of the external magnetic field increase.

7.1 Introduction

The progress of describing disordered systems is connected with the development of electron theory. Substitution alloys are best described among disordered systems. The traditional knowledge about physical properties of alloys is based on Born approximation in the scattering theory. However, this approach obviously can not be applied in the case of large scattering of potential difference of components which are held for the description of alloys with simple transition and rare-earth elements. The same difficulty is typical for the pseudopotential method [8]. Because of the pseudopotentials non-local nature there is a problem of their transferability. It is impossible to use nuclear potentials determined by the properties of some systems to describe other systems. Significant success in the study of the electronic structure and properties of the systems was achieved recently due to use of the ultra-soft Vanderbilt's pseudopotential [21, 37] and the method of projector-augmented waves in density functional theory proposed by Blohl [2, 16]. This approach was developed further thanks to generalized gradient approximation (GGA) usage in density functional theory of multi-electron systems developed by Perdew in his papers [23–25, 35, 36].The wave function of the valence states of electron (all-electron orbital) is denoted in projector-augmented waves approach by using the conversion through the pseudo-orbital. The pseudo-orbital waves expand to pseudo partial ones in the augment area. Even so, the all-electron orbital in the same area are expanded with the same coefficients via partial waves that are described by Kohn–Sham equation. The expression for the Hamiltonian operator, which we have in equation for pseudo wave function, is derived by minimizing the full energy functional. Using this equation, and expanding pseudo orbital by plane waves, we can derive the set of equations for expansion coefficients. From this system it is possible to get the electron energetic spectrum, wave functions, and the value of the full energy functional. The method for describing the electron structure of crystals is shown in Ref. [35], using VASP program package. With the help of the cluster methods of calculation and the GAUSSIAN program package, this approach could be used for molecule electronic structure description.

It should be noted that recently in papers (e.g. Refs. [6, 7, 9–11, 14, 26, 35]), the simple effective calculation method of electronic structure and properties for big molecules have been proposed. This method is based on the tight-binding model and functional density theory, which includes long-range coulomb interaction of electrons on different sites of crystal lattice. The long-range coulomb interaction of electrons on different sites is described in the local density approximation.

However, mentioned methods [6, 7, 9–11, 14, 23–26, 35, 36] are used only for an ideal ordered crystals and molecules description. In disordered crystals, the effects associated with localized electronic states and lattice vibrations are arise. They can not be described in the model of ideal crystal. In this regard, other approaches are also developed.

Essential achievements in description of the disordered systems properties are connected with application of the tight-binding model in the multi-scattering theory, including approximation of the coherent potential. Starting from Slater's and Koster's work [29, 31] tight-binding model in description of the ideal crystals properties were frequently applied. Later it was generalized for the case of disordered systems.

The method of describing magnetic alloys electronic structure, based on the functional density theory, was proposed by Staunton et al. [32] and Razee et al. [27]. The effective potential in Kohn–Sham equation [13, 15] consists of atomic potential and Pauli's addition, which is expressed through the magnetic field induction. Atomic potential and the magnetic field induction are expressed through variational derivatives of exchange-correlation energy by electronic density and magnetization respectively. Calculations of the electronic structure of the magnetic alloy are based on the mentioned effective potentials and usage of the self-consistent Korringa–Kohn–Rostoker coherent potential approximation. It was more developed in further papers [12, 33, 34]. The method of interatomic pair correlations parameter calculation was proposed by Staunton et al. [32] due to the pair mixing potential, which is expressed through the second derivative of the thermodynamic potential of the alloy concentration [3]. This thermodynamic potential is calculated in the one site coherent potential approximation. It should be noted that the methods developed in papers [12, 29, 31–34] do not include long-range Coulomb interaction of electrons at different lattice sites.

In our work a new method of describing electronic correlations in disordered magnetic crystals based on the Hamiltonian of multi-electron system and the diagram method for finding Green's functions was developed. Electronic correlations in crystals were described in self-consistent multi-band tight-binding model. The wave functions of non-interacting atoms are calculated based upon the Kohn–Sham equation. The effective one-electron potential of the many-atom structure is approximated as a sum of spherical Kohn–Sham potential of neutral non-interacting atoms. Potential of neutral atoms are defined by the meta-generalized gradient approximation (MGGA) [24, 35]. Our model includes wave functions and atomic potentials recalculation, taking into account the electronic density redistribution as the result of atomic interaction. It is also includes long-range Coulomb interaction of the electrons on different sites of the crystal lattice. Electron scattering processes on the ionic core potentials of different sorts and on oscillations of crystal lattice are considered. The two-time Green's function calculations are based on the temperature Green's function [22, 28]. In this operation a known relation between spectral representation for two-time and temperature Green's function was used [1].

Calculation of two-time Green's function of the disordered crystal are based on diagram technique, analogous to diagram technique for homogeneous system [1]. The set of equations for two-time Green's function, expressions for free energy and electrical conductivity of disordered crystals are derived. Calculation accuracy of the energetic spectrum, free energy, and electrical conductivity are determined by accuracy of the electron-electron and electron-phonon mass operators vertex parts renormalization. Energetic spectrum, free energy, conductivity and spin-depended transport of nanotubes doped with Cr atoms calculations are based on this method.

7.2 Hamiltonian of Electron's and Phonon's System for Disordered Crystals

Hamiltonian of the disordered crystal consists of single-particle Hamiltonian of electrons in the ionic cores field, the potential energy of the pair electron-electron interaction, lattice oscillations Hamiltonian and Hamiltonian of electron-phonon interaction.

Hamiltonian of the disordered crystal is represented on the basis of wave functions for free neutral atoms. In the Wannier representation the system Hamiltonian is, [28]

$$H = H_0 + H_{\text{int}}, \tag{7.1}$$

where zero-order Hamiltonian

$$H_0 = H_e^{(0)} + H_{ph}^{(0)}, \tag{7.2}$$

consists from single-particle Hamiltonian of electrons in the ionic cores field

$$H_e^{(0)} = \sum_{\substack{n_1 i_1 \gamma_1 \\ n_2 i_2 \gamma_2}} h^{(0)}_{n_1 i_1 \gamma_1, n_2 i_2 \gamma_2} a^{+}_{n_1 i_1 \gamma_1} a_{n_2 i_2 \gamma_2} \tag{7.3}$$

and atomic nucleus Hamiltonian

$$H_{ph}^{(0)} = \sum_{ni\alpha} \frac{P^2_{ni\alpha}}{2M_i} + \frac{1}{2} \sum_{\substack{n_1 i_1 \alpha_1 \\ n_2 i_2 \alpha_2}} \Phi^{(0)}_{n_1 i_1 \alpha_1, n_2 i_2 \alpha_2} u_{n_1 i_1 \alpha_1} u_{n_2 i_2 \alpha_2}. \tag{7.4}$$

Here in Eq. (7.3), (7.4) different sort ion cores distribution on sites (ni) of crystal lattice is the same as for ideally ordered crystal.

Perturbation Hamiltonian in Eq.(7.1) is

$$H_{\text{int}} = \delta\Phi + H_{ei} + H_{eph} + H_{ee} + H_{phi} + H_{phph}. \tag{7.5}$$

Perturbation Hamiltonian H_{int} consists of different sorts ion core interaction energy $\delta\Phi$. Interaction energy $\delta\Phi$ is measured from energy level for ideally ordered crystal. Addition to single-particle Hamiltonian of electrons in the ionic cores field in Eq. (7.3) (electron-ion interaction Hamiltonian) is defined as

$$H_{ei} = \sum_{\substack{n_1 i_1 \gamma_1 \\ n_2 i_2 \gamma_2}} w_{n_1 i_1 \gamma_1, n_2 i_2 \gamma_2} a^{+}_{n_1 i_1 \gamma_1} a_{n_2 i_2 \gamma_2}. \tag{7.6}$$

Hamiltonian of electron's interaction with oscillations of the crystal lattice (named as electron-phonon interaction) is defined as

$$H_{eph} = \sum_{\substack{n_1 i_1 \gamma_1 \\ n_2 i_2 \gamma_2}} v'_{n_1 i_1 \gamma_1, n_2 i_2 \gamma_2} a^{+}_{n_1 i_1 \gamma_1} a_{n_2 i_2 \gamma_2}. \tag{7.7}$$

Electron-electron pair interaction Hamiltonian is

$$H_{ee} = \frac{1}{2} \sum_{\substack{n_1 i_1 \gamma_1 \\ n_2 i_2 \gamma_2 \\ n_3 i_3 \gamma_3 \\ n_4 i_4 \gamma_4}} v^{(2) n_1 i_1 \gamma_1, n_2 i_2 \gamma_2}_{n_3 i_3 \gamma_3, n_4 i_4 \gamma_4} a^{+}_{n_1 i_1 \gamma_1} a^{+}_{n_2 i_2 \gamma_2} a_{n_3 i_3 \gamma_3} a_{n_4 i_4 \gamma_4}. \tag{7.8}$$

Atomic nucleus Hamiltonian component, caused by disordered distribution of atoms (phonon-impurity interaction Hamiltonian) is defined as

$$H_{phi} = \frac{1}{2} \sum_{\substack{n_1 i_1 \alpha_1 \\ n_2 i_2 \alpha_2}} \Delta M^{-1}_{n_1 i_1 \alpha_1, n_2 i_2 \alpha_2} P_{n_1 i_1 \alpha_1} P_{n_2 i_2 \alpha_2} + \frac{1}{2} \sum_{\substack{n_1 i_1 \alpha_1 \\ n_2 i_2 \alpha_2}} \Delta\Phi_{n_1 i_1 \alpha_1, n_2 i_2 \alpha_2} u_{n_1 i_1 \alpha_1} u_{n_2 i_2 \alpha_2}, \tag{7.9}$$

where

$$\Delta M^{-1}_{n_1 i_1 \alpha_1, n_2 i_2 \alpha_2} = \left(\frac{1}{M_{n_1 i_1}} - \frac{1}{M_{i_1}} \right) \delta_{n_1 n_2} \delta_{i_1 i_2} \delta_{\alpha_1 \alpha_2},$$

$\Delta\Phi_{n_1 i_1 \alpha_1, n_2 i_2 \alpha_2} = \Phi_{n_1 i_1 \alpha_1, n_2 i_2 \alpha_2} - \Phi^{(0)}_{n_1 i_1 \alpha_1, n_2 i_2 \alpha_2}$, $M_{n_1 i_1}$, M_{i_1} – mass of the atom in the site (ni) for disordered and ordered alloy accordingly.

The anharmonic component of atomic nucleus Hamiltonian (phonon-phonon interaction Hamiltonian) is

$$H_{phph} = \frac{1}{3!} \sum_{\substack{n_1 i_1 \alpha_1 \\ n_2 i_2 \alpha_2 \\ n_3 i_3 \alpha_3}} \Phi^{(0)}_{n_1 i_1 \alpha_1, n_2 i_2 \alpha_2, n_3 i_3 \alpha_3} u_{n_1 i_1 \alpha_1} u_{n_2 i_2 \alpha_2} u_{n_3 i_3 \alpha_3}. \tag{7.10}$$

In Eq. (7.10) only anharmonic terms of third order are taken into account.

In the previously written expressions $a^{+}_{ni\gamma}$, $a_{ni\gamma}$ are creation and destruction operators in the state, described by Wannier function $\phi_{ni\gamma}(\xi) = \langle \xi | ni\gamma \rangle$, where $\xi = (\mathbf{r}, \sigma)$. In these notations state's index γ includes as quantum number $\sigma = 1/2$, $-1/2$ which defines the value of spin projection on z axis and a set of other quantum numbers which describes electron spatial movements. Here n is the number of primitive cell, i – sublattice site number in primitive cell, $\mathbf{r}$ is electron's radius-vector, u_{ni} – atoms displacement operator in site (ni); $P_{ni\alpha}$ – operator of α-projection of atom's momentum on orthogonal axes.

The wave functions of an electron in free neutral atom sort λ, which is located at the site (ni), are obtained from Kohn–Sham equation in density functional theory [25, 26]:

$$\left[-\frac{\hbar^2}{2m} \nabla^2 + V^{\lambda}_{ext}(\mathbf{r}) + V^{\lambda i}_{H}(\mathbf{r}) + V^{\lambda i}_{XC,\sigma}(\mathbf{r}) \right] \times \psi^{\lambda}_{i\delta\sigma}(\mathbf{r}) = \varepsilon^{\lambda}_{i\delta\sigma} \psi^{\lambda}_{i\delta\sigma}(\mathbf{r}), \tag{7.11}$$

where σ is quantum number of spin projection on axis z; $\delta = (\tilde{\varepsilon} lm)$, l, m are quantum numbers of angular momentum and $\tilde{\varepsilon}$ is quantum number that describes the value of electron energy. To reduce the length of Eq. (7.11) $(\mathbf{r} - \mathbf{r}_{ni})$ is denoted by $(\mathbf{r})$.

In expression (7.11) the value $V^{\lambda}_{ext}(\mathbf{r})$ is potential energy of an electron in the atom's core field of sort λ at the site (n),

$$V^{\lambda i}_{H}(\mathbf{r}) = \int dv' \frac{e^2}{|\mathbf{r} - \mathbf{r}'|} n_{\lambda i}(\mathbf{r}') \tag{7.12}$$

– the Hartree potential.

In Eq. (7.12) electron density is

$$n_{\lambda i}(\mathbf{r}) = n_{\lambda i\sigma}(\mathbf{r}) + n_{\lambda i -\sigma}(\mathbf{r}). \tag{7.13}$$

The electron density with projection of spin σ is given by expression

$$n_{\lambda i\sigma}(\mathbf{r}) = \sum_{\delta} Z^{\lambda}_{i\delta\sigma} \psi^{\lambda} *_{i\delta\sigma}(\mathbf{r}) \psi^{\lambda}_{i\delta\sigma}(\mathbf{r}), \tag{7.14}$$

where $Z^{\lambda}_{i\delta\sigma}$ – occupation number of electron state $(i\delta\sigma)$, assuming that an atom of sort λ is in the site (ni).

In MGGA obtained by Perdew [24, 35] and based on density functional theory, the exchange-correlation potential $V_{XC,\sigma}(\mathbf{r}) = V^{\mathrm{MGGA}}_{XC,\sigma}(\mathbf{r})$ can be represented as:

$$V^{\mathrm{MGGA}}_{XC,\sigma}(\mathbf{r})\,\psi_{\gamma\sigma}(\mathbf{r}) = V^{\mathrm{GGA}}_{XC,\sigma}(r)\psi_{\gamma\sigma}(\mathbf{r}) - \frac{1}{2}\nabla\cdot\{\mu_{XC,\sigma}(r)\nabla\}\,\psi_{\gamma\sigma}(\mathbf{r}), \tag{7.15}$$

where

$$V^{\mathrm{GGA}}_{XC,\sigma}(r) = \left[\frac{\partial e^{\mathrm{MGGA}}_{XC}}{\partial n_\sigma} - \nabla\cdot\left(\frac{\partial e^{\mathrm{MGGA}}_{XC}}{\partial \nabla n_\sigma}\right)\right], \quad \mu_{XC,\sigma}(r) = \frac{\partial e^{\mathrm{MGGA}}_{XC}}{\partial \tau_\sigma},$$

$e^{\mathrm{MGGA}}_{XC}(2n_\sigma)/2$ is exchange-correlational energy density, $\tau_\sigma = \sum_\delta |\nabla\psi_{\delta\sigma}|^2/2$ is kinetic energy density.

Wave functions of the basis set $\phi_{ni\gamma\sigma}(\mathbf{r})$, on which Hamiltonian of the system is represented as in Eq. (7.1), are defined from Kohn–Sham Eq. (7.11) for atom of sort λ_i and equals to $\phi_{ni\gamma\sigma}(\mathbf{r}) = R^{\lambda_i}_{i\tilde{\varepsilon}l\sigma}(|\mathbf{r}-\mathbf{r}_{ni}|)\,Y^{v}_{lm}(\theta,\phi)$, where $R^{\lambda_i}_{i\tilde{\varepsilon}l\sigma}(|\mathbf{r}-\mathbf{r}_{ni}|)$ – radial part of wave function in Eq. (7.11), λ_i – sort of atom which located in site (ni) of ideal ordered crystal. Real spherical functions $Y^{v}_{lm}(\theta,\phi)$, are related with complex spherical functions $Y_{lm}(\theta,\phi)$ by equations

$$Y^{c}_{lm}(\theta,\phi) = \frac{1}{\sqrt{2}}[Y_{lm}(\theta,\phi) + Y^{*}_{lm}(\theta,\phi)],$$

$$Y^{s}_{lm}(\theta,\phi) = \frac{1}{i\sqrt{2}}[Y_{lm}(\theta,\phi) - Y^{*}_{lm}(\theta,\phi)], \quad m \neq 0. \tag{7.16}$$

The potential energy operator of electron in the field of different sort ionic cores can be expressed

$$V(\mathbf{r}) = \sum_{ni} v^{ni}\left(\mathbf{r}-\mathbf{r}'_{ni}\right), \quad \mathbf{r}'_{ni} = \mathbf{r}_{ni} + \mathbf{u}^{s}_{ni} + \mathbf{u}_{ni},$$

where $\mathbf{r}$ – electron's radius vector, $\mathbf{r}_{ni} = \mathbf{r}_n + \boldsymbol{\rho}_i$ – radius-vector of atom's equilibrium position in site of crystal lattice (ni), $\mathbf{u}^{s}_{ni}$ – is vector of atom's static displacement from equilibrium position in site (ni).

Potentials of ionic core $v^{\lambda_i}(\mathbf{r}-\mathbf{r}_{ni})$ are found from Eqs. (7.11), (7.12), (7.13), (7.14) and (7.15). While finding potentials $v^{\lambda_i}(\mathbf{r}-\mathbf{r}_{ni})$ summation in Eq. (7.14) must be done by electronic states of ionic cores.

Values $h^{(0)}_{n_1i_1\gamma_1,n_2i_2\gamma_2}$ in Eq. (7.3) are the matrix elements of the kinetic and potential $\sum_{ni} v^{\lambda_i}(\mathbf{r}-\mathbf{r}_{ni})$ energy operators of electron in the field of ionic core, where λ_i – sort of ion which located in site (ni) of ideal ordered crystal. Matrix elements are calculated by Slater–Koster method [11, 35].

Matrix element of electron-ion interaction Hamiltonian in Eq. (7.6) is

$$w_{n_1 i_1 \gamma_1, n_2 i_2 \gamma_2} = \sum_{ni} w^{ni}_{n_1 i_1 \gamma_1, n_2 i_2 \gamma_2}, \tag{7.17}$$

where

$$w^{ni}_{n_1 i_1 \gamma_1, n_2 i_2 \gamma_2} = \sum_{\lambda} c^{\lambda}_{ni} w^{\lambda ni}_{n_1 i_1 \gamma_1, n_2 i_2 \gamma_2},$$

$$w^{\lambda ni}_{n_1 i_1 \gamma_1, n_2 i_2 \gamma_2} = v^{\lambda ni}_{n_1 i_1 \gamma_1, n_2 i_2 \gamma_2} + \Delta v^{\lambda ni}_{n_1 i_1 \gamma_1, n_2 i_2 \gamma_2} - v^{\lambda_i ni}_{n_1 i_1 \gamma_1, n_2 i_2 \gamma_2},$$

λ_i is sort of ion which located in site (ni) of ideal ordered crystal. Here c^{λ}_{ni} – random numbers, taking value 1 or 0 depending on whether the atom of sort λ is at site (ni) or not.

Hamiltonian of electron-phonon interaction defined in Eq. (7.7) is expressed through derivatives of potential energy of electron in ions core field by projections of displacement vector $\mathbf{u}_{ni}$ of atom. In Eq. (7.7) the value of $v'_{n_1 i_1 \gamma_1, n_2 i_2 \gamma_2}$ is given by $v'_{n_1 i_1 \gamma_1, n_2 i_2 \gamma_2} = \sum_{ni\alpha} v'^{ni\alpha}_{n_1 i_1 \gamma_1, n_2 i_2 \gamma_2} u_{ni\alpha}$, where $v'^{ni\alpha}_{n_1 i_1 \gamma_1, n_2 i_2 \gamma_2} = \sum_{\lambda} c^{\lambda}_{ni} v'^{\lambda ni\alpha}_{n_1 i_1 \gamma_1, n_2 i_2 \gamma_2}$. Value $v'^{\lambda ni\alpha}_{n_1 i_1 \gamma_1, n_2 i_2 \gamma_2}$ is matrix element of operator $-\mathrm{e}_{ni\alpha} \frac{d}{d|\mathbf{r}-\mathbf{r}_{ni}|} v^{\lambda} \left(|\mathbf{r} - \mathbf{r}_{ni}|\right)$, where $\mathrm{e}_{ni} = \frac{\mathbf{r}-\mathbf{r}_{ni}}{|\mathbf{r}-\mathbf{r}_{ni}|}$.

Values $\Delta v^{\lambda ni}_{n_1 i_1 \gamma_1, n_2 i_2 \gamma_2}$ in Eq. (7.17) describe electron scattering on static displacement of atoms and defined by equation

$$\Delta v^{\lambda ni}_{n_1 i_1 \gamma_1, n_2 i_2 \gamma_2} = \sum_{\alpha} v'^{\lambda ni\alpha}_{n_1 i_1 \gamma_1, n_2 i_2 \gamma_2} u^{s}_{ni\alpha}.$$

The matrix of force constant part which is caused by the direct coulomb interaction of ionic cores has the form:

$$\begin{aligned} \Phi_{ni\alpha, n'i'\alpha'} = & -\frac{Z_{ni} Z_{n'i'} e^2}{4\pi\varepsilon_0 \left|\mathbf{r}_n + \boldsymbol{\rho}_i - \mathbf{r}_{n'} - \boldsymbol{\rho}_{i'}\right|^5} \\ \times & \Big[3\left(r_{n\alpha} + \rho_{i\alpha} - r_{n'\alpha} - \rho_{i'\alpha}\right)\left(r_{n\alpha'} + \rho_{i\alpha'} - r_{n'\alpha'} - \rho_{i'\alpha'}\right) \\ & -\left|\mathbf{r}_n + \boldsymbol{\rho}_i - \mathbf{r}_{n'} - \boldsymbol{\rho}_{i'}\right|^2 \delta_{\alpha\alpha'} \Big], \quad (ni) \neq (n'i'), \end{aligned} \tag{7.18}$$

where Z_{ni} – valence of ion cores which located in the site (ni).

Diagonal by number site (ni) elements of the matrix of force constants is determined from the condition:

$$\sum_{n'i'} \Phi_{ni\alpha, n'i'\alpha'} = 0.$$

Values $\Phi^{(0)}_{n_1 i_1 \alpha_1, n_2 i_2 \alpha_2}$, $\Phi^{(0)}_{n_1 i_1 \alpha_1, n_2 i_2 \alpha_2, n_3 i_3 \alpha_3}$ in Eq. (7.4), (7.10) are force constants.

Matrix $\Phi^{(0)}_{ni\alpha, n'i'\alpha'}$ is derived from expression for matrix $\Phi_{ni\alpha, n'i'\alpha'}$ in which $Z_{ni} = Z^{\lambda_i}$, where Z^{λ_i} is valence of ion which is located in the site (ni) of ideal ordered crystal.

Matrix elements $v^{(7.2) n_1 i_1 \gamma_1, n_2 i_2 \gamma_2}_{n_3 i_3 \gamma_3, n_4 i_4 \gamma_4}$ of Hamiltonian in Eq. (7.8) can be calculated by integrating angular variables. Integrals from product of three spherical functions (Gaunt integral) can be represented using the Clebsch–Gordan coefficients [31]. As a result matrix elements $v^{(2)\tilde{\varepsilon}_1 l_1 m_1, \tilde{\varepsilon}_2 l_2 m_2}_{\tilde{\varepsilon}_3 l_3 m_3, \tilde{\varepsilon}_4 l_4 m_4}$ are obtained:

$$v^{(2)\tilde{\varepsilon} l m, \tilde{\varepsilon}_2 l_2 m_2}_{\tilde{\varepsilon}_1 l_1 m_1, \tilde{\varepsilon}' l' m'} = e^2 \sum_{\substack{|l-l'| \leqslant l_3 \leqslant l+l' \\ |l_2 - l_1| \leqslant l_3 \leqslant l_2 + l_1 \\ l + l' + l_3 = 2k, k_1 \in \mathbb{N} \\ l_2 + l_1 + l_3 = 2k_1, k_1 \in \mathbb{N}}} \frac{1}{2l_3 + 1}$$
$$\times \left[\frac{(2l_3+1)(2l'+1)(2l_3+1)(2l_1+1)}{(2l+1)(2l_2+1)} \right]^{1/2}$$
$$\times c(l_3 l' l; 0, 0) c(l_3 l' l; m' - m, m') c(l_3 l_1 l_2; 0, 0) c(l_3 l_1 l_2; m_2 - m_1, m_1)$$
$$\times \left[\int_0^\infty dr_1 r_1^2 R_{\tilde{\varepsilon} l}(r_1) R_{\tilde{\varepsilon}' l'}(r_1) \int_0^{r_1} dr_2 r_2^2 R_{\tilde{\varepsilon}_2 l_2}(r_2) R_{\tilde{\varepsilon}_1 l_1}(r_2) \frac{r_2^{l_3}}{r_1^{l_3+1}} \right.$$
$$\left. + \int_0^\infty dr_2 r_2^2 R_{\tilde{\varepsilon}_2 l_2}(r_2) R_{\tilde{\varepsilon}_1 l_1}(r_2) \int_0^{r_2} dr_1 r_1^2 R_{\tilde{\varepsilon} l}(r_1) R_{\tilde{\varepsilon}' l'}(r_1) \frac{r_1^{l_3}}{r_2^{l_3+1}} \right], \tag{7.19}$$

where l, m are orbital and magnetic quantum numbers, respectively, $c(l'' l' l; m'', m')$ – Clebsch–Gordan coefficients [31], $R_{\tilde{\varepsilon} l}(r)$ – radial part of wave function, $\tilde{\varepsilon}$ – main quantum number.

Matrix elements on the basis of real wave functions [35] $v^{(2) n_1 i_1 \gamma_1, n_2 i_2 \gamma_2}_{n_3 i_3 \gamma_3, n_4 i_4 \gamma_4}$ for each site when $(n_1, i_1) = (n_2, i_2) = (n_3, i_3) = (n_4, i_4)$ are expressed by linear combinations of matrix elements $v^{(2)\tilde{\varepsilon}_1 l_1 m_1, \tilde{\varepsilon}_2 l_2 m_2}_{\tilde{\varepsilon}_3 l_3 m_3, \tilde{\varepsilon}_4 l_4 m_4}$. This procedure of matrix elements calculation can be easily programmed.

Matrix elements on the basis of real wave functions $v^{(2) n_1 i_1 \gamma_1, n_2 i_2 \gamma_2}_{n_3 i_3 \gamma_3, n_4 i_4 \gamma_4}$ for different sites (ni) can be approximately represented in the form similar to formula (7.19), if we describe radial part of the wave function by Gaussian function (namely Gaussian orbital), as this is done in the method of molecular orbitals – linear combinations of atomic orbitals [29]. In this approximation the multi-center integrals $v^{(2) n_1 i_1 \gamma_1, n_2 i_2 \gamma_2}_{n_3 i_3 \gamma_3, n_4 i_4 \gamma_4}$ have the form of one-center integrals, as the product of two Gaussian orbitals that are localized at different centers can be reduced to the product of orbitals that are localized at the joint center.

7.3 Green's Functions of Electrons and Phonons System

In order to calculate the energy spectrum of electrons and phonons, free energy and electrical conductivity of disordered crystal we introduce two-time Green's function. We define two-time retarded $G_r^{AB}(t,t')$ and accelerated $G_a^{AB}(t,t')$ Green's functions as follows[41]:

$$G_r^{AB}(t,t') = -\frac{i}{\hbar}\theta(t-t')\langle[\tilde{A}(t),\tilde{B}(t')]\rangle,$$
$$G_a^{AB}(t,t') = \frac{i}{\hbar}\theta(t'-t)\langle[\tilde{A}(t),\tilde{B}(t')]\rangle. \tag{7.20}$$

Operator in the Heisenberg representation reads

$$\tilde{A}(t) = e^{i\mathsf{H}t/\hbar}Ae^{-i\mathsf{H}t/\hbar},$$

where $\hbar$ – Planck's constant, $\mathsf{H} = H - \mu_e N_e$, μ_e is chemical potential of electron subsystem and N_e is the operator of electrons number:

$$N_e = \sum_{ni\gamma} a_{ni\gamma}^{+} a_{ni\gamma}.$$

In Eq. (7.20)

$$[A,B] = AB - \eta BA, \tag{7.21}$$

where for Bose operators A, B $\eta = 1$ and for Fermi operators $\eta = -1$, $\theta(t)$ is Heaviside's step function.

Brackets $\langle \ldots \rangle$ in Eq. (7.20) denote averaging

$$\langle A\rangle = \mathrm{Sp}(\rho A), \quad \rho = e^{(\Omega-\mathsf{H})/\Theta}, \tag{7.22}$$

where Ω is thermodynamic potential of the system, $\Theta = k_b T$, T – temperature.

Calculation of two-time retarded and accelerated Green's functions defined in Eq. (7.20) is based on the calculation of temperature Green's functions. Known relation between the spectral representation for retarded, accelerated and temperature Green's functions is used.

The temperature Green's function defined as

$$G^{AB}(\tau,\tau') = -\langle T_\tau \tilde{A}(\tau)\tilde{B}(\tau')\rangle, \tag{7.23}$$

where operator $\tilde{A}(\tau)$ is derived from $\tilde{A}(t)$ in Eq. (7.20) by replacing $t = -i\hbar\tau$,

$$\tilde{A}(\tau) = e^{\mathsf{H}\tau} A e^{-\mathsf{H}\tau},$$

$$T_\tau \tilde{A}(\tau)\tilde{B}(\tau') = \theta(\tau - \tau')\tilde{A}(\tau)\tilde{B}(\tau') + \eta\theta(\tau' - \tau)\tilde{B}(\tau')\tilde{A}(\tau),$$

here for Bose operators A, B $\eta = 1$ and for Fermi operators $\eta = -1$.

We introduce the operator

$$\sigma(\tau) = e^{\mathsf{H}_0\tau} e^{-\mathsf{H}\tau}, \tag{7.24}$$

where $\mathsf{H} = \mathsf{H}_0 + H_{\text{int}}$, $\mathsf{H}_0 = H_0 - \mu_e N_e$.

Differentiating the expression $\sigma(\tau)$ in Eq. (7.24) by τ parameter and consequently integrating by the condition$\sigma(0) = 1$, we obtain:

$$\sigma(\tau) = T_\tau \exp\left[-\int\limits_0^\tau H_{\text{int}}(\tau')d\tau'\right], \tag{7.25}$$

where $H_{\text{int}}(\tau) = e^{\mathsf{H}_0\tau} H_{\text{int}}\, e^{-\mathsf{H}_0\tau}$.

Taking into account expression (7.24) we can write

$$\tilde{A}(\tau) = \sigma^{-1}(\tau)A(\tau)\sigma(\tau). \tag{7.26}$$

Taking to account Eqs. (7.25), (7.26) for temperature Green's function in Eq. (7.23) we can obtain equation

$$G^{AB}(\tau, \tau') = -\frac{\langle T_\tau A(\tau)B(\tau')\sigma(1/\Theta)\rangle_0}{\langle\sigma(1/\Theta)\rangle_0} \tag{7.27}$$

where

$$\langle A\rangle_0 = \text{Sp}(\rho_0 A), \quad \rho_0 = e^{(\Omega_0 - \mathsf{H}_0)/\Theta}.$$

Expanding the exponent $\sigma(\tau)$ in expression (7.25) in a series of powers $H_{\text{int}}(\tau)$, substituting the result in Eq. (7.27) and using Wick's theorem for calculating the temperature Green's functions of disordered crystals [28] it is possible to build a diagram technique similar to a homogeneous system [1]. The denominator in Eq. (7.27) reduces from the same factor in the numerator. So Green's function can be expressed in a series only connected diagrams. Using the relation between the spectral representations of temperature and time Green's functions [1], by analytic continuation to real axis we obtain the following system of equations for retarded Green functions (hereinafter index r will be omitted) [28]:

$$G^{aa^+}(\varepsilon) = G_0^{aa^+}(\varepsilon) + G_0^{aa^+}(\varepsilon)\left(w + \Sigma_{eph}(\varepsilon) + \Sigma_{ee}(\varepsilon)\right) G^{aa^+}(\varepsilon)$$

$$\times G^{uu}(\varepsilon) = G_0^{uu}(\varepsilon) + G_0^{uu}(\varepsilon)\left(\Delta\Phi + \Sigma_{phe}(\varepsilon) + \Sigma_{phph}(\varepsilon)\right)$$

$$\times G^{uu}(\varepsilon) + G_0^{uP}(\varepsilon)\Delta M^{-1} G^{Pu}(\varepsilon),$$

$$
\begin{aligned}
G^{PP}(\varepsilon) &= G_0^{PP}(\varepsilon)G_0^{PP}(\varepsilon)\Delta M^{-1}G^{PP}(\varepsilon) + G_0^{Pu}(\varepsilon)\\
&\times\left(\Delta\Phi + \Sigma_{phe}(\varepsilon) + \Sigma_{phph}(\varepsilon)\right)G^{uP}(\varepsilon),\\
G^{uP}(\varepsilon) &= G_0^{uP}(\varepsilon) + G_0^{uP}(\varepsilon)\Delta M^{-1}G^{PP}(\varepsilon) + G_0^{uu}(\varepsilon)\\
&\times\left(\Delta\Phi + \Sigma_{phe}(\varepsilon) + \Sigma_{phph}(\varepsilon)\right)G^{uP}(\varepsilon),\\
G^{Pu}(\varepsilon) &= G_0^{Pu}(\varepsilon) + G_0^{Pu}(\varepsilon)\left(\Delta\Phi + \Sigma_{phe}(\varepsilon) + \Sigma_{phph}(\varepsilon)\right)\\
&\times G^{uu}(\varepsilon) + G_0^{PP}(\varepsilon)\Delta M^{-1}G^{Pu}(\varepsilon),
\end{aligned}
\tag{7.28}
$$

where $\varepsilon = \hbar\omega$.

Here $G^{aa^+}(\varepsilon)$, $G^{uu}(\varepsilon)$, $G^{PP}(\varepsilon)$, $G^{uP}(\varepsilon)$, $G^{Pu}(\varepsilon)$ are spectral representation of one-particle Green's function of the electrons subsystem and Green's functions "shift-shift", "impulse-impulse", "shift-impulse" "impulse-shift" of phonons subsystem; $\Sigma_{eph}(\varepsilon)$, $\Sigma_{phe}(\varepsilon)$, $\Sigma_{ee}(\varepsilon)$, $\Sigma_{phph}(\varepsilon)$ – actual energetic parts (mass operators) which describe the electron-phonon, phonon-electron, electron-electron and phonon-phonon interactions.

The spectral decomposition for two-time (7.20) and temperature (7.23) Green's functions is defined as:

$$
G_{r,a}^{AB}(t) = \frac{1}{2\pi}\int_{-\infty}^{\infty} G_{r,a}^{AB}(\omega)\,e^{-i\omega t}d\omega, \quad G_{r,a}^{AB}(\omega) = \int_{-\infty}^{\infty} G_{r,a}^{AB}(t)\,e^{i\omega t}dt,
$$

$$
G^{AB}(\tau) = \Theta\sum_{\omega_n} G^{AB}(\omega_n)\,e^{-i\omega_n\tau}, \quad G^{AB}(\omega_n) = \frac{1}{2}\int_{-1/\Theta}^{1/\Theta} G^{AB}(\tau)\,e^{i\omega_n\tau}d\tau,
$$

$$
\omega_n = \begin{cases} 2n\pi\,\Theta & \text{for Bose particles,}\\ (2n+1)\pi\,\Theta & \text{for Fermi particles,}\end{cases}
$$
$$
n = 0,\ \pm1,\ \pm2,\ldots
$$

Green's functions definitions Eq. (7.20), (7.23) leads to relation between the spectral representations of temperature and time Green's functions:

$$
G^{AB}(\omega_n) = \begin{cases} G_r^{AB}(i\omega_n/\hbar),\ \omega_n > 0,\\ G_a^{AB}(i\omega_n/\hbar),\ \omega_n < 0.\end{cases}
$$

To obtain Eq. 7.28 we used the above expression.

Accordingly to the indices $(ni\gamma)$ and $(ni\alpha)$ Green's functions of the electrons and phonons subsystems are matrices respectively.

From the motion equations for Green's functions of zero approximation can be obtained [41]:

$$G_0^{aa^+}(\varepsilon) = \left[\varepsilon - H_0^{(7.1)}\right]^{-1}, \quad H_0^{(7.1)} = \left\| h^{(0)}_{ni\gamma,n'i'\gamma'} \right\|,$$
$$G_0^{uu}(\varepsilon) = \left[\omega^2 \mathrm{M}^{(0)} - \Phi^{(0)}\right]^{-1}, \quad \Phi^{(0)} = \left\| \Phi^{(0)}_{ni\alpha,n'i'\alpha'} \right\|,$$
$$M^{(0)} = \left\| M_i \delta_{nn'} \delta_{ii'} \delta_{\alpha\alpha'} \right\|. \tag{7.29}$$

Providing that

$$\frac{\left(\frac{\varepsilon^2}{\hbar^2}\Delta M + \Delta\Phi + \Sigma_{phe}(\varepsilon) + \Sigma_{phph}(\varepsilon)\right)_{ni\alpha,n'i'\alpha'}}{\Phi^{(0)}_{ni\alpha,n'i'\alpha'}} \ll 1, \tag{7.30}$$

solution of the system in Eq. 7.28 has the form:

$$G^{aa^+}(\varepsilon) = \left[\left[G_0^{aa^+}(\varepsilon)\right]^{-1} - \left(w + \Sigma_{eph}(\varepsilon) + \Sigma_{ee}(\varepsilon)\right)\right]^{-1},$$
$$G^{uu}(\varepsilon) = \left[\left[G_0^{uu}(\varepsilon)\right]^{-1} - \left(\frac{\varepsilon^2}{\hbar^2}\Delta M + \Delta\Phi + \Sigma_{phe}(\varepsilon) + \Sigma_{phph}(\varepsilon)\right)\right]^{-1},$$
$$G^{PP}(\varepsilon) = \frac{\varepsilon^2}{\hbar^2}(M^{(0)})^2 G^{uu}(\varepsilon), \tag{7.31}$$

where $\Delta M = \|(M_i - M_{ni})\delta_{nn'}\delta_{ii'}\delta_{\alpha\alpha'}\|$, $\varepsilon = \hbar\omega$.

For solution of system in Eq. 7.28 members proportional to the second and higher power of small parameter (7.30) were neglected.

Using the mentioned above diagram technique, in Ref. [28], has been found explicit expressions for the mass operator of Green's functions that describe the many-particle interactions in the system.

The mass operator of Green's function of electrons for the electron-phonon interaction $\Sigma_{eph}(\tau, \tau')$ is described by the diagram in Fig. 7.1.

Solid lines in Fig. 7.1 correspond to the Green's function of electrons $G^{aa^+}_{ni\gamma,n'i'\gamma'}(\tau, \tau')$ and dashed lines correspond to the Green's function of phonons $G^{uu}_{ni\alpha,n'i'\alpha'}(\tau, \tau')$. Vertex part $\Gamma^{n_2 i_2 \alpha_2}_{ni\gamma,n_1 i_1 \gamma_1}(\tau_2, \tau, \tau_1)$ is described by diagrams in Fig. 7.2.

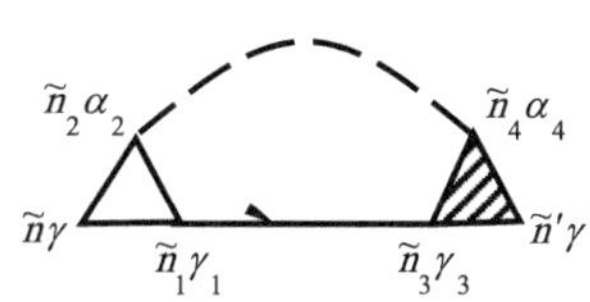

Fig. 7.1 Diagram for $\Sigma_{eph\ ni\gamma,n'i'\gamma'}(\tau, \tau') = \Sigma_{eph\ \tilde{n}\gamma,\tilde{n}'\gamma'}$. Here $\tilde{n} = (ni\tau)$

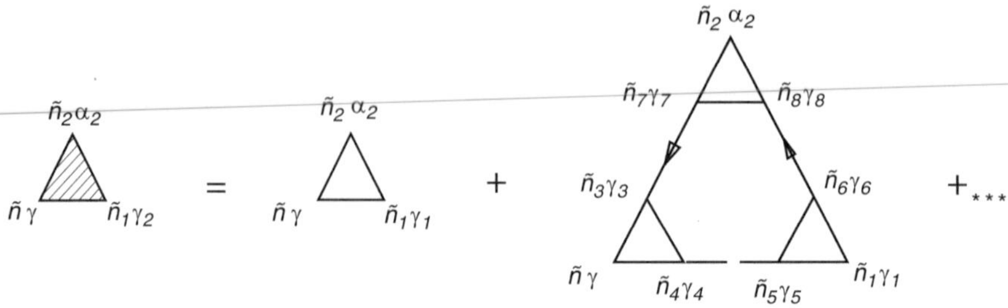

Fig. 7.2 Diagrams for the vertex part $\Gamma^{n_2 i_2 \alpha_2}_{ni\gamma,n_1 i_1 \gamma_1}(\tau_2,\tau,\tau_1) = \Gamma^{\tilde{n}_2\alpha_2}_{\tilde{n}\gamma,\tilde{n}_1\gamma_1}$. Here $\tilde{n} = (ni\tau)$

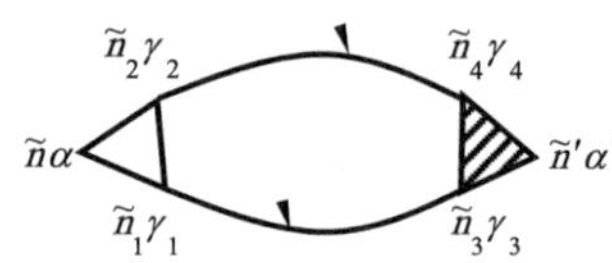

Fig. 7.3 Diagram for $\Sigma_{phe\ ni\alpha,n'i'\alpha'}(\tau,\tau') = \Sigma_{phe\ \tilde{n}\alpha,\tilde{n}'\alpha'}$. In Fig. 7.3 $\tilde{n} = (ni\tau)$

No shaded triangle in Fig. 7.2 corresponds to equation

$$\Gamma^{n_2 i_2 \alpha_2}_{0ni\gamma,n_1 i_1 \gamma_1}(\tau_2,\tau,\tau_1) = v'^{n_2 i_2 \alpha_2}_{ni\gamma,n_1 i_1 \gamma_1}\,\delta(\tau-\tau_2)\,\delta(\tau-\tau_1).$$

In Fig. 7.1 and in Fig. 7.2 summation for internal points $\tilde{n}\gamma$ is carried out. Summation of $\tilde{n}\gamma$ provides summation of $ni\gamma$ and integration over τ. Expressions that correspond instead each diagram attribute multiplier $(-1)^{n+F}$, where n is diagram's order (namely number of vertices Γ_0 in the diagram), and F is the number of lines for the Green's function of electrons G^{aa^+}. This function goes out and goes into in the same vertices.

For the mass operator that describes electron-phonon interaction, we obtained equation

$$\Sigma_{eph\,ni\gamma,n'i'\gamma'}(\varepsilon) = -\frac{1}{4\pi i}\int_{-\infty}^{\infty} d\varepsilon' \coth\left(\frac{\varepsilon'}{2\Theta}\right) v'^{n_1 i_1 \alpha_1}_{ni\gamma,n_3 i_3 \gamma_3}$$
$$\times\left[G^{uu}_{n_1 i_1 \alpha_1,n_2 i_2 \alpha_2}(\varepsilon') - G^{uu\,*}_{n_1 i_1 \alpha_1,n_2 i_2 \alpha_2}(\varepsilon')\right] G^{aa^+}_{n_3 i_3 \gamma_3,n_4 i_4 \gamma_4}$$
$$\times(\varepsilon-\varepsilon')\Gamma^{n_2 i_2 \alpha_2}_{n_4 i_4 \gamma_4,n'i'\gamma'}\left(\varepsilon-\varepsilon',\varepsilon;\varepsilon'\right). \qquad (7.32)$$

For repeated indices summation is conducted.

Phonon-electron interaction is described by the diagram in Fig. 7.3.

Designation in Fig. 7.3 corresponds to designations in Figs. 7.1 and 7.2.

Phonon-electron interaction is described by mass operator

$$\Sigma_{phenia,n'i'\alpha'}(\varepsilon) = \frac{1}{2\pi i}\int_{-\infty}^{\infty} d\varepsilon' f(\varepsilon')v'^{ni\alpha}_{n_2i_2\gamma_2,n_1i_1\gamma_1}$$
$$\times\left\{\left[G^{aa+}_{n_1i_1\gamma_1,n_3i_3\gamma_3}(\varepsilon+\varepsilon') - G^{aa+\,*}_{n_1i_1\gamma_1,n_3i_3\gamma_3}(\varepsilon+\varepsilon')\right]\right.$$
$$\times G^{aa+\,*}_{n_4i_4\gamma_4,n_2i_2\gamma_2}(\varepsilon') + G^{aa+}_{n_1i_1\gamma_1,n_3i_3\gamma_3}(\varepsilon+\varepsilon')$$
$$\left.\times\left[G^{aa+}_{n_4i_4\gamma_4,n_2i_2\gamma_2}(\varepsilon') - G^{aa+\,*}_{n_4i_4\gamma_4,n_2i_2\gamma_2}(\varepsilon')\right]\right\}$$
$$\times\Gamma^{\;n'i'\alpha'}_{n_3i_3\gamma_3,n_4i_4\gamma_4}\left(\varepsilon+\varepsilon',\varepsilon;-\varepsilon'\right). \tag{7.33}$$

Diagrams for the mass operator $\Sigma_{ee}(\tau,\tau')$ that describes electron-electron interaction, are shown in Fig. 7.4.

Vertex part $\Gamma^{n_2i_2\gamma_2,n_1i_1\gamma_1}_{ni\gamma,n'i'\gamma'}(\tau_2,\tau_1\tau,\tau')$ are shown on diagrams in Fig. 7.5.

Not shaded triangle in Fig. 7.5 corresponds to equation

$$\Gamma^{n_2i_2\gamma_2,n_1i_1\gamma_1}_{0ni\gamma,n'i'\gamma'}(\tau_2,\tau_1\tau,\tau') = \tilde{v}^{(2)ni\gamma,n_2i_2\gamma_2}_{n_1i_1\gamma_1,n'i'\gamma'}\delta$$
$$\times(\tau-\tau_2)\delta(\tau-\tau_1)\delta(\tau-\tau'),$$
$$\tilde{v}^{(2)ni\gamma,n_2i_2\gamma_2}_{n_1i_1\gamma_1,n'i'\gamma'} = v^{(2)ni\gamma,n_2i_2\gamma_2}_{n_1i_1\gamma_1,n'i'\gamma'} - v^{(2)ni\gamma,n_2i_2\gamma_2}_{n'i'\gamma',n_1i_1\gamma_1}.$$

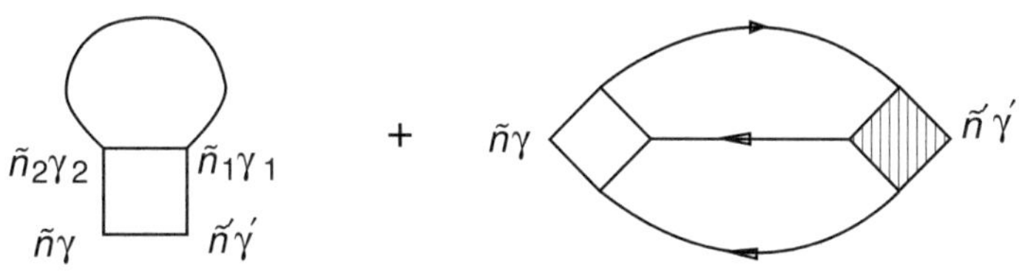

Fig. 7.4 Diagrams for $\Sigma_{eeni\gamma,n'i'\gamma'}(\tau,\tau') = \Sigma_{ee\tilde{n}\gamma,\tilde{n}'\gamma'}$. Here $\tilde{n} = (ni\tau)$

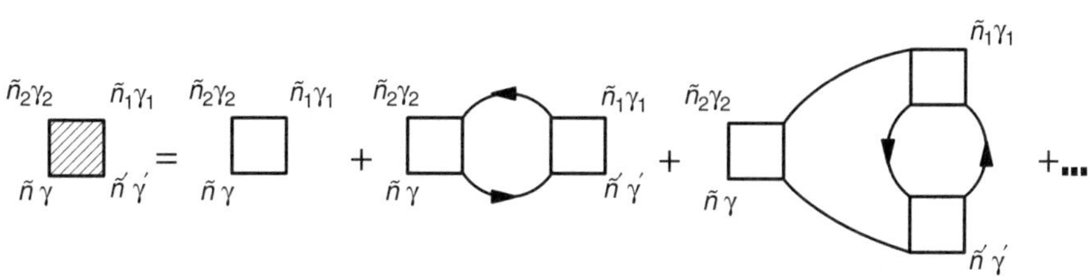

Fig. 7.5 Diagrams for vertex part $\Gamma^{n_2i_2\gamma_2,n_1i_1\gamma_1}_{ni\gamma,n'i'\gamma'}(\tau_2,\tau_1\tau,\tau') = \Gamma^{\tilde{n}_2\gamma_2,\tilde{n}_1\gamma_1}_{\tilde{n}\gamma,\tilde{n}'\gamma'}$. Here $\tilde{n} = (ni\tau)$

The mass operator that describes electron-electron interaction is:

$$\Sigma_{eeni\gamma,n'i'\gamma'}(\varepsilon) = \Sigma^{(1)}_{eeni\gamma,n'i'\gamma'} + \Sigma^{(2)}_{eeni\gamma,n'i'\gamma'}(\varepsilon),$$

$$\Sigma^{(1)}_{een,n'} = -\frac{1}{4\pi i}\int_{-\infty}^{\infty} d\varepsilon' f(\varepsilon')\, \tilde{v}^{(2)n,n_2}_{n_1,n'} \left[G^{aa+}_{n_1,n_2}(\varepsilon') - G^{aa+\,*}_{n_1,n_2}(\varepsilon')\right],$$

$$\Sigma^{(2)}_{een,n'}(\varepsilon) = -\left(\frac{1}{2\pi i}\right)^2 \int_{-\infty}^{\infty} d\varepsilon_1 \int_{-\infty}^{\infty} d\varepsilon_2 \tilde{v}^{(2)n,n_3}_{n_2,n_1} \Big\{ f(\varepsilon_1) f(\varepsilon_2)$$

$$\times \left[G^{aa+\,*}_{n_2,n_5}(\varepsilon - \varepsilon_1 + \varepsilon_2) G^{aa+}_{n_1,n_4}(\varepsilon_1) - G^{aa+}_{n_2,n_5}(\varepsilon - \varepsilon_1 + \varepsilon_2) G^{aa+\,*}_{n_1,n_4}(\varepsilon_1)\right]$$

$$\times \left[G^{aa+}_{n_6,n_3}(\varepsilon_2) - G^{aa+\,*}_{n_6,n_3}(\varepsilon_2)\right] \Gamma^{n_5,n_6}_{n_4,n'}(\varepsilon_1, \varepsilon - \varepsilon_1 + \varepsilon_2; \varepsilon_2, \varepsilon)$$

$$+ f(\varepsilon_1) f(\varepsilon_1 + \varepsilon_2 - \varepsilon) \left[G^{aa+}_{n_2,n_5}(\varepsilon_2) - G^{aa+\,*}_{n_2,n_5}(\varepsilon_2)\right]$$

$$\times \left[G^{aa+}_{n_1,n_4}(\varepsilon_1) G^{aa+}_{n_6,n_3}(\varepsilon_1 + \varepsilon_2 - \varepsilon) - G^{aa+\,*}_{n_1,n_4}(\varepsilon_1) G^{aa+\,*}_{n_6,n_3}(\varepsilon_1 + \varepsilon_2 - \varepsilon)\right]$$

$$\times \Gamma^{n_5,n_6}_{n_4,n'}(\varepsilon_1, \varepsilon_2; \varepsilon_1 + \varepsilon_2 - \varepsilon, \varepsilon) \Big\}, \qquad (7.34)$$

$\tilde{v}^{(7.2)n,n_2}_{n_1,n'} = v^{(7.2)n,n_2}_{n_1,n'} - v^{(7.2)n,n_2}_{n',n_1}$, $(n \equiv ni\gamma)$.

In Ref. [28] expressions for the mass operator $\Sigma_{phph}(\varepsilon)$ that describe phonon-phonon interaction similarly were obtained.

The relations arising from the theory of functions of complex variable are used upon receipt of expressions (7.32), (7.33) and (7.34).

$$\Theta \sum_{\omega_n} \phi(i\omega_n) = \frac{1}{4\pi i} \oint_C dz \coth\left(\frac{z}{2\Theta}\right) \phi(z) \quad (\omega_n = 2n\pi\Theta),$$

$$\Theta \sum_{\omega_n} \phi(i\omega_n) = -\frac{1}{2\pi i} \oint_C dz f\left(\frac{z}{\Theta}\right) \phi(z) \quad (\omega_n = (2n+1)\pi\Theta),$$

$$f\left(\frac{z}{\Theta}\right) = \frac{1}{\exp\left(\frac{z}{\Theta}\right) + 1},$$

where $\phi(z)$ is analytic function of complex z in the region covered by a contour C.

Between spectral representations of the Green's functions $G_\mu(\varepsilon)$ defined at constant chemical potential μ_e by Eq. (7.20), and $G_N(\varepsilon)$, defined at a constant number of electrons N_e a relation $G_\mu(\varepsilon) = G_N(\varepsilon + \mu)$ exists.

Renormalization of diagrams vertex parts can be neglected.

$$\Gamma^{\lambda_2 n_2 i_2 \alpha_2}_{n_4 i_4 \gamma_4, n' i' \gamma'}(\varepsilon - \varepsilon', \varepsilon; \varepsilon') = v'^{\lambda_2 n_2 i_2 \alpha_2}_{n_4 i_4 \gamma_4, n' i' \gamma'},$$

$$\Gamma^{n_5,n_6}_{n_4,n'}(\varepsilon_1, \varepsilon - \varepsilon_1 + \varepsilon_2; \varepsilon_2, \varepsilon) = \tilde{v}^{(2)n_4,n_5}_{n_6,n'}.$$

In expressions (7.33) and (7.34) $f(\varepsilon)$ is Fermi function. Fermi level $\varepsilon_{\mathrm{F}} \equiv \mu_e$ of system is determined by the equation:

$$\langle Z \rangle = \int_{-\infty}^{\infty} f(\varepsilon, \varepsilon_{\mathrm{F}}) g_e(\varepsilon) d\varepsilon, \tag{7.35}$$

where $\langle Z \rangle$ – average number of electrons per atom, $g_e(\varepsilon)$ – density of electron states.

$$g_e(\varepsilon) = -\frac{1}{\pi \nu N} \mathrm{ImSp} \left\langle G^{aa^+}(\varepsilon) \right\rangle_c . \tag{7.36}$$

In Eq. (7.36) brackets $\langle \ldots \rangle_c$ denote configurational averaging, N is a number of primitive cells, ν – number of atoms in the primitive cell. To further reduce of recordings the index c near brackets $\langle \ldots \rangle_c$ will be omitted. In the formula (7.35) $\langle Z \rangle$ – average number of electrons per atom.

Equation (7.36) follows from the definition of operator number of electrons in Eq. (7.20) and Green's function $G^{AB}(\tau, \tau')$ (7.23), $A = a^+_{ni\gamma}$, $B = a_{ni\gamma}$.

It should be noted that the first term $\Sigma^{(1)}_{eeni\gamma,n'i'\gamma'}$ in Eq. (7.34) for the mass operator of electron-electron interaction describes the Coulomb and exchange electron-electron interactions in the Hartree–Fock approximation. The second term $\Sigma^{(2)}_{eeni\gamma,n'i'\gamma'}(\varepsilon)$ which is caused by output beyond the Hartree–Fock approximation, describes the electronic correlation. In contradiction to Refs. [6, 7, 9–11, 14, 26] long-range Coulomb interaction of electrons located at the different sites of the crystal lattice is described taking into account an arbitrary number of energy bands.

Expression in Eq. (7.31) differs from corresponding expressions for the Green's function of the single-particle Hamiltonian of disordered system only by form of mass operators. Therefore to calculate the Green's function given by Eq. (7.31) we use well known methods of the disordered systems theory [28].

7.4 Localized Magnetic Moments

Obtained results give possibilities to investigate the influence of electronic correlations on the electronic structure and properties of the system. To perform this, the heterogeneous distribution of electron density is taken into account. We assume that distribution of electron density corresponds to the minimum of free energy. The mass operator of electron-electron interaction in Eq. (7.34) can be defined through the occupation number $Z^{\lambda m_{\lambda i}}_{ni\gamma\sigma}$ of electron's states $(ni\gamma\sigma)$. The value $Z^{\lambda m_{\lambda i}}_{ni\gamma\sigma}$ is determined by Eq. (7.35), where density of electronic states $g_e(\varepsilon)$ is replaced by conditional partial density of states $g^{\lambda m_{\lambda i}}_{ni\gamma\sigma}(\varepsilon)$ for energy band γ and spin projection σ.

The occupation number of electron states $Z_{ni\gamma\sigma}^{\lambda m_{\lambda i}}$ and conditional partial density of states $g_{ni\gamma\sigma}^{\lambda m_{\lambda i}}(\varepsilon)$ are

$$Z_{ni\gamma\sigma}^{\lambda m_{\lambda i}} = \int_{-\infty}^{\infty} f(\varepsilon, \varepsilon_{\mathrm{F}})\, g_{ni\gamma\sigma}^{\lambda m_{\lambda i}}(\varepsilon) d\varepsilon,$$

$$g_{ni\gamma\sigma}^{\lambda m_{\lambda i}}(\varepsilon) = -\frac{1}{\pi}\, \mathrm{Im} \left\langle G_{ni\gamma\sigma, ni\gamma\sigma}^{aa^{+}}(\varepsilon) \right\rangle \Big|_{(ni)\in\lambda m_{\lambda i}}. \tag{7.37}$$

In the last formula averaging is done along with condition that atom of sort λ is located at the site (ni), and projection of localized magnetic moment equals to $m_{\lambda i}$. The probability of such event is $P_{ni}^{\lambda m_{\lambda i}}$ and we can write $\sum_{\lambda, m_{\lambda i}} P_{ni}^{\lambda m_{\lambda i}} = 1$.

Thus, we consider the localized magnetic moments which are distributed not homogeneously on the crystal lattice sites (static magnetization fluctuations).

We can also write equations

$$Z_{ni\gamma}^{\lambda m_{\lambda i}} = Z_{ni\gamma\sigma}^{\lambda m_{\lambda i}} + Z_{ni\gamma,-\sigma}^{\lambda m_{\lambda i}}, \quad m_{\lambda i\gamma} = Z_{ni\gamma\sigma}^{\lambda m_{\lambda i}} - Z_{ni\gamma,-\sigma}^{\lambda m_{\lambda i}}. \tag{7.38}$$

From Eq. (7.38) we obtain

$$Z_{ni\gamma\sigma}^{\lambda m_{\lambda i}} = \frac{Z_{ni\gamma}^{\lambda m_{\lambda i}} + m_{\lambda i\gamma}}{2}, \quad Z_{ni\gamma,-\sigma}^{\lambda m_{\lambda i}} = \frac{Z_{ni\gamma}^{\lambda m_{\lambda i}} - m_{\lambda i\gamma}}{2}.$$

Number of electrons and magnetic moment projection for the atom of sort λ at the site (ni) are, respectively:

$$Z_{ni}^{\lambda m_{\lambda i}} = \sum_{\gamma} Z_{ni\gamma}^{\lambda m_{\lambda i}}, \quad m_{\lambda i} = \sum_{\gamma} m_{\lambda i\gamma}.$$

In Eq. (7.31) by introducing the mass operator as the sum of one site operators and selecting as a zero one site approximation of the effective medium Green's function cluster expansion for Green's functions $G^{aa^{+}}(\varepsilon)$, $G^{uu}(\varepsilon)$ is performed. Specified expansion is a generalization of the cluster expansion for the Green's function $G^{aa^{+}}(\varepsilon)$ of single-particle Hamiltonian.

Green's functions in Eq. (7.31) satisfy this equation:

$$G(\varepsilon) = \tilde{G}(\varepsilon) + \tilde{G}(\varepsilon) T(\varepsilon) \tilde{G}(\varepsilon),$$

where T is matrix of scattering and could be represented expansion in series, where which terms describe scattering of the clusters with different numbers of nodes

$$T = \sum_{(n_1 i_1)} t^{n_1 i_1} + \sum_{(n_1 i_1)\neq(n_2 i_2)} T^{(2) n_1 i_1, n_2 i_2} + \ldots$$

Here $T^{(2)n_1i_1,n_2i_2} = \left[I - t^{n_1i_1}\tilde{G}t^{n_2i_2}\tilde{G}\right]^{-1} t^{n_1i_1}\tilde{G}t^{n_2i_2}\left[I + \tilde{G}t^{n_1i_1}\right]$, where $t^{n_1i_1}$ is scattering operator on the same site is determined by equation $t^{n_1i_1} = \left[I - (\Sigma^{n_1i_1} - \sigma^{n_1i_1})\tilde{G}\right]^{-1}(\Sigma^{n_1i_1} - \sigma^{n_1i_1})$. This is $\sigma^{n_1i_1}$ potential of the effective medium (coherent potentials).

By using Eqs. (7.36) and (7.37) and performing averaging over the distribution of atoms of different sort and localized magnetic moments projection at the sites of the crystal lattice and neglecting the contribution of processes of electron scattering in clusters consisting of three or more atoms that are small by the small parameter for the density of electronic states we obtain:

$$g_e(\varepsilon) = \frac{1}{v}\sum_{i,\gamma,\sigma,\lambda,m_{\lambda i}} P_{0i}^{\lambda m_{\lambda i}} g_{0i\gamma\sigma}^{\lambda m_{\lambda i}}(\varepsilon),$$

$$g_{0i\gamma\sigma}^{\lambda m_{\lambda i}}(\varepsilon) = -\frac{1}{\pi}\mathrm{Im}\Big\{\tilde{G} + \tilde{G}t_{0i}^{\lambda m_{\lambda i}}\tilde{G} + \sum_{\substack{(lj)\neq(0i)\\ \lambda', m_{\lambda' j}}} P_{lj\,0i}^{\lambda' m_{\lambda' j}/\lambda m_{\lambda i}}$$

$$\times\tilde{G}\left[t_{lj}^{\lambda' m_{\lambda' j}} + T^{(2)\lambda m_{\lambda i}0i,\lambda' m_{\lambda' j}lj} + T^{(2)\lambda' m_{\lambda' j}lj,\lambda m_{\lambda i}0i}\right]\tilde{G}\Big\}^{0i\gamma\sigma,0i\gamma\sigma},$$

$$T^{(2)\lambda m_{\lambda i}0i,\lambda' m_{\lambda' j}lj} = \left[I - t^{\lambda m_{\lambda i}0i}\tilde{G}t^{\lambda' m_{\lambda' j}lj}\tilde{G}\right]^{-1}$$

$$\times t^{\lambda m_{\lambda i}0i}\tilde{G}t^{\lambda' m_{\lambda' j}lj}\left[I + \tilde{G}t^{\lambda m_{\lambda i}0i}\right], \tag{7.39}$$

where $\tilde{G} = \tilde{G}^{aa^+}(\varepsilon)$.

Energy spectrum calculation is done by iterative procedure. The first step of iterative procedure is described above.

On the next steps of iterative procedure values of occupation number of electron basis states in Eq. (7.14) is calculated by expression:

$$Z_{i\delta\sigma}^{\lambda} = \frac{1}{2}\left(Z_{i\delta c\sigma}^{\lambda} + Z_{i\delta s\sigma}^{\lambda}\right), \tag{7.40}$$

where $Z_{i\delta c\sigma}^{\lambda}, Z_{i\delta s\sigma}^{\lambda}$ defined by expression (7.37) for $Z_{ni\gamma\sigma}^{\lambda m_{\lambda i}}$, $\gamma = (\delta c), (\delta s)$ and $\delta = (\tilde{\varepsilon} lm)$. Expression (7.40) obtained using Eq. (7.16).

Above mentioned procedure of energy spectrum calculation is self-consistent.

7.5 Free Energy

Thermodynamic potential of the system is

$$\Omega = -\Theta \ln \mathrm{Sp}(e^{-H/\Theta}).$$

Free energy F as a function of volume V, temperature T, number of electrons N_e and parameters of interatomic correlations $(\varepsilon_{n_1 i_1, n_2 i_2}, \eta)$ to the thermodynamic potential Ω is expressed via expression $F = \Omega + \mu_e \langle N_e \rangle$. Free energy F in the weak dependence of the mass operators on the energy of electrons and phonon approximation can be represented in the form [17–19, 28]

$$F = \langle \delta \Phi \rangle - \Theta S_c + \Omega_e + \Omega_{ph} + \mu_e \langle Z \rangle, \tag{7.41}$$

where Ω_e, Ω_{ph} are given by

$$\Omega_e = -\Theta \int\limits_{-\infty}^{\infty} \ln\left(1 + e^{(\mu_e - \varepsilon)/\Theta}\right) g_e(\varepsilon) d\varepsilon,$$

$\Omega_{ph} = \Theta \int_{-\infty}^{\infty} \ln\left(1 - e^{-\varepsilon/\Theta}\right) g_{ph}(\varepsilon) d\varepsilon$ – the thermodynamic potential of atomic nucleus in field of ionic cores.

In Eq. (7.41) the values F, S_c, Ω_e, Ω_{ph} are calculated per one atom.

7.6 Electrical Conductivity

For electrical conductivity tensor calculation Kubo's formula is used [20]

$$\sigma_{\alpha\beta}(\omega) = \int\limits_0^{1/\Theta} \int\limits_0^{\infty} e^{i\omega t - \delta t} \left\langle \tilde{J}_\beta(0) \tilde{J}_\alpha(t + i\hbar\tau) \right\rangle d\tau dt, \tag{7.42}$$

where J_α is the operator of the α is projection of the current density.

From Eq. (7.42) it follows:

$$\operatorname{Re} \sigma_{\alpha\beta}(\omega) = \frac{i}{2\omega} \left[G_r^{J_\alpha J_\beta}(\omega) - G_a^{J_\alpha J_\beta}(\omega) \right].$$

To calculate the spectral representations $G_r^{J_\alpha J_\beta}(\omega)$ and $G_a^{J_\alpha J_\beta}(\omega)$ of the retarded and the advanced Green's functions the expression for current-density operator is used

$$J_\alpha(t) = e \int \Psi^+(\xi, t)\, v_\alpha \Psi(\xi, t) d\xi,$$

where $\Psi^+(\xi, t)$ and $\Psi(\xi, t)$ are field operators of electron's creation and annihilation respectively; v_α is the operator of the α – projection of the velocity; e is the electron charge; by integrating over ξ we mean integrating over the crystal volume and summing over the projection of spin σ onto the z axis; and the crystal's volume is assumed to be equal to unity.

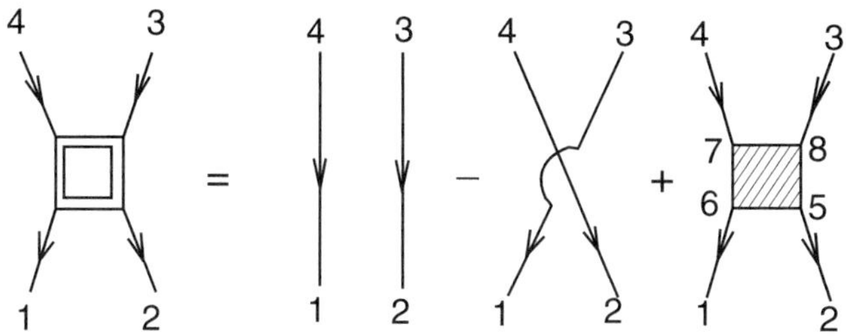

Fig. 7.6 Diagrams for the two-particle Green's function

In this case the temperature Green's function is

$$G^{J_\alpha J_\beta}(\tau,\tau') = \frac{e^2}{NV_1}\sum_{n_1n_2n_3n_4} v_{\alpha n_4 n_2} v_{\beta n_3 n_1} G''\left(n_1\tau', n_2\tau, n_3\tau', n_4\tau\right),$$

where V_1 – volume of primitive cell, the two-particle Green's function is

$$\begin{gathered} G''(n_1\tau', n_2\tau, n_3\tau', n_4\tau) \\ = \left\langle T_\tau a_{n_1}\left(\tau'\right) a_{n_2}(\tau)\, a^+_{n_3}\left(\tau'\right) a^+_{n_4}(\tau)\,\sigma\left(1/\theta\right)\right\rangle_0 \left\langle \sigma\left(1/\theta\right)\right\rangle_0^{-1}, \\ (n = ni\gamma). \end{gathered}$$

The two-particle Green's function is described by the diagram in Fig. 7.6. Numbers of Fig. 7.6 correspond to point numbers, e.g., 1 corresponds to $(n_1 i_1 \gamma_1 \tau_1)$.

Using the diagram technique for two-particle temperature Green's function and neglecting the contributions of scattering processes on clusters of three or more sites for the static conductivity tensor we can get

$$\begin{aligned} \sigma_{\alpha\beta} = \frac{e^2\hbar}{4\pi V_1}\Bigg\{ &\int_{-\infty}^{\infty} d\varepsilon_1 \frac{\partial f}{\partial \varepsilon_1} \sum_{s,s'=+,-} (2\delta_{ss'}-1) \sum_{\sigma\gamma,i} \Bigg\{ \left[v_\beta \tilde{K}\left(\varepsilon_1^s, v_\alpha, \varepsilon_1^{s'}\right)\right] \\ &+ \sum_{\lambda,m_{\lambda i}} P_{0i}^{\lambda m_{\lambda i}} \tilde{K}(\varepsilon_1^{s'}, v_\beta, \varepsilon_1^s)(t_{0i}^{\lambda m_{\lambda i}}(\varepsilon_1^s)\tilde{K}(\varepsilon_1^s, v_\alpha, \varepsilon_1^{s'}) t_{0i}^{\lambda m_{\lambda i}}(\varepsilon_1^{s'}) \\ &+ \sum_{\lambda,m_{\lambda i}} P_{0i}^{\lambda m_{\lambda i}} \sum_{\substack{lj\neq 0i,\\ \lambda', m_{\lambda' j}}} P_{lj0i}^{\lambda' m_{\lambda' j}/\lambda m_{\lambda i}} \Big[[\tilde{K}(\varepsilon_1^{s'}, v_\beta, \varepsilon_1^s) v_\alpha \tilde{G}(\varepsilon_1^{s'})] \\ &\qquad \times T^{(2)\lambda m_{\lambda i} 0i, \lambda' m_{\lambda' j} lj}(\varepsilon_1^{s'}) \\ &\quad + [\tilde{K}(\varepsilon_1^{s'}, v_\beta, \varepsilon_1^s) v_\alpha \tilde{G}(\varepsilon_1^{s'})] T^{(2)\lambda' m_{\lambda' j} lj, \lambda m_{\lambda i} 0i}(\varepsilon_1^{s'}) \\ &\quad + [\tilde{K}(\varepsilon_1^s, v_\alpha, \varepsilon_1^{s'}) v_\beta \tilde{G}(\varepsilon_1^s)] T^{(2)\lambda m_{\lambda i} 0i, \lambda' m_{\lambda' j} lj}(\varepsilon_1^s) \end{aligned}$$

$$+[\tilde{K}(\varepsilon_1^s, v_\alpha, \varepsilon_1^{s'})v_\beta \tilde{G}(\varepsilon_1^s)]T^{(2)\lambda' m_{\lambda'j} lj, \lambda m_{\lambda i} 0i}(\varepsilon_1^s)$$

$$+\tilde{K}(\varepsilon_1^{s'}, v_\beta, \varepsilon_1^s)\Big[(t_{lj}^{\lambda' m_{\lambda' j}}(\varepsilon_1^s)\tilde{K}(\varepsilon_1^s, v_\alpha, \varepsilon_1^{s'})t_{0i}^{\lambda m_{\lambda i}}(\varepsilon_1^{s'})$$

$$+t_{lj}^{\lambda' m_{\lambda' j}}(\varepsilon_1^s)\tilde{K}(\varepsilon_1^s, v_\alpha, \varepsilon_1^{s'})T^{(2)\lambda m_{\lambda i} 0i, \lambda' m_{\lambda' j} lj}(\varepsilon_1^{s'})$$

$$+T^{(2)\lambda' m_{\lambda' j} lj, \lambda m_{\lambda i}\, 0i}(\varepsilon_1^s)\tilde{K}(\varepsilon_1^s, v_\alpha, \varepsilon_1^{s'})t_{0i}^{\lambda m_{\lambda i}}(\varepsilon_1^{s'})$$

$$+T^{(2)\lambda' m_{\lambda' j} lj, \lambda m_{\lambda i} 0i}(\varepsilon_1^s)\tilde{K}(\varepsilon_1^s, v_\alpha, \varepsilon_1^{s'})T^{(2)\lambda m_{\lambda i} 0i, \lambda' m_{\lambda' j} lj}(\varepsilon_1^{s'})$$

$$+T^{(2)\lambda' m_{\lambda' j} lj, \lambda m_{\lambda i} 0i}(\varepsilon_1^s)\tilde{K}(\varepsilon_1^s, v_\alpha, \varepsilon_1^{s'})T^{(2)\lambda' m_{\lambda' j} lj, \lambda m_{\lambda i} 0i}(\varepsilon_1^{s'}))\Big]\Big]\Big\}^{0i\gamma\sigma, 0i\gamma\sigma}$$

$$+\int_{-\infty}^{\infty}\int_{-\infty}^{\infty} d\varepsilon_1 d\varepsilon_2 f(\varepsilon_1) f(\varepsilon_2) \left\langle \Delta G_{\alpha\beta}^{II}(\varepsilon_1, \varepsilon_2)\right\rangle\Bigg\}, \tag{7.43}$$

where

$$\tilde{K}\left(\varepsilon_1^s, v_\alpha, \varepsilon_1^{s'}\right) = \tilde{G}^{aa^+}\left(\varepsilon_1^s\right) v_\alpha \tilde{G}^{aa^+}\left(\varepsilon_1^{s'}\right),$$

$$\tilde{G}^{aa^+}\left(\varepsilon_1^+\right) = \tilde{G}_r^{aa^+}(\varepsilon_1),$$

$$\tilde{G}^{aa^+}\left(\varepsilon_1^-\right) = \tilde{G}_a^{aa^+}(\varepsilon_1) = \left(\tilde{G}_r^{aa^+}\right)^*(\varepsilon_1).$$

In Eq. (7.43) component of two-particle Green's function $\Delta G_{\alpha\beta}^{II}(\varepsilon_1, \varepsilon_2)$ is caused by the interaction and has the form:

$$\Delta G_{\alpha\beta}^{II}(\varepsilon_1, \varepsilon_2) = \frac{i}{2\pi} v_{\alpha n_4 n_2} v_{\beta n_3 n_1}\Big\{\Big[G_{rn_1n_6}^{aa^+}(\varepsilon_1) - G_{an_1n_6}^{aa^+}(\varepsilon_1)\Big]$$

$$\times\Big[G_{rn_2n_5}^{aa^+}(\varepsilon_2) - G_{an_2n_5}^{aa^+}(\varepsilon_2)\Big]\Big[G_{an_7n_4}^{aa^+}(\varepsilon_2)\, G_{rn_8n_3}^{aa^+}(\varepsilon_1)$$

$$- G_{rn_7n_4}^{aa^+}(\varepsilon_2)\, G_{an_8n_3}^{aa^+}(\varepsilon_1)\Big] + G_{an_1n_6}^{aa^+}(\varepsilon_1)\Big[G_{rn_2n_5}^{aa^+}(\varepsilon_2) - G_{an_2n_5}^{aa^+}(\varepsilon_2)\Big]$$

$$\times G_{an_7n_4}^{aa^+}(\varepsilon_2)\Big[G_{rn_8n_3}^{aa^+}(\varepsilon_1) - G_{an_8n_3}^{aa^+}(\varepsilon_1)\Big] - G_{rn_1n_6}^{aa^+}(\varepsilon_1)$$

$$\times\Big[G_{rn_2n_5}^{aa^+}(\varepsilon_2) - G_{an_2n_5}^{aa^+}(\varepsilon_2)\Big]\, G_{rn_7n_4}^{aa^+}(\varepsilon_2)\Big[G_{rn_8n_3}^{aa^+}(\varepsilon_1) - G_{an_8n_3}^{aa^+}(\varepsilon_1)\Big]$$

$$+\Big[G_{an_1n_6}^{aa^+}(\varepsilon_1)\, G_{rn_2n_5}^{aa^+}(\varepsilon_2) - G_{rn_1n_6}^{aa^+}(\varepsilon_1)\, G_{an_2n_5}^{aa^+}(\varepsilon_2)\Big]$$

$$\times\Big[G_{rn_7n_4}^{aa^+}(\varepsilon_2) - G_{an_7n_4}^{aa^+}(\varepsilon_2)\Big]\Big[G_{rn_8n_3}^{aa^+}(\varepsilon_1) - G_{an_8n_3}^{aa^+}(\varepsilon_1)\Big]$$

$$+\Big[G_{rn_1n_6}^{aa^+}(\varepsilon_1) - G_{an_1n_6}^{aa^+}(\varepsilon_1)\Big]\, G_{rn_2n_5}^{aa^+}(\varepsilon_2)\Big[G_{rn_7n_4}^{aa^+}(\varepsilon_2) - G_{an_7n_4}^{aa^+}(\varepsilon_2)\Big]$$

$$+\left[G^{aa+}_{rn_1n_6}(\varepsilon_1)-G^{aa+}_{an_1n_6}(\varepsilon_1)\right]G^{aa+}_{rn_2n_5}(\varepsilon_2)\left[G^{aa+}_{rn_7n_4}(\varepsilon_2)-G^{aa+}_{an_7n_4}(\varepsilon_2)\right]$$
$$\times G^{aa+}_{rn_8n_3}(\varepsilon_1)-\left[G^{aa+}_{rn_1n_6}(\varepsilon_1)-G^{aa+}_{an_1n_6}(\varepsilon_1)\right]G^{aa+}_{an_2n_5}(\varepsilon_2)$$
$$\times\left[G^{aa+}_{rn_7n_4}(\varepsilon_2)-G^{aa+}_{an_7n_4}(\varepsilon_2)\right]G^{aa+}_{an_8n_3}(\varepsilon_1)\Big\}\,\Gamma^{n_6n_7}_{n_5n_8}(\varepsilon_1,\varepsilon_2;\varepsilon_2,\varepsilon_1) \qquad (7.44)$$

$(n \equiv ni\gamma\sigma)$.

Operator α-projection of the electron velocity in Eq. (7.43) is:

$$v_\alpha(\mathbf{k})=\frac{1}{\hbar}\frac{\partial H_0^{(1)}(k)}{\partial k_\alpha}.$$

To simplify the formula (7.44) we use approximate expression $\left\langle\Delta G^{II}_{\alpha\beta}(\varepsilon_1;\varepsilon_2)\right\rangle \approx \Delta\tilde{G}^{II}_{\alpha\beta}(\varepsilon_1;\varepsilon_2)$, where $\Delta\tilde{G}^{II}_{\alpha\beta}(\varepsilon_1;\varepsilon_2)$ is derived from the Eq. (7.43) by replacing the $G^{aa^+}(\varepsilon)$ with $\tilde{G}^{aa^+}(\varepsilon)$.

7.7 Spin-Dependent Transport of Carbon Nanotubes with Chromium Atoms

In this section the results of calculation of the energy spectrum of electrons and phonons and conductivity of carbon nanotubes doped with chromium are represented. Renormalization of vertex parts diagrams for the mass operators in Eqs. (7.32), (7.33) and (7.34) and contribution of the static displacements of atoms in the calculations are neglected. The contribution of matrix elements on the basis of real wave functions $v^{(2)n_1i_1\gamma_1,n_2i_2\gamma_2}_{n_3i_3\gamma_3,n_4i_4\gamma_4}$ in Eq. (7.34) for different sites (ni) is neglected. As the basis $2s$, $2p$-states wave functions of the neutral carbon atoms and 3d, 4s-states wave functions of the neutral chromium atoms were chosen. The initial ion core's valence of C and Cr atoms Z^{λ_i} are 4 and 6 respectively. To simplify calculations in the above-mentioned self-consistent iterative procedure was performed only first step. In expression (7.14) $Z^{\lambda}_{i\delta\sigma} = 1$ for occupied electronic states. The off-diagonal matrix elements (ni) by site index of Hamiltonian in Eq. (7.1) were calculated by taking into account the first three coordination spheres. For Green function calculation 10^3 points for Brillouin zone are used. For mass operator calculation in Eqs. (7.32), (7.33) and (7.34) numerical integration is applied. Calculations were performed for the temperature $T = 300$ K.

We performed geometry optimization of the crystal structure of carbon nanotube with (3, 0) chirality doped by Cr atoms. Geometric optimization of the crystal structure was achieved by minimization the free energy F, defined in Eq. (7.41). Carbon nanotube doped with Cr has a one-dimensional crystal structure. Primitive cell contains 18 non-equivalent atom positions. Carbon atoms are located in 12

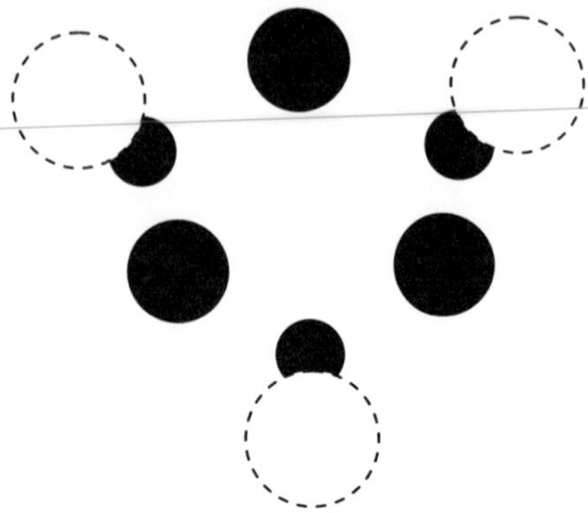

Fig. 7.7 The crystal structure of carbon nanotube with (3,0) chirality doped by Cr atoms. The top view. *Black* – C atoms, *white* – Cr atoms

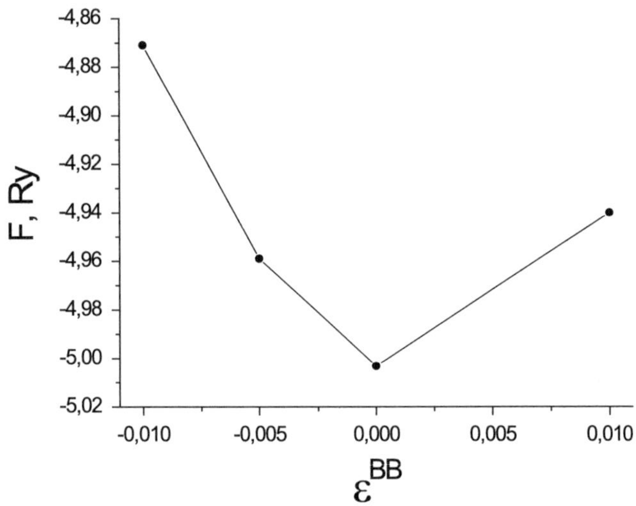

Fig. 7.8 Dependence of free energy F for carbon nanotubes with 5 atoms of Cr per primitive cell on parameter of pair correlations in the arrangement of Cr impurities on lattice sites ε^{BB}

positions on the surface of the inner cylinder. The distance between the carbon atoms is 0.142 nm. Cr atoms are located in 6 positions on the outer surface of the cylinder opposite the center of a hexagon, the vertices of which are carbon atoms. The distance between carbon atoms and Cr is 0.22 nm. The crystal structure of carbon nanotube with (3,0) chirality with Cr impurity is showed at Fig. 7.7.

Through the study of free energy minimum found that Cr atoms are randomly located on the surface of nanotubes. In Fig. 7.8 points shows the dependence of the free energy F in Eq. (7.41) on the parameter of pair correlations in the arrangement of Cr impurities on lattice sites $\varepsilon^{BB} = \varepsilon^{BB}_{lj0i}$ in Eq. (7.33) for the first coordination sphere. Atom of Cr is denoted as atom of sort B. The dependence $F(\varepsilon^{BB})$ is shown in the region of free energy minimum.

As shown in Fig. 7.8, the free energy F minimum corresponds to $\varepsilon^{BB} = 0$. This is a result of Cr atoms are randomly located on the surface of carbon nanotube.

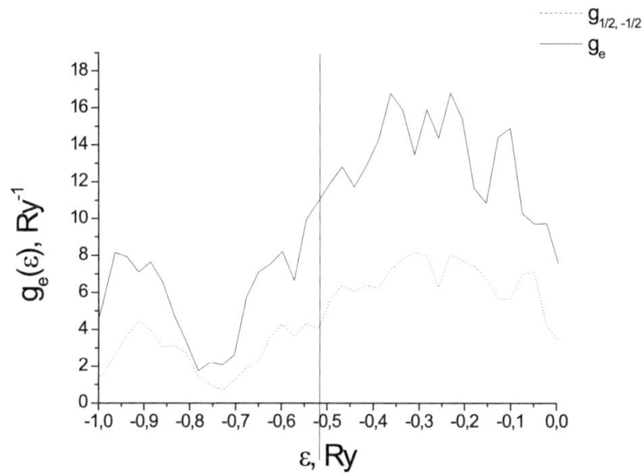

Fig. 7.9 Densities of electron states of carbon nanotube with an admixture of Cr

The relative position of carbon atoms and Cr is similar to the location of atoms of transition metals on the surface of carbon nanotubes of large diameter, which are described in Ref. [39, 40] by ultrasoft pseudopotential method.

The value of localized magnetic moment projection of the atom Cr and induced localized magnetic moment of an atom C in the direction of the magnetic field increases with the size of the field. For carbon nanotubes of 5 Cr atoms in primitive cell value of projection of magnetic moment of the atom Cr varies within $m_{\mathrm{Cr}} = (1.02; 2,24)\mu_B$, and the magnetic moment of the atom C – within $m_{\mathrm{C}} = (0,0036; 0,02)\mu_B$ with increasing values of the magnetic field from zero to $H = 200$ A/m. The magnetic field is oriented along the axis of the carbon nanotube. Parameter of pair correlations in the orientation of localized magnetic moments on lattice sites for the first coordination sphere in the absence of magnetic field equals to $\varepsilon^m = 0.235$. The value ε^m for the second and third coordination spheres are close to zero. A positive value of ε^m for the first coordination sphere indicates that the localized magnetic moment given carbon atom is oriented in the same direction as the magnetic moment of the nearest Cr atom.

Figure 7.9 shows a partial $g_{e\sigma}(\varepsilon) = \frac{1}{v}\sum_{i,\gamma,\lambda} P_{0i}^{\lambda} g_{0i\gamma\sigma}^{\lambda}(\varepsilon)$ and full $g_e(\varepsilon) = \sum_{\sigma} g_{e\sigma}(\varepsilon)$ densities of electron states of carbon nanotube with an admixture of Cr in the absence of external magnetic field calculated by Eq. (7.32). In the absence of a magnetic field $g_{1/2}(\varepsilon) = g_{-1/2}(\varepsilon)$. Vertical line shows the Fermi level ε_{F}.

Figure 7.10 shows partial $g_{e\sigma}(\varepsilon)$ and full $g_e(\varepsilon)$ densities of electron states of carbon nanotube with 5 atoms of Cr per primitive cell in external magnetic field $H = 100$ A/m.

In Fig. 7.10 the part of energy spectrum that is close to the Fermi level is showed.

As shown in Fig. 7.10, line of partial $g_{e\sigma}(\varepsilon)$ density of electron states for spin $\sigma = 1/2$ is shifted relative to line for spin $\sigma = -1/2$. The results presented in Fig. 7.10, are qualitatively consistent with results obtained by another method

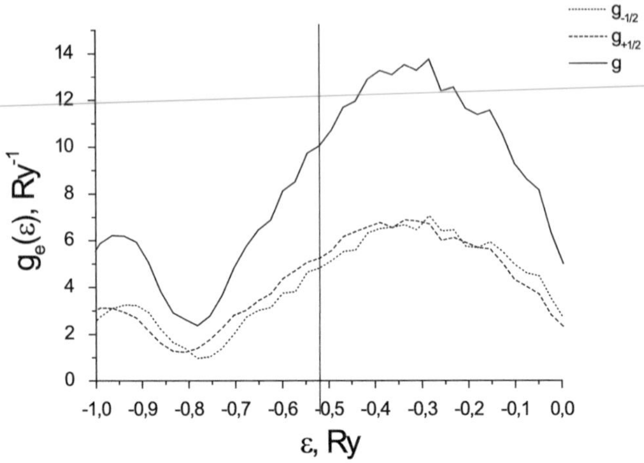

Fig. 7.10 Densities of electron states of carbon nanotube with 5 atoms of Cr per primitive cell in external magnetic field $H = 100$ A/m

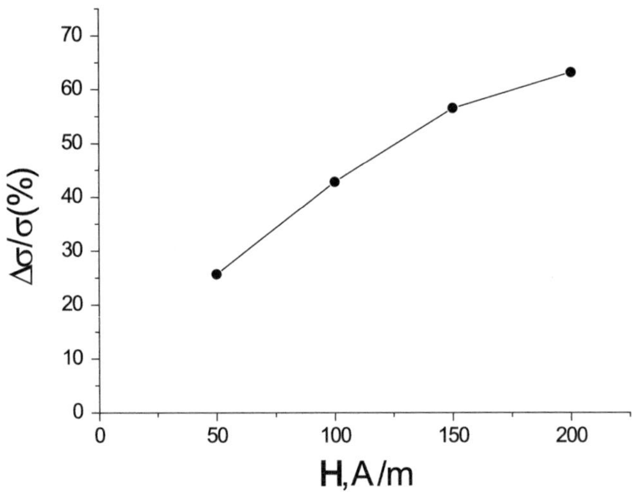

Fig. 7.11 The dependence of spin polarized electric current $\Delta\sigma/\sigma$ of carbon nanotube on the magnitude of the external magnetic field H

in Ref. [39]. In contradiction to our research in Ref. [39] carbon nanotubes with (9;0) chirality doped by Co atoms located inside of the nanotube were described. Quantitative differences in results derived in our research and results of Ref. [39] are determined by this.

In Fig. 7.11 the dependence of the spin polarization electric current $\Delta\sigma/\sigma = (\sigma_{1/2} - \sigma_{-1/2})$ of carbon nanotube with chirality (3.0) and 5 atoms of Cr per primitive cell on the magnitude of the external magnetic field calculated by Eq. (7.43) for temperature 300 K is shown.

7.8 Conclusion

The previously mentioned methods [4, 6, 7, 9–11, 14, 23–26, 30, 35, 36, 38] are used only for the description of ideal ordered crystals and molecules. However, a new method of describing electronic correlations in disordered magnetic crystals based on the Hamiltonian of multi-electron system and diagram method for finding Green's functions is developed.

The nature of spin-dependent electron transport of carbon nanotubes with chromium atoms, which were adsorbed on the surface was found. The phenomenon of spin-dependent electron transport in a carbon nanotube is the result of strong electron correlations caused by the presence of chromium atoms. The value of the spin polarization of electron transport in carbon nanotubes with chromium atoms, which were adsorbed on the surface, is determined by the difference of the partial densities of electron's states (see Fig. 7.10) with opposite spin projection at the Fermi level. This value is also determined by the difference between the relaxation times which arise from different occupation numbers of single-electron states $Z^{\lambda}_{n_1 i_1 \gamma_1 \sigma_1}$ of atoms of carbon and chromium [see Eq. (54)]. The spin polarization of the electric current value increases with Cr atoms concentration and magnitude of the external magnetic field increase. The electronic structure and properties of carbon nanotubes with transition metal chains, adsorbed on the surface, based on the density functional method [23–25, 35, 36] using ultra-soft pseudopotential was calculated [5]. Our results are qualitatively consistent with the results of Ref. [5]. In this paper electron density functional method showed that the chains of transition metals adsorbed on the surface of carbon nanotubes. They open a gap in the electrons states with a certain spin value. Contradictory to Ref.[5] it is shown that atoms of Cr are randomly placed on the carbon nanotube surface.

References

1. Abrikosov AA, Gorkov LP, Dzyaloshinski IE (1963) Methods of quantum field theory in statistical physics (edited by Silverman RA). Prentice-Hall, Englewood Cliffs
2. Blochl PE (1994) Projector augmented-wave method. Phys Rev B 50:17953
3. Chepulskii RV, Butler WH (2005) Temperature and particle-size dependence of the equilibrium order parameter of FePt alloys. Phys Rev B 72:134205
4. Ducastelle F (1974) Derivation of the coherent potential approximation from the atomic limit in the disordered alloy problem. J Phys C 7:1795
5. Durgun E, Ciraci S (2006) Spin-dependent electronic structure of transition-metal atomic chains adsorbed on single-wall carbon nanotubes. Phys Rev B 74:125404
6. Elstner M, Porezag D, Jungnickel G, Elsner J, Haugk M, Frauenheim T, Suhai S, Seifert G (1998) Self-consistent-charge density-functional tight-binding method for simulations of complex materials properties. Phys Rev B 58:7260
7. Enyaschin A, Gemming S, Heine T, Seifert G, Zhechkov L (2006) C_{28} fullerites-structure, electronic properties and intercalates. Phys Chem Chem Phys 8:3320
8. Harrison WA (1966) Pseudopotentials in the theory of metals. Benjamin, New York

9. Ivanovskaya VV, Seifert G (2004) Tubular structures of titanium disulfide TiS_2. Solid State Commun 130:175
10. Ivanovskaya VV, Heine T, Gemming S, Seifert G (2006) Structure, stability and electronic properties of composite $Mo_{1-x}Nb_xS_2$ nanotubes. Phys Status Solidi B 243:1757
11. Ivanovskaya VV, Köhler C, Seifert G (2007) 3d metal nanowires and clusters inside carbon nanotubes: structural, electronic, and magnetic properties. Phys Rev B 75:075410
12. Johnson DD, Nicholson DM, Pinski FJ, Gyorffy BL, Stocks GM (1990) Total-energy and pressure calculations for random substitutional alloys. Phys Rev B 41:9701
13. Jones RO, Gunnarsson O (1989) The density functional formalism, its applications and prospects. Rev Mod Phys 61:689
14. Köhler C, Seifert G, Gerstmann U, Elstner M, Overhof H, Frauenheim T (2001) Approximate density-functional calculations of spin densities in large molecular systems and complex solids. Phys Chem Chem Phys 3:5109
15. Kohn W, Sham LJ (1965) Self-consistent equations Including exchange and correlation fffects. Phys Rev 140:A1133
16. Kresse G, Joubert D (1999) From ultrasoft pseudopotentials to the projector augmented-wave method. Phys Rev B 59:1758
17. Kruchinin SP (1995) Functional integral of antiferromagnetic spin fluctuations in high-temperature superconductors. Modern Phys Lett B 9:205–215
18. Kruchinin SP (2014) Physics of high-Tc superconductors. Rev Theor Phys 2:124–145
19. Kruchinin S, Nagao H, Aono S (2010) Modern aspects of superconductivity: theory of Superconductivity. World Scientific, Singapore, p 220
20. Kubo R (1957) Statistical-mechanical theory of irreversible processes. 1. General theory and simple applications to magnetic and conduction problems. J Phys Soc Jpn 12:570
21. Laasonen K, Car R, Lee C, Vanderbilt D (1991) Implementation of ultrasoft pseudopotentials in ab initio molecular dynamics. Phys Rev B 43:6796
22. Los' VF, Repetsky SP (1994) A theory for the electrical conductivity of an ordered alloy. J Phys Condens Matter 6:1707
23. Perdew JP, Burke K, Ernzerhof M (1996) Generalized gradient approximation made simple. Phys Rev Lett 77:3865
24. Perdew JP, Kurth S, Zupan A, Blaha P (1999) Accurate density functional with correct formal properties: a step beyond the generalized gradient approximation. Phys Rev Lett 82:2544
25. Perdew JP, Ruzsinszky A, Csonka GI, Constantin LA, Sun J (2009) Workhorse semilocal density functional for condensed matter physics and quantum chemistry. Phys Rev Lett 103:026403
26. Porezag D, Frauenheim T, Köhler T, Seifert G, Kascher R (1995) Construction of tight-binding-like potentials on the basis of density-functional theory: application to carbon. Phys Rev B 51:2947
27. Razee SSA, Staunton JB, Ginatempo B, Bruno E, Pinski FJ (2001) The effects of magnetic annealing of transition metal alloys deduced from ab initio electronic structure calculations. J Phys: Condens Matter 13:8565
28. Repetsky SP, Shatnii TD (2002) Theor Math Phys 131:456
29. Sharma RR (1979) General expressions for reducing the Slater-Koster linear combination of atomic orbitals integrals to the two-center approximation. Phys Rev B 19:2813
30. Slater JC (1963) Quantum theory of molecules and solids: electronic structure of molecules, vol 1. McGraw-Hill, New York
31. Slater JC, Koster GF (1954) Simplified LCAO method for the periodic potential problem. Phys Rev 94:1498
32. Staunton JB, Razee SSA, Ling MF, Johnson DD, Pinski FJ (1998) Magnetic alloys, their electronic structure and micromagnetic and microstructural models. J Phys D: Appl Phys 31:2355
33. Stocks GM, Winter H (1982) Self-consistent-field-Korringa-Kohn-Rostoker-coherent-potential approximation for random alloys. Z Phys B 46:95

34. Stocks GM, Temmerman WM, Gyorffy BL (1978) Complete solution of the Korringa-Kohn-Rostoker coherent-potential-approximation equations: Cu-Ni alloys. Phys Rev Lett 41:339
35. Sun J, Marsman M, Csonka GI, Ruzsinszky A, Hao P, Kim Y-S, Kresse G, Perdew JP (2011) Self-consistent meta-generalized gradient approximation within the projector-augmented-wave method. Phys Rev B 84:035117
36. Tao J, Perdew JP, Staroverov VN, Scuseria GE (2003) Climbing the density functional ladder: nonempirical meta-generalized gradient approximation designed for molecules and solids. Phys Rev Lett 91:146401
37. Vanderbilt D (1985) Soft self-consistent pseudopotentials in a generalized eigenvalue formalism. Phys Rev B 41:7892
38. Wigner EP (1959) Group theory. Academic Press, New York/London
39. Yang C, Zhao J, Lu JP (2003) Magnetism of transition-metal/carbon-nanotube hybrid structures. Phys Rev Lett 90:257203
40. Yang C, Zhao J, Lu JP (2004) Complete spin polarization for a carbon nanotube with an adsorbed atomic transition-metal chain. Nano Lett 4:561
41. Zubarev DN (1974) Nonequilibrium statistical thermodynamics (edited by Gray P, Shepherd PJ). Consultants Bureau, New York

Chapter 8
Cell Monolayer Functioning Detection Based on Quantum Polarization Effects in Langmuir–Blodgett Multi-Walled Carbon Nanotubes Films

H.V. Grushevskaya, N.G. Krylova, I.V. Lipnevich, T.I. Orekhovskaja, and B.G. Shulitski

Abstract By means of Langmuir–Blodgett (LB) technique the multi-walled carbon nanotubes were decorated by organometallic complexes of high-spin Fe(II). Fabricated LB-nanoheterostructures screen efficiently the double electrically charged layer in the near-electrode region. It was shown that these polarization effects are spin-dependent. The proposed capacitive sensor on spin-dependent polarization effects was used to detect phenomena of CGJC-network activity.

8.1 Introduction

At present, investigations of cooperative dynamics of living cell monolayer have shown that there exist a dynamical network of gap junction channels (GJCs), which penetrate cellular membranes [1–5]. The formation of communicative gap junction channel network (CGJC network) occurs in the result of both an selfassembly/selfdisassambly of individual GJCs and an establishment of bistable channels in given state: "open" or "closed". The process of self-assembly/self-disassambly has been studied a lot and GJC inhibitors were found. However, the establishing of threshold signal transduction principles for opening/closing of GJCs in the gap junction network is the challenge because of multifactor and complicated of signaling pathways with large number of regulatory molecules [6].

Any signal of living cell leads to change of potential difference on boundary living cell/phisiological medium. An interphase boundary is a double electrically charged layer. It is known [7, 8] that an impedance analysis allows to detect a

H.V. Grushevskaya (✉) • N.G. Krylova • I.V. Lipnevich • T.I. Orekhovskaja
Physics Department, Belarusan State University, 4 Nezalezhnasti Ave., 220030 Minsk, Belarus
e-mail: grushevskaja@bsu.by

B.G. Shulitski
Belarusan State University of Informatics and Radioelectronics, 6 P. Brovki Str, 220013 Minsk, Belarus

J. Bonča, S. Kruchinin (eds.), *Nanomaterials for Security*, NATO Science for Peace and Security Series A: Chemistry and Biology, DOI 10.1007/978-94-017-7593-9_8

capacity of the double layer and, respectively, a dielectric polarization in it, for example, to study a membrane polarization of living cells in suspension (see [9]). The biocompatible graphene-like materials with high conductance and large spin relaxation times for charged carrier density are perspective transducers [10]. The use of carbon nanotubes (CNT) or graphene as sensitive elements is hampered due to the lack of free charge carriers at the Fermi level in these materials [11–14]. However, when the materials were decorated with conducting particles, one can obtain an excitonic dielectric with non-zero electric charge density at the Fermi level [15].

In this paper we will utilize the LB technique to fabricate new biocompatible layered nanoheterostructures consisting of multi-walled CNT (MWCNT) bundles, which were deposited on ultrathin LB-films of organometallic compounds. A electrochemical technique to detect functional changes proceeded in living cell monolayers immediately after their stimulation or inhibition, has been developed in [16, 17]. In our paper the technique will be used to establish principles of CGJC-network fine-tuning. Understanding of these principles would allow to program the CGJC network and, hence, to correct of pathological states.

The goal is to reveal quantum polarization effects on interface between water and nanoheterostructures from MWCNT bundles decorated by organometallic complexes, and to developed a sensor that functions based on these effects and detects an interface polarization in a gap junction network of a living cell monolayer.

8.2 Materials and Methods

8.2.1 Reagents

MWCNTs with diameters ranging from 2.0 to 5 nm and length of ~2.5 μm were obtained by the method of chemical vapor deposition (CVD-method) [18]. A TEM-image and a Raman spectrum of MWCNTs are shown in Fig. 8.1. MWCNTs were covalently modified by carboxyl groups and non-covalently functionalized by stearic acid molecules.

Salts $Fe(NO_3)_3 \cdot 9H_2O$, $Ce_2(SO_4)_3$ (Sigma, USA), hydrochloric acid, deionized water were used to preparate subphases. Iron-containing films were fabricated from an amphiphilic oligomer of thiophene derivatives with chemically bounded hydrophobic 16-link hydrocarbon chain: 3-hexadecyl-2,5-di(thiophen-2-yl)-1H-pyrrole (H-DTP, H-dithionilepyrrole). H-dithionilepyrrole was synthesized by a method proposed in [19]. Working solution of H-dithionilepyrrole, 1.0×10^{-3} M, was prepared by dissolving precisely weighed substances in hexane. All salt solutions have been prepared with deionized water with resistivity 18.2 MΩ·cm.

C6 rat glioma cells obtained from culture collection of Institute of Epidemiology and Microbiology (Minsk, Belarus) were grown on aligned MWCNT arrays and Fe-containing LB-DTP-films in a Dulbecco's modified Eagle medium (DMEM) (Sigma, USA) supplemented with 10 % fetal bovine serum and 1×10^{-4} g/ml

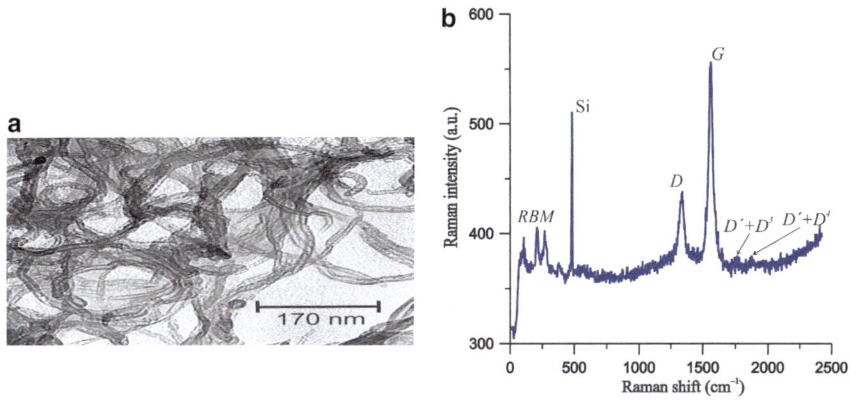

Fig. 8.1 (**a**) TEM-image of original carboxylated multi-walled CNTs; (**b**) Raman spectrum of original carboxylated MWCNTs on Si

gentamycin at 37 °C in a humidified 5 % CO_2 atmosphere. Menadione and GJC inhibitor carbenoxolone sodium sald (CSS) (Sigma, USA) were utilized to alter cellular communication activity.

All used materials belong to class of analytical pure reagents.

8.2.2 Methods

Langmuir–Blodgett Technique

Langmuir–Blodgett monolayer formation was carried out on an automated handmade Langmuir trough with controlled deposition on a substrate, and with computer user interface working under Microsoft Windows operational system. Control of the surface tension has been performed by a highly sensitive resonant inductive sensor. The Y-type transposition of monolayers on supports was performed by their vertical dipping. The complexes Fe(II)DTP_3 of high-spin Fe(II) with the dithionilepyrrole ligands were synthesized by LB-method at compression of H-dithionilepyrrole molecules on the surface of subphase with salts of three-valence Fe [20].

Horizontally and vertically arranged LB-MWCNT-bundles can be fabricated from the multi-walled CNTs [21]. An image of the LB-MWCNT-bundle and its microdiffraction pattern are represented in Fig. 8.2.

We use the Langmuir–Blodgett technique to fabricate new biocompatible layered nanoheterostructures consisting of two MWCNT LB-monolayers which are deposited on five-monolayer LB-film of the organometallic complexes.

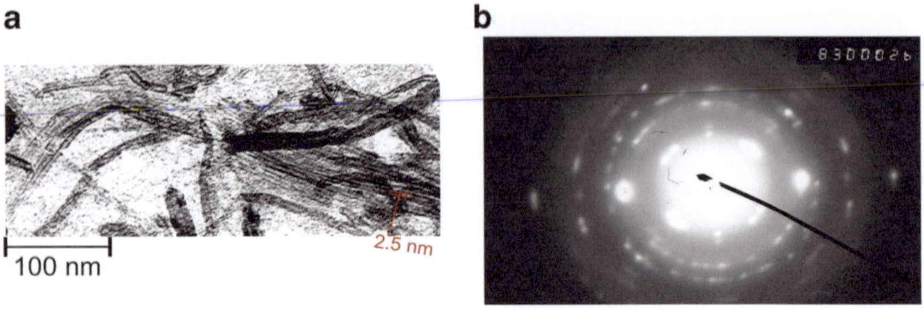

Fig. 8.2 A TEM-image (**a**) of LB-MWCNT-bundle deposited on the Fe-containing dithionilepyrrole LB-film and (**b**) a microdiffraction pattern of two LB-monolayers of MWCNTs

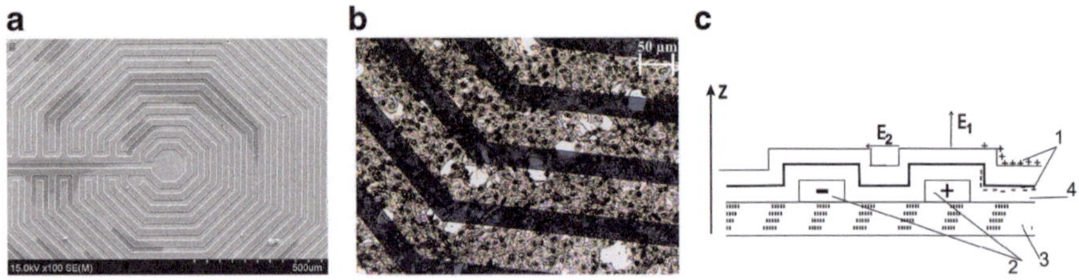

Fig. 8.3 (**a**) A SEM image of pure sensor. (**b**) A confocal microscopic image of tumor (C6 glioma) cells grown on sensor surface coated by MWCNT-Fe-containing dithionilepyrrole LB-film. (**c**) A model of i-th "open type" capacitor: *1* – double electrically charged layer, *2* – positive and negative electrodes, *3* – dielectric substrate of sensor, *4* – dielectric barrier layer of AOA; $\boldsymbol{E}_1$, $\boldsymbol{E}_2$ are electric field strengths near the electrodes

Impedance Measurements

For electrochemical studies, we use a planar capacitive sensor of interdigital-type on pyroceramics support. N pairs, $N = 20$, of aluminum electrodes of the sensors in Fig. 8.3a, b are arranged in an Archimedes-type spiral configuration. Every such pair is an "open type" capacitor. A dielectric coating of the electrodes represents itself nanoporous anodic alumina layer (AOA) with a pore diameter of 10 nm. The obtained LB-nanoheterostructures were suspended on the interdigital electrode system covered by dielectric nanoporous anodic alumina. The sensor was placed into aqueous salt solutions, or culture medium, or deionized water. A double electrically charged layer forms on the interface shown in Fig. 8.3c.

To excite harmonic auto-oscillations of electric current (charge–discharge processes in the capacitors), the sensor was connected as the capacitance C into the relaxation RC-generator (self-excited oscillator) [22]. Operating of such RC-generator is based on the principle of self-excitation of an amplifier with a positive feedback on the quasi-resonance frequency. The capacitance C of the sensor entered in measuring RC-oscillating circuit has been calculated by the formula $C = 1/(2\pi Rf)$, where R is the measuring resistance, f is the frequency of quasi-resonance.

Structural and Spectroscopy Studies

Scanning electron microscopy (SEM) images were taken on LEO 1455 VP (Carl Zeiss, Gernamy) JEM-100CX. The accelerating voltage was 20 kV. Signals of reflected and secondary electrons were detected simultaneously. Microdiffraction patterns and transmission electron microscopic images were obtained by means of transmission electron microscope JEM-100CX (JEOL, Japan) (TEM) at accelerating voltage of 100 kV. After the objects were previously deposited on a copper grid with a formvar polymer coating, structural analysis has been performed. Spectral studies in visible range were carried out using a confocal micro-Raman spectrometer Nanofinder HE ("LOTIS-TII", Tokyo, Japan–Belarus) by laser excitation at wavelength 532 nm with power 20 mW.

8.3 A Dielectric Polarization in Metal- and MWCNT-Containing LB-Monolayers and Effects of Screened Electrically Charged Double Layer

8.3.1 Spontaneous Dielectric Polarization of Helicoidal MWCNT Bundles

The conducting multi-walled carbon nanotubes can form helicoidal LB-bundles shown in Fig. 8.4a. Such helicoidal inclusions can make a medium to be right(left)-active one. As shown in Fig. 8.4b, current-voltage characteristics of the pure sensor (without LB-films) and of a four monolayer LB-film of stearic acid are linear when this characteristic for three monolayers LB-film from MWCNTs and steric acid is nonlinear one. The nonlinearity is due to that LB-bundles can be helicoidal ones and, hence, are polarized spontaneously [23]. A hysteresis cyclic volt-ampere

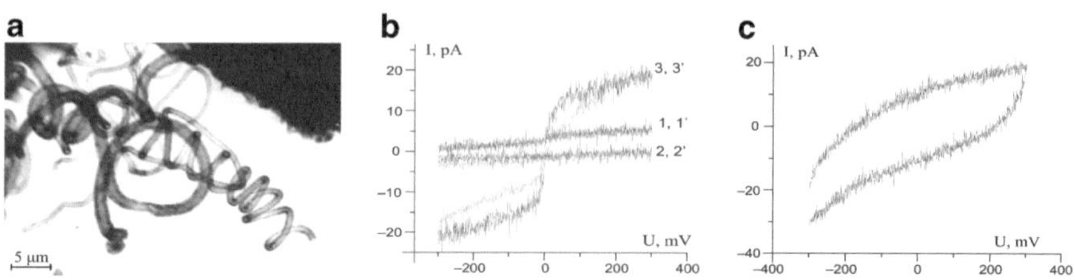

Fig. 8.4 (**a**) A TEM-image of helicoidal LB-MWCNT-bundle deposited on three LB-monolayers of stearic acid; (**b**) Current-voltage characteristics of sensor without LB-film (curves 1 and 1′), with four LB-monolayers of stearic acid (curves 2 and 2′), and with a three monolayer LB-film from MWCNTs and steric acid (curves 3 and 3′). Prime denotes a repeat measurement; (**c**) Cyclic current-voltage characteristics of the three monolayer LB-film

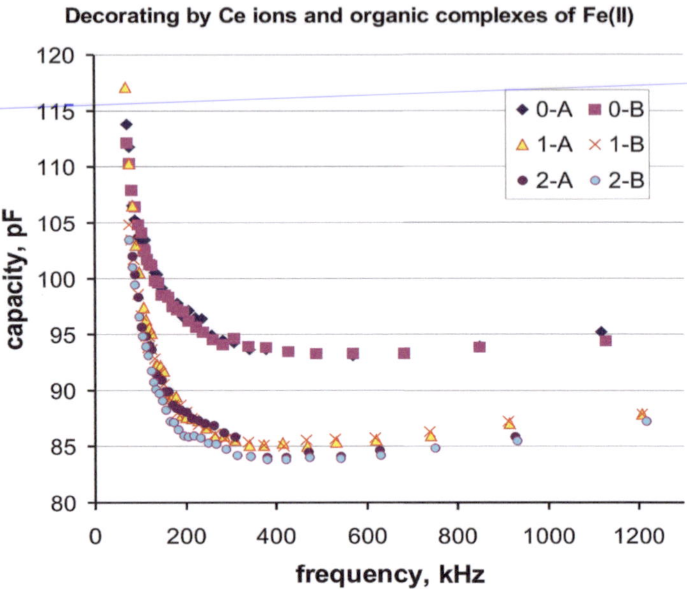

Fig. 8.5 Cyclic frequency dependencies of capacity for sensors without (curve 0) and with dithionilepyrrole LB-film (curve 1) and MWCNTs (hysteresis curve 2) in deionized water. *A* and *B* denote direct and reverse branches, respectively

characteristic (VAC) for the three monolayer LB-film from MWCNTs and steric acid is observed, as shown in Fig. 8.4c. A hysteresis frequency dependency of capacity in Fig. 8.5 is due to the spontaneous dipole polarization of the helicoidal MWCNT bundles, which interact with the organometallic complexes.

So, medium with helicoidal MWCNTs-bundles is right (left)-active one. For such mediums we observe a hysteresis frequency dependence of capacity as well as for VAC.

The MWCNT-monolayer, which is deposited on the Fe(II)-dithionilepyrrole complex LB-film, are decorated by Fe(II) atoms. As known [15], an impurity metal atom (adhesive atom, adatom) near the graphene sheet causes a local spin-orbit field, and, respectively, the scattering of current components with positive (negative) angular momentum is enhanced (suppressed) for charge carriers with a spin projection $s_z = 1$ ($s_z = -1$). Moreover, a spin-orbit splitting of the band dispersion occurs by bringing heavy metallic atoms in close contact to graphene. Under the action of Coulomb field of adatoms there appears a gap E_g in band structure of graphene and metallic graphene-like materials. The helicity prevents the spin-dependent polarization. Since the helicoidal structure diminishes the spin-dependent spontaneous polarization of the sample. In further investigations the nanocomposite LB-coatings without hysteresis properties have been used.

8.3.2 Screening Effects

Screening of Electrically Charged Double Layer by Metal- and MWCNTs-Containing LB-Monolayers in Water

The near-electrode double electrically charged layer is similar to plane capacitor. Because of a distance between plates of such capacitor is very small, its electric field strength is large. A high value of low-frequency dielectric permeability of water is due to the ionization of water molecules and impurities in water with following formation of hydrated complexes of ions OH^-, H^+ and of impurity ions. The existence of such complexes impedes the ion recombination. According to frequency dependencies of capacity (curves 0 in Fig. 8.6), there is a frequency of oscillating alternative electrical field (a plasmon resonance frequency) of the near-electrode layer. This frequency coincides with an eigenfrequency of hydrated-complex oscillations (ion vibrations). The plasmon resonance enhances a process of decay of hydrated complexes and, respectively, following recombination of ions into neutral molecules H_2O, that results in decreasing of dielectric permeability of the medium. The density of charge carriers in the double electrically charged layer, which does not modified by ultra-thin LB-film, is moderate because this resonance is observed as a wide band in dielectric spectrum with maximum in frequency range from 200 to 600 kHz. The conducting Fe-containing dithionilepyrrole LB-films add their free charged carriers to the electrical density of near-electrode layer, as the screening effect is observed according curves 1 in Fig. 8.6a,b. Since the decorated MWCNTs possess non-zero electron density on the Fermi level, MWCNTs screen the electrical field of electrodes according curves 2 in Fig. 8.6a,b. Because of these screening effects stipulate sharp increasing intensity of the double layer electrical field, the probability of hydrated complexes decay with following recombination into neutral molecules increases sharply, narrowing the dielectric band with explicit extremum about 220 kHz for MWCNTs decorated Fe(II) (curves 2 in Fig. 8.6a) or 270 kHz for MWCNTs decorated complexes of Fe(II) and Ce atoms. The large

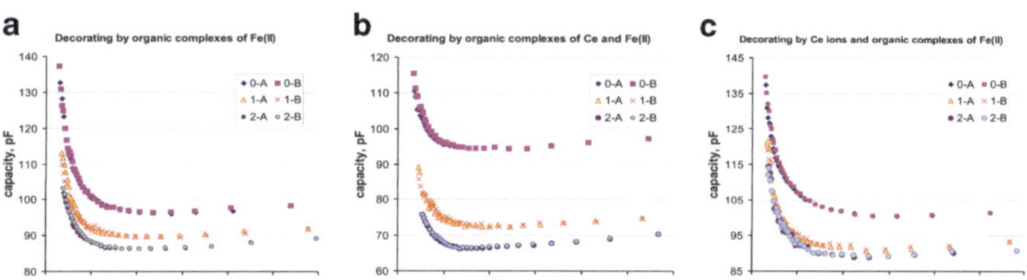

Fig. 8.6 Cyclic frequency dependencies of capacity for sensors without (0) and with dithionilepyrrole LB-films (1) and MWCNTs (2) in deionized water. *A* and *B* denote direct and reverse branches, respectively. (**a**) Decorating by organic complexes of Fe(II). (**b**) Decorating by organic complexes of Ce and Fe(II). (**c**) Decorating by Ce ions and organic complexes of Fe(II)

screening effect is due to decoration of all surface of MWCNTs by the complexes of Fe(II) and Ce atoms. An addition of impurity ions Ce^{3+} into water decreases sharply the screening effect, according the comparison of displacements of curves 1 and 2 in Fig. 8.6, because, as known [15], decoration with small doses of adatoms reduces and one with large doses suppresses the charge carrier mobilities of graphene devices.

So, if helicoidal MWCNT bundles are absent, then the plasmon resonance enhances the ion recombination process in the electrically charged double layer, and, respectively, causes a sharp drop in capacitance value of the sensor at the resonance frequency.

Screening by Living Cell Monolayers

Culture mediums are salt solutions with a large ionic strength. Fig. 8.7 demonstrates a dependence of the sensor capacity on concentration of ions in the solution at presence of the LB-coatings and in the case of the pure sensors. The LB-coatings screen an electric field of the double layer significantly decreasing the capacitance value and, therefore, not allowing appearance of break-down voltages. This dependence has at least three inflection points. The stepwise dependence of the sensor capacity is due to a quantized low-frequency Maxwell-Wagner polarization of the spin-polarized LB-coating. The large spin relaxation time and the spin-polarization of graphene-like materials allow us to detect the neutral molecules of hydrogen peroxide H_2O_2 and hypochlorous acid HOCl with lone-electron pairs, co-called, reactive oxygen species (ROS). Addition of ROS in the medium increases the capacitance value, as shown in Fig. 8.8.

Comparing concentration dependencies of double layer capacity in salt solution without and/or with living cell monolayer in Fig. 8.9a, b one gets that the cellular monolayer screens the electric field. It is known [1–5] that glucose and ions

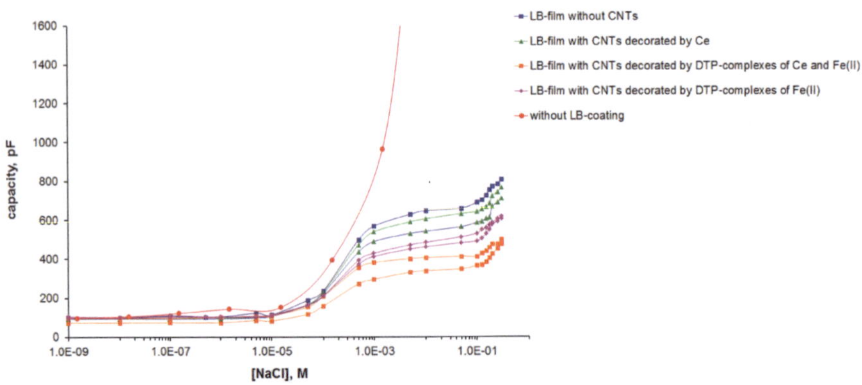

Fig. 8.7 Concentration dependencies of capacity of sensors without and with Fe-containing dithionilepyrrolle LB-film and MWCNTs at frequency 150 kHz

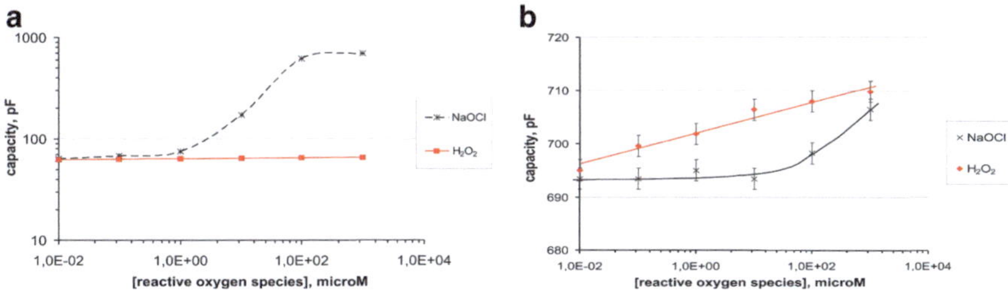

Fig. 8.8 The dependencies of sensor capacity on H_2O_2 and HOCl concentrations in the deionized water (**a**) and in Earle balanced salt solution (**b**)

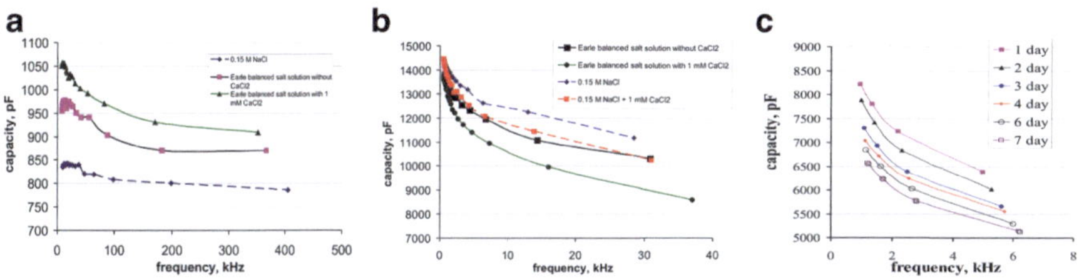

Fig. 8.9 Frequency dependencies of sensor without (**a**) and with tumor (C6 glioma) cell monolayer (**b**) in different mediums and a frequency dependencies of sensor with tumor cell monolayer for different days of growth (**c**)

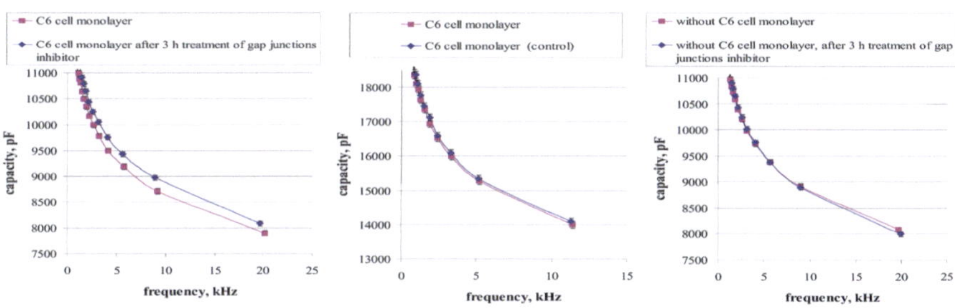

Fig. 8.10 Frequency dependencies of capacity of sensors with or without tumor cell monolayer after 3 h treatment of gap junction inhibitor (0.1 mM CSS in DMSO) and control (after 3 h treatment of the same volume of solvent (DMSO))

Ca^{2+} transport between cells in monolayer occurs through GJCs. Because of this, one can assume that the screening is a phenomenon of CGJC network activity. The extent of screening rises with growth of cellular monolayer, as Fig. 8.9c demonstrates, because the size of the CGJC network increases. Further we will reveal characteristic features of intercellular contact network functioning in the cellular monolayer.

According to the results in Fig. 8.10, the gap junction inhibitor CSS, disrupting gap junction channels, decreases the density of electrical charge carriers in the double layer that leads to the decrease of screening and, respectively, to the increase

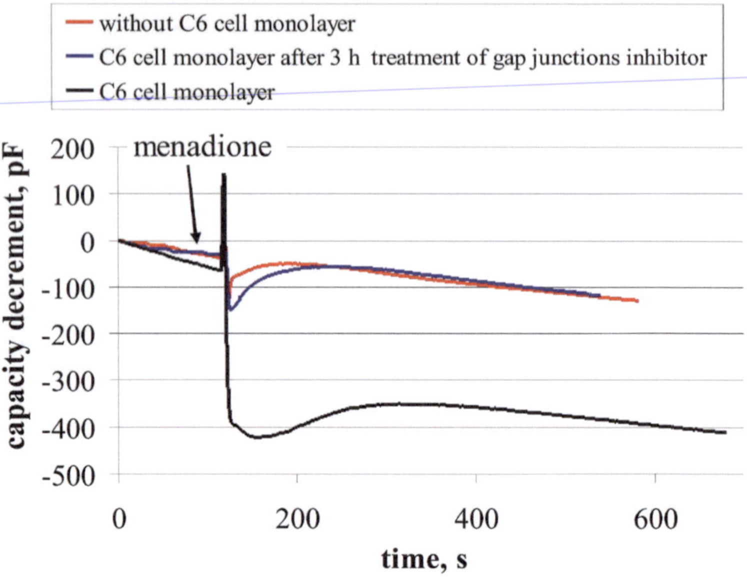

Fig. 8.11 Kinetic dependencies of capacity decrement of sensors without or with tumor cell monolayer at action of 10 μM menadione

of sensor capacity. On the contrary, the addition of menadione decreases sensor capacity on 400 pF in Fig. 8.11. It is not due to extracellular ROS generation, because ROS increases capacitance value in Fig. 8.8. Menadione does not participate in self-assembly processes, because according Fig. 8.11, its addition to the cellular monolayer with completely uncoupled GJC network does not change the sensor capacity.

Therefore, menadione is a regulatory molecule that opens the already existing GJCs.

8.4 Conclusion

So, it have been found out that MWCNTs, decorated by organometalic complexes of high-spin Fe(II), screen efficiently the double electrically charged layer in the near-electrode region. It was shown that these polarization effects are spin-dependent. The proposed capacitive sensor based on the spin-dependent polarization effects has been used to detect phenomena of CGJC-network activity.

References

1. Kurtenbach S, Kurtenbach S, Zoidl G (2014) Gap junction modulation and its implications for heart function. Front Physiol 5:82
2. Mese G, Richard G, White TW (2007) Gap junctions: basic structure and function. J Invest Derm 127:2516
3. Salameh A, Dhein S (2005) Pharmacology of gap junctions. New pharmacological targets for treatment of arrhythmia, seizure and cancer? Biochim Biophys Acta 1719:36
4. Dere E (ed) (2013) Gap junctions in the brain. Elsevier, San Diego
5. Moinfar Z, Dambach H, Faustmann PM (2014) Influence of drugs on gap junctions in glioma cell lines and primary astrocytes in vitro. Front Physiol 5:186
6. Klotz LO, Hou X, Jacob C (2014) 1,4-Naphthoquinones: from oxidative damage to cellular and inter-cellular signaling. Molecules 19:14902
7. Delachei P (1967) Double layer and kinetics of electrode processes (ed. by AN Frumkin), vol 351. Mir, Moscow
8. Chelidze TL, Derevjanko AI, Kurilenko OD (1977) Electric spectroscopy of heterogenous systems. Naukova dumka, Kiev
9. Bonincontro A, Risuleo G (2015) Electrorotation: a spectroscopic imaging approach to study the alterations of the cytoplasmic membrane. Adv Mol Imag 5:1
10. Zomer PJ, Guimares MHD, Tombros N, van Wees BJ (2012) Long-distance spin transport in high-mobility graphene on hexagonal boron nitride. Phys Rev B 86:161416
11. Morozov SV, Novoselov KS, Geim AK (2008) Electron transport in graphene. Phys – Usp 178:776
12. Grushevskaya HV, Krylov G (2014) Partially Breaking Pseudo-Dirac Band Symmetry in Graphene. J Nonlin Phen Complex Syst 17:86
13. Grushevskaya HV, Krylov GG (2015) Graphene: beyond the massless Dirac's fermion approach. In: Bonča J, Kruchinin S (eds) Nanotechnology in the security systems. NATO science for peace and security series C: environmental security. Graphene: Beyond the Massless Dirac's Fermion Approach. Springer, Dordrecht, p 21
14. Kruchinin SP, Repetsky SP, Vyshyvana IG (2016) Spin-depent transport of carbon nanotubes with chromium atoms. In: Bonca J, Kruchinin S (eds) Nanomaterials for security. Springer, Berlin
15. Ferreira A, Rappoport TG, Cazalilla MA, Castro Neto AH (2014) Extrinsic spin hall effect induced by resonant skew scattering in graphene. Phys Rev Lett 112:066601
16. Semenkova GN, Hrushevsky VV, Krylova HV, Krylova NG, Kulahava TA, Lipnevich IV, Orekhovskaya TI, Shulitsky BG (2011) Monitoring of blood phagocyte functional state by cyclic dielectric spectroscopy. J Nonlinear Phen Complex Syst 14:295
17. Grushevskaya HV, Krylova NG, Lipnevich IV, Orekhovskaya TI, Egorov AS, Egorova VP, Govorov MI, Kulahava TA, Semenkova GN, Shulitsky BG, Ulashchik VS, Vasiljev NV (2013) Cellular adhesion to carbon nanotubes-anchored oligonucleotide: electro-chemical detection. Sci J Proc Int Conf Nanomater: Apl Prop 2:04NABM22
18. Labunov V, Shulitski B, Prudnikava A, Shaman YP, Basaev AS (2010) Composite nanostructure of vertically aligned carbon nanotube array and planar graphite layer obtained by the injection CVD method. Quantum Electron Optoelectron 13:137
19. Kel'in A, Kulinkovich O (1995) A new synthesic approach to the conjugated five membertd heterocycles with ling-chain substituents. Folia Pharm Univ Carol (supplementum) 18:96
20. Grushevskaya HV, Lipnevich IV, Orekhovskaya TI (2013) Coordination interaction between rare earth and/or transition metal centers and thiophene series Oligomer derivatives in ultrathin Langmuir–Blodgett films. J Mod Phys 4:7

21. Egorov AS, Krylova HV, Lipnevich IV, Shulitsky BG, Baran LV, Gusakova SV, Govorov MI (2012) A structure of modified multi-walled carbon nanotube clusters on conducting organometallic Langmuir–Blodgett films. J Nonlinear Phen Complex Syst 15:121
22. Abramov II, Hrushevsky VV, Krylov GG, Krylova HV, Lipnevich IV, Orekhovskaya TI (2012) Method to calculate impedance of interdigital capacity sensor. Petersburg J Electronics 4(73):59
23. Krylova HV, Shulitsky BG, Hrushevsky VV (2010) Electron diffusion and spontaneous dielectric polarization in modified carbon nanotube clusters fabricated by Langmuir–Blodgett method on metal-containing nanocomposites. J Nonlinear Phen Complex Syst 13:334

Chapter 9
Magnetic Properties of Cobalt and Nitrogen Co-modified Titanium Dioxide Nanocomposites

N. Guskos, J. Typek, G. Zolnierkiewicz, E. Kusiak-Nejman, S. Mozia, and A.W. Morawski

Abstract A short review of papers devoted to study the magnetic properties of cobalt doped titanium dioxide (TiO_2) in thin film or nanoparticles form is given. Besides, cobalt and nitrogen co-modified titanium dioxide nanocomposites, nCo,N-TiO_2 (where $n = 1$, 5 and 10 wt % of Co) have been prepared and investigated by dc magnetization measurements in 2–300 K range and in external magnetic fields up to 7 T. Complex magnetic structure of investigated nanocomposites have been revealed. The samples are ferromagnetic with Curie temperature above room temperature and no superparamagnetic behavior was registered in the studied temperature range. Comparison of obtained magnetic parameters (saturation magnetization, remanent magnetization and coercive field) of nCo,N-TiO_2 with previously studied similar nanocomposites, nFe,N-TiO_2 and nNi,N-TiO_2, is presented.

9.1 Introduction

In the last five decades a large amount of research on TiO_2 nanoparticles has been done and many papers published, including reviews, describing their many interesting physical and chemical properties, see e.g. [1–3]. Titanium dioxide, belonging to a large family of transition metal oxides, has three main polymorphs found in nature: rutile (tetragonal), anatase (tetragonal), and brookite (orthorhombic). The thermodynamically favorable phase is rutile with 6 atoms in the units cell forming a slightly deformed octahedron. Anatase TiO_2 has also the tetragonal crystal structure, but its octahedron is more deformed in comparison to rutile phase. On the other hand, brookite has a larger cell and belongs to the orthorhombic crystal system. Its unit cell contains 8 TiO_2 units forming edge-sharing octahedra. After the TiO_2

N. Guskos (✉) • J. Typek • G. Zolnierkiewicz
Department of Physics, West Pomeranian University of Technology, Al. Piastow 48, 70-311 Szczecin, Poland
e-mail: nikos.guskos@zut.edu.pl

E. Kusiak-Nejman • S. Mozia • A.W. Morawski
Department of Chemical and Environmental Engineering, West Pomeranian University of Technology, Al. Piastow 17, 70-310 Szczecin, Poland

J. Bonča, S. Kruchinin (eds.), *Nanomaterials for Security*, NATO Science for Peace and Security Series A: Chemistry and Biology, DOI 10.1007/978-94-017-7593-9_9

nanoparticles reach a certain size (depending on preparation method a crossover size is roughly 30 nm) anatase and brookite phases transform to rutile structure which becomes prevailing for larger nanoparticles. Also the heat treatment during synthesis of particles plays a vital role in this phase transformation. An anatase to rutile phase transformation undergoes usually in the range from In the last four decades a large amount of research on TiO_2 nanoparticles has been done and many papers published, including reviews, describing their many interesting physical and chemical properties, see e.g. [1–3]. Titanium dioxide belongs to a large family of transition metal oxides, has three main polymorphs found in nature: rutile (tetragonal), anatase (tetragonal), and brookite (orthorhombic). The thermodynamically favorable phase is rutile with 6 atoms in the units cell forming a slightly deformed octahedron. Anatase TiO_2 has also the tetragonal crystal structure, but its octahedron is more deformed in comparison to rutile phase. On the other hand, brookite has a larger cell and belongs to the orthorhombic crystal system. Its unit cell contains 8 TiO_2 units, forming edge-sharing octahedra. After the TiO_2 nanoparticles reach a certain size (depending on preparation method, a crossover size is roughly 30 nm) anatase and brookite phases transform to rutile structure which becomes prevailing for larger nanoparticles. Also the heat treatment during synthesis of particles plays a vital role in this phase transformation. An anatase to rutile phase transformation undergoes usually in the range from 600–700 °C, but it is influenced by preparation conditions, impurities, precursors, oxygen vacancies, particle sizes etc. It has been reported that this transformation temperature is increased by dopants such as Si, Al, and Zr, while it is decreased by addition of Co, Mn and V. Photocatalytic activity of TiO_2 is strongly depended on this phase transformation.

In order to achieve various required properties needed for specific applications, TiO_2 has been synthesized in many different forms: powders, crystals, thin films, nanotubes, nanorods, using a multitude of experimental conditions [4]. The most commonly used method of obtaining TiO_2 nanoparticles is the liquid phase processing. Small pure anatase nanoparticles (>30 nm) are generally prepared from titanium (IV) isopropoxide and acetic acid, but larger anatase particles are difficult to obtain because of transformation to rutile phase. The synthesis of small brookite nanoparticles (>10 nm) is achieved by thermolysis of $TiCl_4$ in aqueous HCl solution, while quite large brookite particles (<300 nm) have been prepared using amorphous titania as the starting material and hydrothermal treatment with NaOH. Phase-pure rutile nanoparticles have been synthesized from $TiCl_4$ or $TiCl_3$ in HCl solution or from titanium(IV) isopropoxide in nitric acid.

Among many possible applications of TiO_2 in nanoparticle form the most important seems to be as a photocatalyst because it possesses suitable band gap energy, is cost effective, photostable in aqueous media and nontoxic. It is used in antiseptic and antibacterial compositions, in degrading organic contaminants and microorganisms. Hydroxyl radicals and reactive oxygen species generated on the illuminated TiO_2 surface play a role in inactivating microorganism by oxidising the polyunsaturated phospholipid components of the cell membrane of the microbes [5]. This property might be very useful in food packaging applications. TiO_2

nanoparticles could be also applied as a UV-resistant material, in manufacture of printing ink, self-cleaning ceramics and glass, coating, in making of cosmetic products such as sunscreen creams, whitening creams, morning and night creams, skin milks, etc. In the paper industry it may be used for improving the opacity of paper.

As the electric properties of TiO_2 are concerned, it is a semiconductor with a rather large band gap (3.02 eV in case of rutile). The conduction band is formed solely by 3d orbitals of Ti, while the valence band is a mixture of 2p orbitals of oxygen and 3d orbitals of titanium. There is a band bending due to a surface in titanium dioxide and this effect is stronger in anatase than in rutile. This bending causes surface holes and electrons trapping in anatase and only holes trapping in rutile. As the size of a particle decreases, the charge carriers show more quantum mechanical properties and in consequence the band gap increases and the band edges shift producing larger redox potentials.

A quite wide band gap of different phases of TiO_2 (<3 eV) means that it can only be activated by UV light and this prevents its extensive technological applications. Another drawback is a rapid electron–hole pairs recombination resulting in a low electron concentration in the conduction band of TiO_2 which diminishes the photocatalytic efficiency of this semiconductor. To remedy these limitations and to advance efficient visible light activation, two main approaches have been developed: dye sensitization [6] and doping with metal (Fe, Ni, Cr, Co, Cu, V, Mn, Ag, Ce, Gd) and nonmetal (F, P, C, N, B) ions to reduce band gap energy [7]. By doping, new electronic states are formed in the gap, and the electronic structure of the host material is changed by interaction of the host electrons with the dopant electrons. More recently, the simultaneous doping of two kinds of atoms into TiO_2, e.g. Co and N, has attracted significant interest, since it can result in a higher photocatalytic activity compared with single element doping. An extra advantage of the transition metal doping is the improved trapping of electrons to inhibit electron-hole recombination during irradiation. Doping of transition metals in TiO_2 not only increases its photoactivity but also introduces magnetic properties not previously possessed by the host material. These doped semiconductor materials may even exhibit ferromagnetism at room temperature (RT) while still preserving their semiconducting properties. Because of these enriched properties the doped TiO_2 has found applications in spintronic devices, where both spin and charge degrees of freedom can be controlled. In conventional electronic devices only electronic charges are subjected to manipulation.

Ferromagnetism with Curie temperature above RT was first discovered in Co doped TiO_2 thin films. This discovery has given an impulse in study of semiconductors with metal ions introduced in cationic sites and the so called diluted magnetic semiconductors (DMS) have brought promise of obtaining new materials with very broad potential applications in spintronic and magneto-optical devices. Although it has been found that Co, Fe, Cr, Ni, and Mn doped TiO_2 with anatase and rutile structures in thin film form are ferromagnetic well above RT, the origin of that ferromagnetism remains still a very controversial topic [8–11]. A number of studies has claimed that for the ferromagnetic behaviour the precipitation of magnetic clusters or the secondary magnetic phases are responsible, so it is an external effect. On the other hand many papers supported the intrinsic sources of RT ferromagnetism (ferromagnetism between localized spins is mediated by delocalized carriers with long range interaction). A majority of recent papers is in favour of the later point of view and claims that ferromagnetism is connected to the presence of defects such as oxygen or metal interstitials. Cobalt doped TiO_2 in different forms (thin film, nanoparticles, nanorods) is the most studied system from the point of view of magnetic properties. In Table 9.1 a short presentation of papers devoted to that subject is given. Brief information about the preparation method and the form of the investigated samples is presented as well as the used magnetic characterization method and the main conclusions drawn from their study.

The aim of this work is to present a short review of magnetic investigations of Co doped TiO_2 (in various forms, like e.g. nanoparticles, thin films, nanorods, nanotubes, etc.) and to study the temperature and magnetic field dependence of dc magnetisation in ZFC (zero field cooling) and FC (field cooling) modes of three co-modified nCo,N-TiO_2 ($n = 1, 5, 10$ wt % of Co) nanocomposites. The obtained magnetic results will be compared with previous similar studies done on related nFe,N-TiO_2 and nNi,N-TiO_2 nanocomposites.

9.2 Experimental

A method of preparation of three nanocomposites of the starting cobalt titanium dioxides modified by various concentrations of cobalt ions and their characterization was described in the paper [39]. In brief, a fixed amount of amorphous titanium dioxide (Chemical Factory Police S.A., Poland) was mixed with $Co(NO_3)_2{\cdot}5H_2O$ aqueous solution and stirred for 48 h. The amount of Co introduced to the beaker was of 1, 5, or 10 wt % relatively to TiO_2 content. After stirring samples were dried at 80 °C for 24 h in an oven and subsequently annealed at 800 °C in the NH_3 flow. The obtained samples are denoted as 1 % Co,N-TiO_2, 5 % Co,N-TiO_2, and 10 % Co,N-TiO_2, or in general as nCo,N-TiO_2, where the cobalt concentration index $n =$ 1 wt %, 5 wt %, or 10 wt %.

X-ray diffraction (XRD) patterns of the three synthesized samples have shown the presence of 11 main reflections attributed to the TiO_2 in rutile phase and one additional, weak peak from the metallic cobalt in 5 % Co,N-TiO_2, and 10 % Co,N-TiO_2 samples. The intensity of Co line, as expected, increases with an increase of index n. X-ray photoelectron spectroscopy (XPS) has shown that the cobalt ions are on the second level of oxidation.

DC magnetisation measurements were carried out by using a MPMS-7 SQUID magnetometer in the temperature range 2–300 K and in magnetic fields up to 70 kOe in the zero-field-cooling (ZFC) and field cooling (FC) modes.

Table 9.1 Short review of papers devoted to the magnetic study of TiO_2 doped with Co

Ref.	Sample preparation/form	Magnetic method of characterization	Main results
[12]	Anatase thin films with 0–8 % Co deposited on substrates by laser molecular beam epitaxy	Magnetic images by SQUID microscope	• Existence of magnetic domains • RT ferromagnetism • Magnetic moment 0.32 μ_B/Co • No clustering of Co
[13]	Co-doped anatase TiO_2 in thin film form	• Magnetic circular dichroism • Magnetisation	Ferromagnetism is induced by a small amount of clustered Co
[14]	Pulsed laser deposited films of Co-doped anatase TiO_2	SQUID magnetometry	• Co clusters as a source of ferromagnetism • Full matrix incorporation of Co after annealing
[15]	Review paper	Review paper	• A homogeneous Co distribution seems to be the most important factor in RT ferromagnetism • Smaller concentration of Co (<10 at%) avoids formation of metal clusters • Samples should be thoroughly characterized
[16]	Nanorods adapted from a low-temperature hydrolytic route	• Magnetic circular dichroism instrument • SQUID magnetometry	Creation of grain boundary defects, proposed to be oxygen vacancies at nanocrystal fusion interfaces is the most important factor
[17]	Highly insulating thin films	SQUID magnetometry	• RT ferromagnetism is not carrier mediated • Co not metallic, but as Co^{2+} substitutes for Ti • Ferromagnetism is intrinsic, $M_S = 1.1\mu_B$/Co at RT
[18]	Heteroepitaxy of Co doped anatase TiO_2 on $LaAlO_3$ by molecular beam epitaxy	SQUID magnetometry	• No Co segregation or secondary phases • Despite good quality films very weak magnetism $M_S = 0.1\mu_B$/Co • Strong ferromagnetism in faster-grown samples is due to extended structural defects

(continued)

Table 9.1 (continued)

[19]	Co-doped (4 %) anatase titanate nanotubes synthesized under hydrothermal conditions	• EPR • dc magnetisation	• Co was incorporated into the titanate lattice as Co^{2+} and substituted for Ti cations sites • Paramagnetic behaviour with weakly ferromagnetic ordering at room temperature • Ferromagnetism is due to the exchange interaction between Co^{2+} mediated by oxygen vacancies, not being caused by metallic Co
[20]	• ($Co_xTi_{1-x}O_2$) films on Si (0 0 1) substrates • Annealed (at 600 °C) in air and in hydrogen	• Alternating gradient magnetometer (at RT) • SQUID magnetometer in 2–300 K range	• No ferromagnetism in as-prepared $Co_{0.1}Ti_{0.9}O_2$ and $Co_{0.2}Ti_{0.8}O_2$ films • $Co_xTi_{1-x}O_2$ films acquire RT ferromagnetism after hydrogenation • Metal Co may be produced by hydrogenation of $Co_{0.2}Ti_{0.8}O_2$ films, leading to extrinsic RT ferromagnetism
[21]	Co-doped anatase TiO_2 thin films	SQUID magnetometry	• Co atoms are dispersed and occupy preferred interstitial sites • Oxygen vacancies (V_O), introduced by annealing, anchor the Co atoms into Co-Ti^{+3}-V_O complexes • No superparamagnetism, no nanometer-scale metallic Co clusters
[22]	Nanoparticles of pure TiO_2 anatase with four different Co contents $x = 0.03, 0.07, 0.08$, and 0.10 at%)	SQUID magnetometry; ac and dc susceptibility in 4–300 K range; isothermal magnetisation up to 3 T	• Oxygen vacancies play an important role in promoting long-range FM order in bulk Co:TiO_2 anatase phase • Results support the bound magnetic polaron model
[23]	• Cobalt doped rutile TiO_2 epitaxial thin films grown on sapphire	• SQUID magnetometry • Magnetic circular dichroism	• Cobalt as Co^{2+} in high-spin state ruling out metal segregation • Charge carriers induce ferromagnetism
[24]	• Anatase TiO_2 cluster films having Co concentrations of 0, 6, and 8 at% produced by inert-gas condensation • Co cluster sizes in 26–30 nm range	SQUID magnetometry; hysteresis loops from 10 to 400 K	• An initial increase and subsequent decrease of the magnetic moment with doping level • The film with 8 % Co contains CoO, as evidenced by the occurrence of inverted hysteresis loops • The enhancement of moment for low Co concentration is due to the Co atoms occupying interstitial sites

[25]	Polycrystalline anatase $Ti_{1-x}Co_xO_2$ ($x = 0-0.06$) films fabricated by sol-gel spin coating	SQUID magnetometry	• RT ferromagnetism in Co-doped TiO_2 films is not an intrinsic property • Saturated magnetisation is around 1.8° μ_B/Co
[26]	• Sol-gel technique • 53–60 nm crystallite size	SQUID magnetometry	• Ferromagnetism at RT • Peak at $T = 40$ K due to AFM $CoTiO_3$ • Ferromagnetism is due to the exchange interaction between Co^{2+}
[27]	• Sol-gel technique • 12–15 nm crystallite size • Anatase structure	Vibrating sample magnetometer at RT	• Ferromagnetism not from cobalt oxide phases • A weak ferromagnetism at RT • Increase in Co concentration decreases magnetic moment
[28]	• Co doped TiO_2 nanoparticles grown by a ball milling method with two different Co concentrations (3 % and 8 %) • Nanoparticles of sizes 35–50 nm identified as the anatase TiO_2	Vibrating sample magnetometer	• Ferromagnetism is due to oxygen vacancies created by the incorporation of Co into the TiO_2 crystal lattice • Bound magnetic polaron model could not explain RT ferromagnetism
[29]	• Sol-gel technique • 8 nm crystallite size • Anatase structure	Vibrating sample magnetometer	• Ferromagnetism only in vacuum annealed samples • Oxygen vacancies are responsible for ferromagnetism • The presence of AFM Co_3O_4 phase
[30]	• Cobalt doped rutile TiO_2 epitaxial thin films grown on sapphire	SQUID magnetometry • Magnetic circular dichroism	• Electron carriers mediate the ferromagnetic exchange interaction • An increase of electron density causes ferromagnetism to appear
[31]	Thin films (anatase structure) deposited by pulsed laser deposition on Si substrate	SQUID magnetometry	• No clustering of Co is responsible for RT ferromagnetism • Phase and crystallinity plays a major role in deciding magnetic moment

(continued)

Table 9.1 (continued)

[32]	• Two-year-aged $Ti_{1-x}Co_xO_{2-\delta}$ reduced anatase nanopowders with different Co contents ($0.03 \leqslant x \leqslant 0.10$)	• SQUID magnetometry • Dc and ac susceptibility in 2–300 K range	• RT ferromagnetic order of reduced $Ti_{1-x}Co_xO_{2-\delta}$ anatase nanopowders is fairly stable in time, in particular for $x \leqslant 0.07$ • Some degree of metastability of the diluted Co doped anatase structure could be inferred in the case of the sample with higher Co content ($x = 0.10$) • High Co contents ($x = 10$ at%) should be avoided whenever stable ferromagnetic samples are required
[33]	• Hydrothermal method with post-annealing • Crystallite size 15.2 nm • Anatase phase	Vibrating sample magnetometer	• Paramagnetic at RT • Magnetism comes from secondary phase $CoTiO_3$
[34]	Co(II) doped on TiO_2 with varying concentrations using impregnation method	EPR at RT	Strong EPR signal from high spin Co^{2+} on the surface of TiO_2
[35]	Films on Si substrates by sol-gel method	Vibrating sample magnetometer; measurements at RT	• Ferromagnetic behavior of samples • Magnetisation decrease with increasing dopant content • Anatase-rutile phase junction destroys magnetic polarons
[36]	Co-doped TiO_2 (3at% Co in rutile TiO_2) prepared by the solid state reaction of milling with subsequent annealing (800 °C and 1000 °C)	SQUID magnetometer at RT	• The oxygen vacancies only around the doped Co ions due to charge neutrality • No metallic Co clusters • The Co ion occupies the Ti site with some oxygen vacancies • Support of the oxygen vacancy mediated ferromagnetic exchange mechanism
[37]	Polycrystalline samples $Ti_{1-x}Co_xO_2$ (where $x = 0.00$, 0.05 and 0.10) prepared by the solid-state reaction	SQUID magnetometry in 2–300 K range, in magnetic fields up to 7 T	• RT ferromagnetism in doped samples • Ferromagnetism is the result of the intrinsic properties of the doped extended defects in TiO_2

[38]	Irradiated Co-doped TiO_2 thin films	SQUID magnetometer, isothermal magnetisation at RT	• M_S decreases exponentially with irradiation fluence • Irradiation induces more magnetic disorder than structural disorder
[39]	• Hydrothermal method with post-annealing at 800 °C • N and Co commodified TiO_2 • 3 different concentrations of cobalt: 1, 5, 10 wt %	EPR/FMR at RT	• Two types of strongly correlated spin systems • Two magnetic sublattices • In one sublattice the magnetic anisotropy did not vary with Co concentration, in the other system it decreased with Co concentration
[40]	Sol-gel method; cobalt concentration 3 and 6 mol %; 30 nm in size nanocrystals	Vibrating sample magnetometer; magnetic hysteresis loops	• Weak ferromagnetic behavior at RT • Ferromagnetism is due to the carriers exchange coupling
[41]	• Hydrothermal method with post-annealing at 800 °C • N and Co commodified TiO_2 • 3 different concentrations of cobalt: 1, 5, 10 wt %	EPR/FMR in 4–300 K range	• Magnetic agglomerates of two types, one derived from metallic cobalt residues and another involving cobalt ions in the TiO_2 lattice • Spectra of Ti^{3+} observed in low temperature range • Metallic cobalt spectrum most intense in 1 wt % Co sample
[42]	• Hydrothermal method with post-annealing at 800 °C • N and Co commodified TiO_2 • 3 different concentrations of cobalt: 1, 5, 10 wt %	SQUID magnetometry in 2–300 K range; isothermal magnetisation in magnetic fields up to 7 T	• Samples show weak ferromagnetism at RT and below • Co clusters and spurious phases could not be responsible for the ferromagnetic properties

9.3 Results and Discussion

Figure 9.1 presents the temperature dependence of dc magnetic susceptibilities (defined as $\chi = M/H$) in ZFC and FC modes of cobalt and nitrogen co-modified TiO_2 nanocomposites, nCo,N-TiO_2 ($n = 1$, 5 and 10 wt %), nanocomposites, registered at different external applied magnetic fields ($H = 10$, 100, 500, 6000, and 70000 Oe). All nanocomposites exhibit a rather similar behavior, independent on Co concentration. The ZFC and FC curves show temperature dependence of susceptibility which is typical for ferromagnetic materials. ZFC and FC branches split at the temperature higher than RT, so no superparamagnetic phase exists below RT and the Curie temperature must be accordingly much higher than 300 K. In external fields smaller than ~1 kOe the FC branch do not shows much temperature change when samples are cooled from RT down to ~30 K. Only below that temperature an abrupt increase of χ_{FC} is registered with temperature decrease. On the other hand χ_{ZFC} branch decreases significantly on cooling the nanocomposites from RT and only at the lowest temperatures an increase is observed. A very strong increase of susceptibility during temperature decrease seen in the low temperature range ($T < 20$ K) might be due to some paramagnetic component present in our samples (e.g. intrinsic defects or unpaired Co^{2+} ions). The difference of susceptibilities, $\chi_{FC} - \chi_{ZFC}$, is a measure of magnetic anisotropy of the system and in case of our nanocomposites the smallest the Co concentration

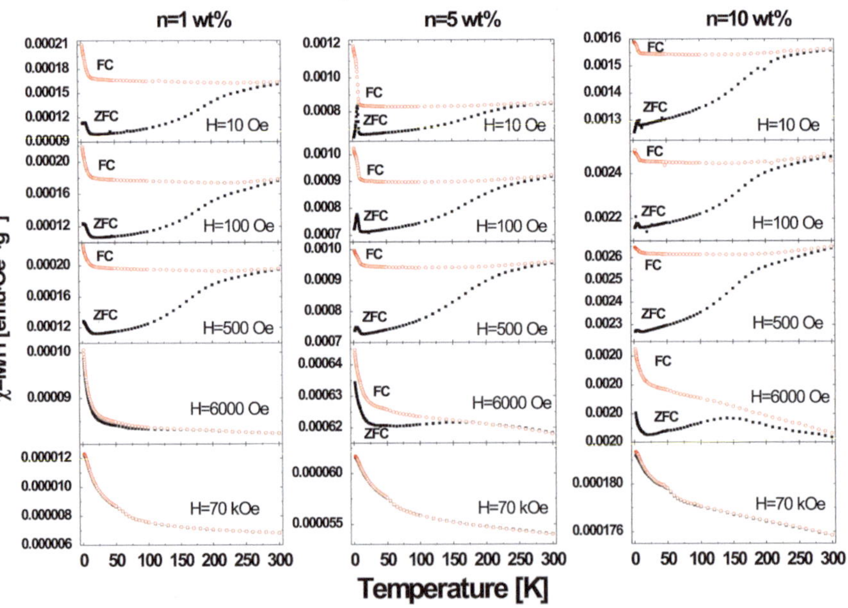

Fig. 9.1 Temperature dependence of dc magnetic susceptibility in different external magnetic fields H in ZFC and FC modes registered for three studied nanocomposites

the largest the anisotropy. In external magnetic fields stronger than few kOe the χ_{ZFC} and χ_{FC} branches acquire similar thermal evolution and in $H = 7\,T$ they are hard to discriminate. Temperature dependence of the susceptibility in strong magnetic fields can be approximated with the Curie–Weiss law, producing a negative Curie–Weiss temperature indicative of an effective antiferromagnetic interaction. The temperature dependence of dc susceptibility presented in Fig. 9.1 for nCo,N-TiO_2 nanocomposites shows many similarities with analogous study of nFe,N-TiO_2 and nNi,N-TiO_2 samples, but in the present case it seems to be more complex. It is obvious that many magnetic components are present in nCo,N-TiO_2 nanocomposites producing very complicated magnetic response to an external magnetic field.

There are two features in Fig. 9.1 that are worth noticing: a low-temperature peak in χ_{ZFC} clearly visible in $n = 5$ and 10 wt % nanocomposites, and a kink or a broad peak in $\chi_{ZFC}(T)$ curve in the middle temperature range ($T \sim 150\,K$) also discerned in the same $n = 5$ and 10 wt % nanocomposites. As these features are visible only for samples with large concentrations of cobalt ions, they must involve these particular ions. Cobalt ions might be assembled in metallic clusters (as evidenced by XRD study) and also confined in different secondary phases containing Co, Ti and O atoms (e.g. $CoTiO_3$). It has been shown that ferromagnetism originating from Co clusters would result in a peak in magnetisation in ZFC mode, which will be a signature of the blocking temperature for that nanocomponent. The narrow peak in $\chi_{ZFC}(T)$ at low temperature seems to be a signal of the phase transition from paramagnetic to antiferromagnetic state of some cobalt secondary phase while the broad peak in intermediate temperature range might be connected with the superparamagnetic transition in Co metallic clusters.

In Figs. 9.2 and 9.3 isothermal magnetisation at $T = 2\,K$ and $T = 300\,K$ for the three studied nanocomposites is shown. The magnetic loops are very narrow evidencing a rather weak ferromagnetism at RT and at low temperature. The three parameters characterizing the magnetic loops (saturation magnetisation, remanent magnetisation and the coercive field) have been calculated and are presented in Figs. 9.4, 9.5, and 9.6. For comparison, the data of similar measurements on previously studied nFe,N-TiO_2 and nNi,N-TiO_2 nanocomposites has been used and are shown in these figures.

The saturation magnetisation of nCo,N-TiO_2 nanocomposites (Fig. 9.4) increases with concentration of Co ions both at low (2 K) and high temperature (300 K). In comparison with the other two nanocomposite, nCo,N-TiO_2 is generally magnetically weaker as nFe,N-TiO_2, but stronger than nNi,N-TiO_2. This is not unexpected as spins of these three magnetic ions are considered. What is surprising, if only $n = 10$ wt % nanocomposites are taken into account, that 10 % Co,N-TiO_2 sample has the largest saturation magnetisation. It could mean that in the other two nanocomposites antiferromagnetic interactions are reducing their magnetism.

The remanent magnetisation (remanence) of nCo,N-TiO_2 sample is roughly independent on concentration at low temperature, but decreases very strongly with concentration increase at high temperature (Fig. 9.5). This behavior is quite the opposite

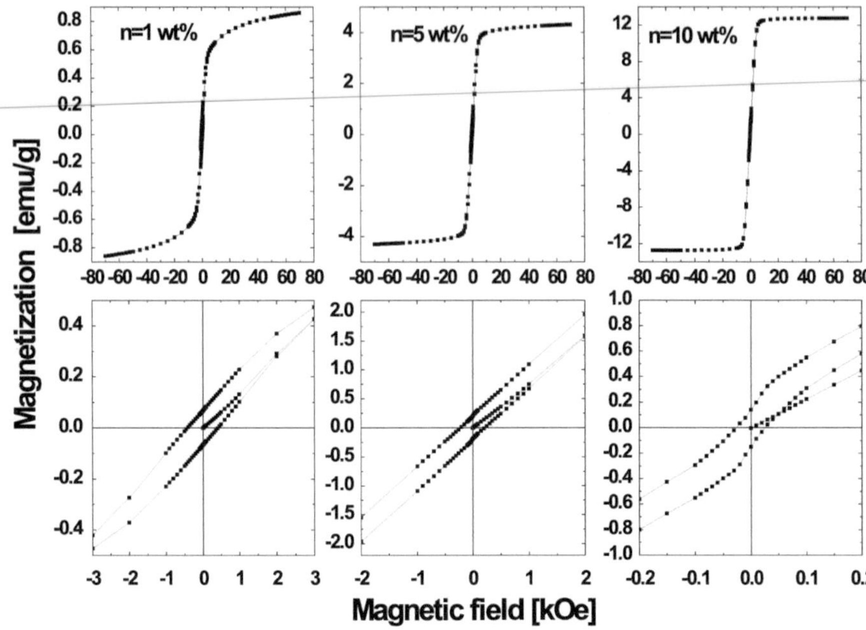

Fig. 9.2 Isothermal magnetization at $T = 2\,\mathrm{K}$ for three studied nanocomposites. *Bottom row* in these figures shows magnified view of the central part of the registered hysteresis loops

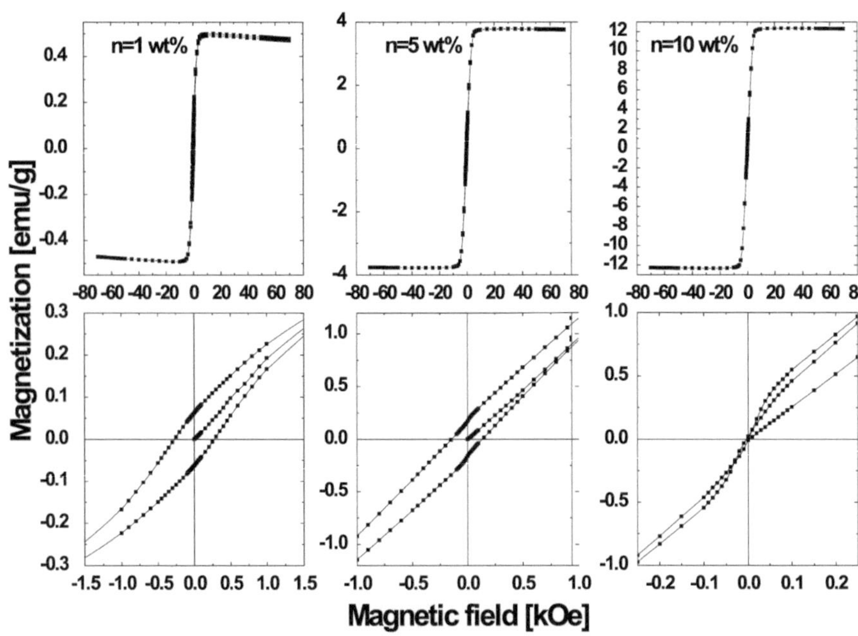

Fig. 9.3 Isothermal magnetization at $T = 300\,\mathrm{K}$ for three studied nanocomposites. *Bottom row* in these figures shows magnified view of the central part of the registered hysteresis loops

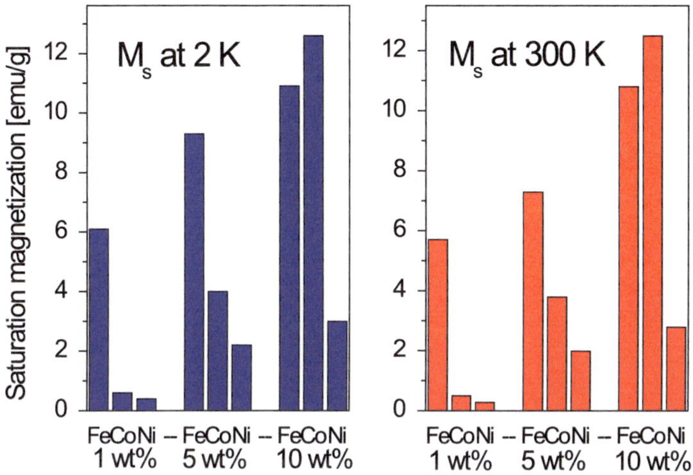

Fig. 9.4 Comparison of the saturation magnetization for Fe, Co and Ni doped and N co-doped TiO_2 nanoparticles measured at $T = 2$ K (*left panel*) and 300 K (*right panel*) for three different concentrations (1, 5, 10 wt %) of doped 3d ions. Data for *n*Ni,N-TiO_2 nanocomposites is taken from Ref. [43], while for *n*Fe,N-TiO_2 samples from Ref. [44]

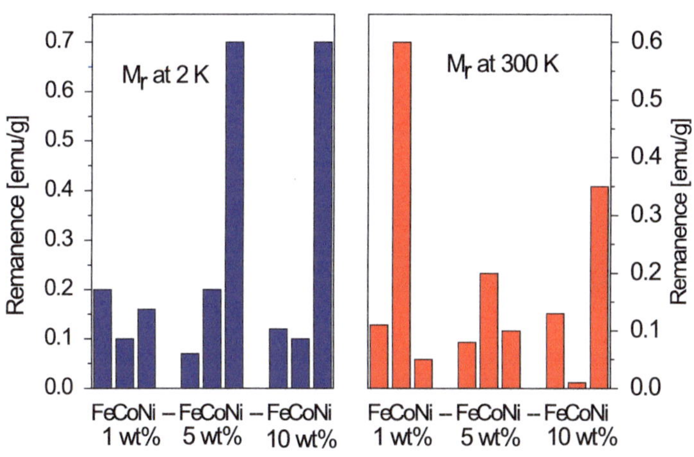

Fig. 9.5 Comparison of the remanent magnetization (remanence) for Fe, Co and Ni doped and N co-doped TiO_2 nanoparticles measured at $T = 2$ K (*left panel*) and 300 K (*right panel*) for three different concentrations (1, 5, 10 wt %) of doped 3d ions. Data for *n*Ni,N-TiO_2 nanocomposites is taken from Ref. [43], while for *n*Fe,N-TiO_2 samples from Ref. [44]

in *n*Ni,N-TiO_2 samples, where an increase of remanence with concentration of Ni ions is observed at $T = 300$ K. For the *n*Fe,N-TiO_2 nanocomposites, $n = 5$ wt % sample has the smallest remanence at both low and high temperatures among other concentrations.

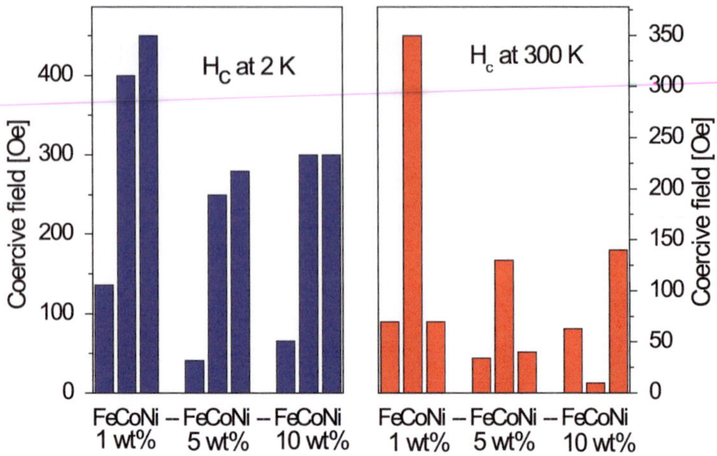

Fig. 9.6 Comparison of the coercive field for Fe, Co and Ni doped and N co-doped TiO_2 nanoparticles measured at $T = 2\,K$ (*left panel*) and 300 K (*right panel*) for three different concentrations (1, 5, 10 wt %) of doped 3d ions. Data for *n*Ni,N-TiO_2 nanocomposites is taken from Ref. [43], while for *n*Fe,N-TiO_2 samples from Ref. [44]

The coercive field of *n*Co,N-TiO_2 samples seems to decrease with the concentration at both low and high temperatures, but at the later the effect is quite impressive. There is no clear concentration dependence of the coercive field for *n*Fe,N-TiO_2 and *n*Ni,N-TiO_2 nanocomposites and the values of H_C for these samples are generally smaller than for *n*Co,N-TiO_2.

9.4 Conclusions

Magnetisation measurements of the three cobalt and nitrogen co-modified titanium dioxide nanocomposites, *n*Co,N-TiO_2 ($n = 1, 5$ and 10 wt %), have revealed a weak ferromagnetic behaviour at RT and at lower temperatures of these samples that is rather independent on the concentration of Co ions. Temperature dependence of dc susceptibility in ZFC and FC modes has indicated on the presence of additional magnetic components, including metallic cobalt clusters and spurious phases. Comparison with similar previously studied systems, *n*Fe,N-TiO_2 and *n*Ni,N-TiO_2 has shown that *n*Co,N-TiO_2 nanocomposites displays a unique, very interesting magnetic characteristics.

Acknowledgements Partially this research was supported by The National Science Center (Poland) under project MAESTRO No. DEC-2012/06/A/ST5/00226.

References

1. Gupta SM, Tripathi M (2011) A review of TiO_2 nanoparticles. Chin Sci Bull 56:1639
2. Chen X, Mao SS (2007) Titanium dioxide nanomaterials: synthesis, properties, modifications, and applications. Chem Rev 107:2891
3. Diebold U (2003) The surface science of titanium dioxide. Sur Sci Rep 48:53
4. Reyes-Coronado D, Rodriguez-Gattorno G, Espinosa-Pesqueira ME, Cab C, de Coss R, Oskam G (2008) Phase-pure TiO_2 nanoparticles: anatase, brookite and rutile. Nanotechnology 19:145605
5. Othman SH, Abd Salam NR, Zainal N, Basha RK, Talib RA (2014) Antimicrobial activity of TiO_2 nanoparticle-coated film for potential food packaging applications. Int J Photoenergy 2014:945930
6. Kisch H, Macyk W (2002) Visible-light photocatalysis by modified titania. Chem Phys Chem 3:399
7. Dozzi MV, Selli E (2013) Doping TiO_2 with p-block elements: effects on photocatalytic activity. J Photochem Photobiol C: Photochem Rev 14:13
8. Tian Y-F, Hu S-J, Yan S-S, Mei L-M (2013) Oxide magnetic semiconductors: materials, properties, and devices. Chin Phys B 22(8): 088505
9. Choudhury B, Verma R, Choudhury A (2014) Oxygen defect assisted paramagnetic to ferromagnetic conversion in Fe doped TiO_2 nanoparticles. RSC Adv 4:29314
10. Kruchinin S, Dzezherya Yu, Annett J (2006) Imteractions of nanoscale ferromagnetic granules in a London superconductors. Supercond Sci Technol 19:381–384
11. Kruchinin S, Nagao H (2012) Nanoscale superconductivity. Inter J Mod Phys B 26:1230013
12. Matsumoto Y, Murakami M, Shono T, Hasegawa T, Fukumura T, Kawasaki M, Ahmet P, Chikyow T, Koshihara S-Y, Koinuma H (2001) Room-temperature ferromagnetism in transparent transition metal-doped titanium dioxide. Science 291:854–856
13. Kim J-Y, Park J-H, Park B-G, Noh H-J, Oh S-J, Yang JS, Kim D-H, Bu SD, Noh T-W, Lin H-J, Hsieh H-H, Chen CT (2003) Ferromagnetism induced by clustered Co in Co-doped anatase TiO_2 thin films. Phys Rev Lett 90:017401
14. Shinde SR, Ogale SB, Sarma SD, Simpson JR, Drew HD, Lofland SE, Lanci C, Buban JP, Browning ND, Kulkarni VN, Higgins J, Sharma RP, Greene RL, Venkatesan T (2003) Ferromagnetism in laser deposited anatase $Ti_{1-x}Co_xO_{2-\delta}$ films. Phys Rev B 67:115211
15. Janisch R, Gopal P, Spaldin N (2005) Transition metal-doped TiO_2 and ZnO–present status of the field. J Phys Condens Matter 17:R657–R689
16. Bryan JD, Santangelo SA, Keveren SC, Gamelin DR (2005) Activation of high-TC ferromagnetism in Co^{2+}:TiO_2 and Cr^{3+}:TiO_2 nanorods and nanocrystals by grain boundary defects. J Am Chem Soc 127:15568–15574
17. Griffin KA, Pakhomov AB, Wang CM, Heald SM, Krishan KM (2005) Intrinsic ferromagnetism in insulating cobalt doped anatase TiO_2. Phys Rev Lett 94:157204
18. Kaspar TC, Droubay T, McCready DE, Nachimuthu P, Heald SM, Wang CM, Lea AS, Shutthanandan V, Chamber SA, Toney MF (2006) Magnetic properties of epitaxial Co-doped anatase TiO_2 thin films with excellent structural quality. J Vac Sci Technol B 24:2012
19. Huang C, Liu X, Kong L, Lan W, Su Q, Wang Y (2007) The structural and magnetic properties of Co-doped titanate nanotubes synthesized under hydrothermal conditions. Appl Phys A 87:781–786
20. Liu LF, Kang JF, Wang Y, Tang H, Kong LG, Sun L, Zhang X, Han RQ (2007) The influence of hydrogen annealing on magnetism of Co-doped TiO_2 films prepared by sol–gel method. J Magn Magn Mater 308:85–89
21. Griffin RK, Varela M, Rashkeev S, Pantelides ST, Pennycook SJ, Krishnan KM (2008) Defect-mediated ferromagnetism in insulating Co-doped anatase TiO_2 thin films. Phys Rev B 78:014409
22. Pereira LCJ, Nunes MR, Monteiro OC, Silvestre AJ (2008) Magnetic properties of Co-doped TiO_2 anatase nanopowders. Appl Phys Lett 93:222502

23. Fukumura T, Toyosai H, Ueno K, Nakano M, Kawasaki M (2008) Role of chargé carriers for ferromagnetism in cobalt-doped rutile TiO_2. New J Phys 10:055018
24. Wei XH, Skomski R, Sellmyer DJ (2009) Structure and magnetism of pure and co-doped TiO_2 clusters. IEEE Trans Magn 45:4089
25. Xu J, Shi S, Li L, Zhang X, Wang Y, Chen X, Wang J, Lv L, Zhang F, Zhong W (2010) Structural, optical, and ferromagnetic properties of Co-doped TiO_2 films annealed in vacuum. J Appl Phys 107:053910
26. Jiang BZ, Phan TL, Yang DS, Lee KW, Yu SC (2010) Magnetism in Co-doped rutile TiO_2 nanoparticles. Solid State Commun 150:1932
27. Karthik K, Pandian SK, Kumar KS, Jaya NV (2010) Influence of dopant level on structural, optical and magnetic properties of Co-doped anatase TiO_2 nanoparticles. Appl Surf Sci 256:4757
28. Santara B, Pal B, Giri PK (2011) Signature of strong ferromagnetism and optical properties of Co doped TiO_2 nanoparticles. J Appl Phys 110:114322
29. Choudhury B, Choudhury A, MaidulIslam AKM, Alagarsamy P, Mukherjee M (2011) Effect of oxygen vacancy and dopant concentration on the magnetic properties of high spin Co^{2+} doped TiO_2 nanoparticles. J Magn Magn Mater 323:440
30. Yamada Y, Fukumura T, Ueno K, Kawasaki M (2011) Control of ferromagnetism at room temperature in $(Ti,Co)O_{2-\delta}$ via chemical doping of electron carriers. Appl Phys Lett 99:242502
31. Mohanty P, Mishra NC, Choudhary RJ, Banerjee A, Shripathi T, Lalla NP, Annapoorni S, Chandana R (2012) Oxygen vacancy induced phase formation and room temperature ferromagnetism in undoped and Co-doped TiO_2 thin films. J Phys D: Appl Phys 45:32530
32. Silvestre AJ, Pereira LC, Nunes MR, Monteiro OC (2012) Ferromagnetic order in aged Co-doped TiO_2 anatase nanopowders. J Nanosci Nanotechnol 12:6850
33. Rashad MM, Elsayed EM, Al-Kotb MS, Shalan AE (2013) The structural, optical, magnetic and photocatalytic properties of transition metal ions doped TiO_2 nanoparticles. J Alloy Compd 581:71–78
34. Sadanandam G, Lalitha K, Kumari VD, Shankar MV, Subrahmanyam M (2013) Cobalt doped TiO_2: a stable and efficient photocatalyst for continuous hydrogen production from glycerol: water mixtures under solar light irradiation. Int J Hydrog Energy 38:9655
35. Tian J, Gao H, Kong H, Yang P, Zhan W, Chu J (2013) Influence of transition metal doping on the structural, optical, and magnetic properties of TiO_2 films deposited on Si substrates by a sol–gel process. Nanoscale Res Lett 8:533
36. Nakai I, Sasano M, Inui K, Korekawa T, Ishijima H, Katoh H (2013) Oxygen vacancy and magnetism of a room temperature ferromagnet Co-doped TiO_2. J Korean Phys Soc 63(3): 532–537
37. Kaushik A, Dalela B, Kumar S, Alvi PA, Dalela S (2013) Role of Co doping on structural, optical and magnetic properties of TiO_2. J Alloy Compd 552:274–278
38. Mohanty P, Singh VP, Mishra NC, Ojha S, Kanjilal D, Rath C (2014) Evolution of structural and magnetic properties of Co-doped TiO_2 thin films irradiated with 100 MeV Ag^{7+} ions. J Phys D: Appl Phys 47:315001
39. Guskos N, Zolnierkiwicz G, Guskos A, Typek J, Berczynski P, Dolat D, Mozia S, Aidinis K, Morawski AW (2015) Magnetic resonance study of co-modified (Co,N)-TiO_2 nanocomposites. Nukleonika 60(3):411–416
40. Wang Q, Liu X, Wei X, Dai J, Li W (2015) Ferromagnetic property of Co and Ni doped TiO_2 nanoparticles. J Nanomater 2015:371582
41. Guskos N, Zolnierkiwicz G, Guskos A, Typek J, Berczynski P, Dolat D, Mozia S, Aidinis A, Kruk K, Morawski AW (2016) Temperature study of magnetic resonance spectra of Co-modified (Co,N)-TiO_2 nanocomposites. Mater Sci-Poland 33:627
42. Guskos N, Typek J, Zolnierkiewicz G, Glenis S, Diamantopoulou A, Mozia S, Morawski AW (2016) Magnetic properties of cobalt and nitrogen co-modified titanium dioxide nanocomposites. In: Bonca J, Kruchinin S (eds) Nanomaterials for Security, NATO Science for Peace and Security Series A: Chemistry and Biology (This publication)

43. Guskos N, Typek J, Zolnierkiewicz G, Glenis S, Diamantopoulou A, Dolat D, Mozia S, Morawski AW (2016) Magnetic properties of co-modified Ni,N-TiO_2 nanocomposites. Rev Adv Mat Sci 44:194
44. Guskos N, Glenis S, Zolnierkiewicz G, Guskos A, Typek J, Berczynski P, Dolat D, Mozia S, Morawski AW (2015) Magnetic properties of Co-modified Fe,N-TiO_2 nanocomposites. Open Phys 13:76–86

Chapter 10
Physical Properties of $(As_2Se_3)_{1-x}:Sn_x$ and $(As_4S_3Se_3)_{1-x}:Sn_x$ Glasses

O.V. Iaseniuc, D.V. Harea, E.E. Harea, G.F. Volodina, and M.S. Iovu

Abstract Experimental results on some physical and optical properties of $(As_2Se_3)_{1-x}$:Sn_x and $(As_4S_3Se_3)_{1-x}$:Sn_x ($x = 0 \div 10$ at %) glasses and amorphous films ($d \sim 2.0\,\mu m$) are presented. The bulk chalcogenide glasses are studied by X-ray diffraction spectroscopy and nanoindentation methods. It is established that the addition of these amounts of tin ($x = 0 \div 10$ at %) does not lead to significant changes in the physical properties of the glass, such as values of stress and Young's modulus related to the modification of the density and compactness. The XRD measurements show that the Sn impurities in the $(As_4S_3Se_3)_{1-x}$:Sn_xdo not significantly change the shape of the first sharp diffraction peak (FSDP) of the X-ray diffraction patterns either; the intensity and the position of the FSDP nonmonotonically depend on the Sn concentration. It has been found that the addition of these amounts of tin in $(As_4S_3Se_3)_{1-x}$:Sn_x does not lead to significant changes in the glass physical properties, such as values of stress and Young's modulus related to the modification of the density and compactness. The study of the photoplastic effect is performed in situ, with illumination of the bulk and thin film samples during indentation as well as their indentation after illumination with a green laser ($\lambda = 532\,nm$) at a power of $P = 50\,mV/cm^2$. The hardness is calculated from load-displacement curves by the Oliver–Pharr method. A sharp increase in hardness is registered if the tin concentration exceeds a value of 3–4 % Sn. The hardness H of $(As_2Se_3)_{1-x}$:Sn_x films varies between 115 and 130 kg/mm^2. It is found that the hardness H of amorphous thin films is generally higher than the hardness of bulk samples with the same chemical composition. In this study, we are focused on the mechanical characteristics of high-purity As_2Se_3:Sn_x thin films.

O.V. Iaseniuc • D.V. Harea (✉) • E.E. Harea • G.F. Volodina • M.S. Iovu
Institute of Applied Physics, Academy of Sciences of Moldova
Academiei str. 5, Chisinau, MD-2028, Republic of Moldova
e-mail: dyanaharea@gmail.com

J. Bonča, S. Kruchinin (eds.), *Nanomaterials for Security*, NATO Science for Peace and Security Series A: Chemistry and Biology, DOI 10.1007/978-94-017-7593-9_10

10.1 Introduction

Chalcogenide glasses (CGs), such as As_2S_3, As_2Se_3, As_2Te_3, As-S-Se, As-Sb-S(Se), As-Ge-S(Se), represent a new class of advanced materials exhibiting attractive combinations of properties, such as high strength/hardness and excellent wear/corrosion resistance [1–4], excellent optical properties (high refractive index, high transmittance in the near IR and IR regions of the spectrum), which make them good candidates for photonic and optoelectronic applications (optical elements and memories, optical sensors, nonlinear optical devices, holographic elements, IR telecommunications, biosensing, signal processing, and photonic applications [5–7]). The effect of light-induced photostructural transformations is characteristic of many amorphous chalcogenides films, and they have served as a base of many applications in photonics and optoelectronics, especially as inorganic photoresists for submicron technology [8, 9].

These distinguished properties are primarily due to a disordered atomic arrangement in the amorphous materials resulting in the absence of grain boundaries and defects in the microstructure. While the nonequilibrium nature of amorphous materials offers outstanding properties, it also presents significant challenges in the processing of these materials.

It is even more interesting to study the chalcogenides doped with metal impurities, which alter the physicochemical, electrical, and optical properties. It was shown that the Sn impurity introduced in the As_2Se_3, AsSe, and Sb_2S_3 glass network reduces the photodarkening effect [10–13]. Tin-containing chalcogenides showed the some deviation in behavior due to modification in the local ordering of the host chalcogenide matrix after the tin incorporation, which was observed in the bulk glasses as well as in the corresponding thin films.

For numerous applications, the mechanical behavior is one of important characteristics of the material [14–16]. Moreover, the photoplastical effect in some amorphous films (As-Se, Ss-S-Se) that occurs under band-gap illumination is described. This effect is considered electronic rather than thermal and consists of two parts: negative (decreasing in viscosity) and positive (increasing in viscosity); it occurs only under illumination. In the opinion of the authors, the first part of the photoplastic effect is general for CGs and causes all photostructural phenomena in amorphous materials.

The photoplastical effect is evident from the hardening of illuminated materials [17] as well as their softening [18]. The photoplastical effect depends on a number of factors, such as radiation power, temperature, and wavelength. Studies of the spectral dependence of photoplasticity showed a maximum effect under illumination of samples with a wavelength close to the band gap value [19]. Photoplasticity in chalcogenide films was studied in [20–22], where the effect was observed during illumination of samples with a wavelength comparable to the band gap value. The nature of the photoplastical effect remains poorly understood. The effect was attributed to thermal expansion of the film due to the absorption of exciting light as well as recombination of electrons and holes under photoexcitation. The properties of glass can be varied and regulated over an extensive range by modifying the composition and production techniques. It was shown that the addition of a tin impurity in amorphous As_2Se_3 films can provide a pronounced effect on electrical, transport properties, optical and photoinduced phenomena [23–27].

In this study, we have combined the experimental investigations of X-ray diffraction (XRD) measurements and the mechanical properties of bulk and thin films of $(As_2Se_3)_{1-x}$:Sn_x and $(As_4S_3Se_3)_{1-x}$:Sn_x ($x = 0\div10$ at % Sn) chalcogenides.

10.2 Experimental

The bulk CGs $(As_2Se_3)_{1-x}$:Sn_x and $(As_4S_3Se_3)_{1-x}$:Sn_x ($x = 0\div10$ at % Sn) glasses were prepared from the elements of 6N (As, S, Se, Sn) purity by a conventional melt quenching method. The starting components $As_4S_3Se_3$ and Sn were mixed in quartz ampoules and then evacuated to a pressure of $P \sim 10^{-5}$ torr, sealed and heated to a temperature of $T = 900\,°C$ at a rate of 1 °C/min. The quartz tubes were held at this temperature for 48 h for homogenization and then slowly quenched in a furnace. For nanoindentation and optical measurements of the bulk samples, planar parallel plates with a thickness of about $d = 2\div3.5$ mm were prepared and polished. Experiments were carried out at room temperature. Indentation was performed with CSM Indentation Testers (Ultra Nano, Nano and Micro; with max load $P = 100$ mN; load resolution of 0.001 μN) with software. Hardness (Hv) and Young's modulus (E) were automatically computed. The X-ray diffraction (XRD) measurements were performed on a DRON-UM1 diffractometer using Fe-Kα radiation ($\lambda = 1.93604$ Å) with a Mn filter by the $\theta/2\theta$ scanning method. Thin film samples with a thickness of $d = 0.5\div3$ μm were prepared by vacuum flash thermal evaporation of the synthesized initial CG onto glass substrates held at $T_{subs} = 100\,°C$.

10.3 Results and Discussion

10.3.1 X-Ray Diffraction Measurements

Crystalline semiconductors are characterized by the ***ordering on distance***, e.g., by the existence of a correlation between two atoms situated at any distance in the network. The noncrystalline semiconductors preserve a ***short range order (SRO)*** in a range of 0.3–0.5 nm and are characterized by the existence of a correlation in the first atomic coordination sphere. In the case of dominantly covalently bounded amorphous solids, SRO is described by local coordination polygons, e.g., pyramidal AsS_3 in As_2S_3. In many amorphous semiconductors, such as CGs, this order is shown at longer distances, the so-called ***medium range order (MRO)*** which

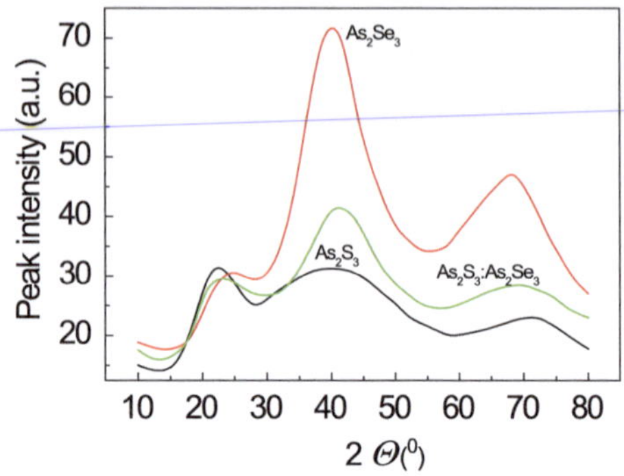

Fig. 10.1 X-ray diffraction patterns for As_2S_3, As_2Se_3, and $As_4S_3Se_3$ glasses

extends up to a range of 0.5–1.0 nm. For disordered semiconductors, a ***long range order (LRO)*** is absent, and this means that the amorphous chalcogenides have no translational symmetry.

The X-ray diffraction technique is a nondestructive method which gives information from the atomic scale range; it is a widely used method for investigations of disordered semiconductors with multilayered structures [28]. Using the X-Ray diffraction method, the diffraction patterns were obtained in the range of 2θ diffraction angles of 10°–80° (θ is the Bragg angle) for CGs As_2S_3, As_2Se_3, $As_4S_3Se_3$, and $As_4S_3Se_3$:Sn_x ($\tilde{o} = 0.01, 0.02, 0.04, 0.06, 0.07$, and 0.10).

Figure 10.1 shows the angular distribution of X-Ray diffraction intensity for As_2S_3, As_2Se_3, and $As_4S_3Se_3$ chalcogenides glasses. The position of the first sharp diffraction peak (FSDP) for As_2S_3 is $2\theta = 22.47°$ and increases to $2\theta = 24.60°$ for As_2Se_3. For the intermediate composition of $As_4S_3Se_3$, the maximum of the FSDP is located at $2\theta = 23.00°$. These spectra represent a sum of diffraction patterns of is structural vitreous As_2S_3 and As_2Se_3 with three broad lines of diffractograms, which are similar to the unvelope of the rounded lines of the spectra of crystalline As_2S_3 and As_2Se_3. It can be assumed that, in the microcrystalline state of the investigated glasses, there are domains with an ordered structure with dimensions of about 15–20 Å. Previously, a analogy between the structure of vitreous and crystalline states of As_2S_3 was vindicated by short-range order investigations – interatomic distances and coordination numbers – with the addition of the radial distribution function [29].

It was established that Van der Waals forces with a reduced covalent component act between As_2S_3 and As_2Se_3 layers. The interaction forces between the layers are hundred times weaker than the binding forces between the layers. According to [28], the structure of the glasses represents an interlinking of As-S_3 and As-Se_3 pyramids that form rings with six units. The arsenic atoms are situated at the top of the pyramid, while the chalcogen atoms form the basis. As was shown, the crystalline semiconductors are characterized by a LRO; for example, there is a good correlation between the position of the each two atoms in the network, which is not found in noncrystalline semiconductors [30, 31]. Noncrystalline semiconductors exhibit only a SRO, which involves individual atoms in the first coordination sphere. Since the range order in CGs can be extended to several interatomic distances, a new concept of the average order was introduced (MRO).

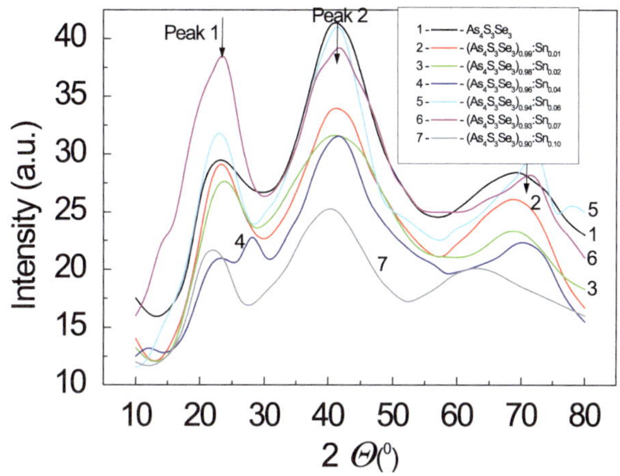

Fig. 10.2 X-ray diffraction patterns for $(As_4S_3Se_3)_{1-x}$: Sn_x glasses; x, at % Sn: (1) 0, (2) 1.0, (3) 2.0, (4) 4.0, (5) 6.0, (6) 7.0, and (7) 10.0

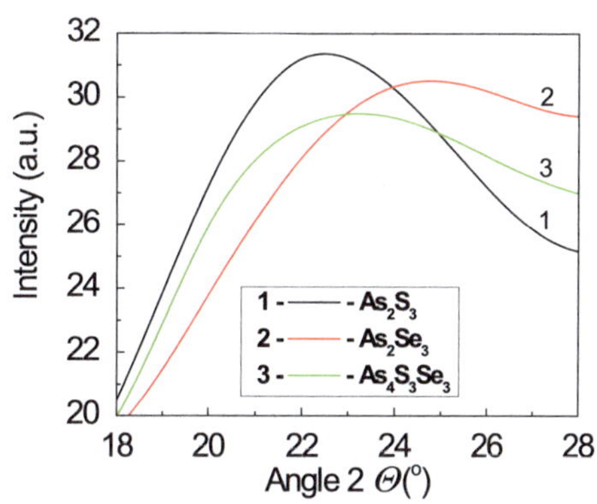

Fig. 10.3 FSDP in the X-ray diffraction patterns of As_2S_3 (1), As_2Se_3 (2), and $As_4S_3Se_3$ (3)

Comprehensive studies of the FSDP of vitreous As_2S_3 and As_2Se_3 show that they have a similar structure [28]. According to [28], the first coordination spheres (first order neighbor position) of the central atom in the structure is $r_1 = 2.414$ Å for As_2Se_3, and $r_1 = 2.306$ Å for As_2S_3, respectively. The second coordination spheres (second order neighbor position) of the central atom in the structure is $r_1 = 3.625$ Å for As_2Se_3, and $r_1 = 3.475$ Å for As_2S_3, respectively.

The Sn concentration in the mixed glasses $As_4S_3Se_3$:Sn_x does not significantly change the shape of the FSDP of the X-ray diffraction patterns (Fig. 10.2). In general, the diffraction patterns of the $As_4S_3Se_3$:Sn_x glasses are similar and form three wide lines, the maxima of which correspond to the interlayer distances $d \sim$ 4.8–2.8 – 1.7 Å. As in the case of As_2Se_3:Sn_x [31], the angular position of the FSDP in $As_4S_3Se_3$:Sn_x slightly depends on the Sn concentration. Figures 10.3 and 10.4 show the angular position and the intensity of the FSDP in the X-ray diffraction patterns of As_2S_3 (1), As_2Se_3 (2), and $As_4S_3Se_3$ (3), and the FSDP in the X-ray diffraction patterns of the $(As_4S_3Se_3)_{1-x}$:Sn_x glasses.

Some parameters of the X-ray diffraction patterns of the investigated CGs are listed in Table 10.1. Figures 10.5 and 10.6 show the dependences of the

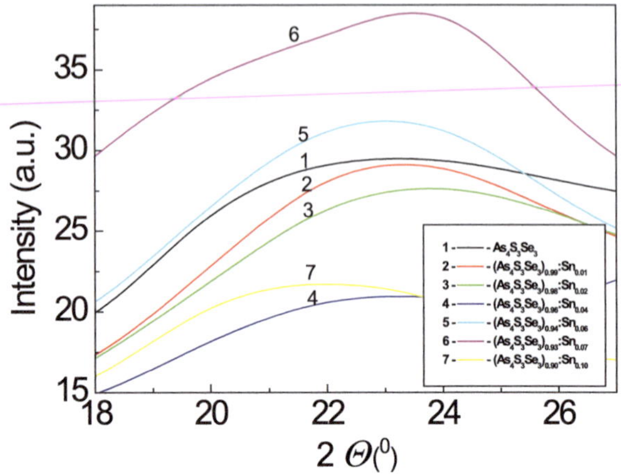

Fig. 10.4 FSDP in the X-ray diffraction patterns of $(As_4S_3Se_3)_{1-x}$:Sn_x; x, at % Sn: (1) 0, (2) 1.0, (3) 2.0, (4) 4.0, (5) 6.0, (6) 7.0, and (7) 10.0

Table 10.1 Parameters of the X-ray diffraction patterns of the studied As_2S_3, As_2Se_3, and $(As_4S_3Se_3)_{1-x}$: Sn_x CGs

N/o	Glass composition	Position of Peak 1, 2θ	Intensity of Peak 1, a.u.	Position of Peak 2, 2θ	Intensity of Peak 2, a.u.	Position of Peak 3, 2θ	Intensity of Peak 3, a.u.
1	As_2S_3	22.32	31.36	40.50	31.07	71.78	22.92
2	As_2Se_3	24.35	30.52	40.50	71.54	68.25	46.81
3	$As_4S_3Se_3$	22.86	29.43	41.01	41.14	69.28	28.37
4	$(As_4S_3Se_3)_{0.99}$:$Sn_{0.01}$	23.39	29.06	41.53	33.92	69.15	26.09
5	$(As_4S_3Se_3)_{0.98}$:$Sn_{0.02}$	23.67	27.54	41.14	31.64	68.61	23.51
6	$(As_4S_3Se_3)_{0.96}$:$Sn_{0.04}$	22.86	20.93	41.53	31.49	70.62	22.38
7	$(As_4S_3Se_3)_{0.94}$:$Sn_{0.06}$	23.00	31.64	41.53	41.14	72.08	30.35
8	$(As_4S_3Se_3)_{0.93}$:$Sn_{0.07}$	23.39	38.33	41.81	39.17	71.55	28.15
9	$(As_4S_3Se_3)_{0.90}$:$Sn_{0.10}$	21.93	21.69	40.34	25.26	62.75	20.17

For the $(As_4S_3Se_3)_{0.96}$:$Sn_{0.04}$ glass composition, an additional peak at $2\theta = 28.20$ was observed

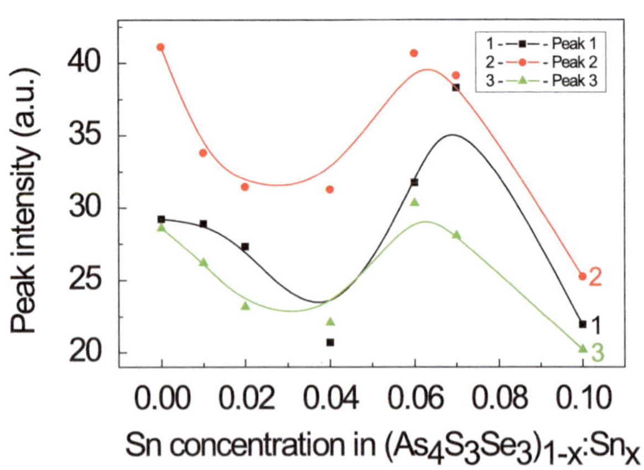

Fig. 10.5 Dependence of the angular position of the diffraction peaks vs. concentration of tin in some $As_4S_3Se_3$:Sn_x CGs

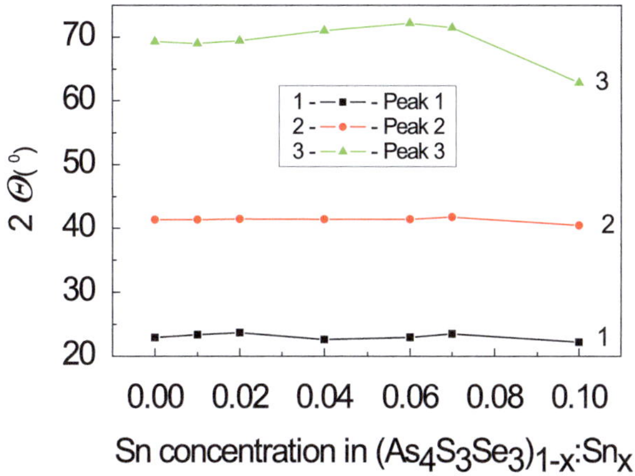

Fig. 10.6 Dependence of the angular position of the diffraction peaks vs. concentration of tin in the some $As_4S_3Se_3:Sn_x$ CGs

peaks intensities vs. concentrations of tin of the three diffraction peaks of CG $As_4S_3Se_3:Sn_x$ situated at $2\theta \sim 22.86 \div 23.67°$, $2\theta \sim 40.34 \div 42°$, and $2\theta \sim 62 \div 70°$, and the dependence of the angular position of the diffraction peaks vs. tin concentration, respectively. The intensity of the FDSP shows a nonlinear behavior with the different amounts of doping with Sn. The maximum intensity is achieved at about 6 at % Sn in $As_4S_3Se_3:Sn_x$, while in the case of $As_2Se_3:Sn_x$ the maximum intensity is situated at the tin concentration of about 2 at % Sn. A similar behavior was found for the second and third diffraction peaks. According to [31], when Sn is added in As_2Se_3 or As_2S_3 ChGs, due to the tetrahedral disposal of the sp3 bonds in the chalcogens, the dopant atom implanted in the network increases the thickness of the layered configuration as revealed by a significant shift of the FSDP towards lower angles. This implantation corresponds to the introduction of the structural units of the $SnSe_2$ or SnS_2 type in the glass network; the same fact was confirmed by the Mössbauer spectroscopy experiments [11].

10.3.2 Physical Properties and Photoplastic Effect in $As_2Se_3:Sn_x$ Glasses

The physical properties (density, hardness, glass transition temperature, electrical conductivity, etc.) of amorphous As-Se films have been previously reviewed by Borisova [32, 33]. The investigation of bulk, thin films, and illuminated thin film samples was performed using a NHT CSM nanohardnes tester. The hardness was calculated from load-displacement curves by the Oliver–Pharr method. In-situ illumination of samples was carried with a green laser ($\lambda = 532$ nm) with a power of $P = 50\,mV/cm^2$. To change the direction of the incident laser beam, an optical glass prism was used (Fig. 10.7).

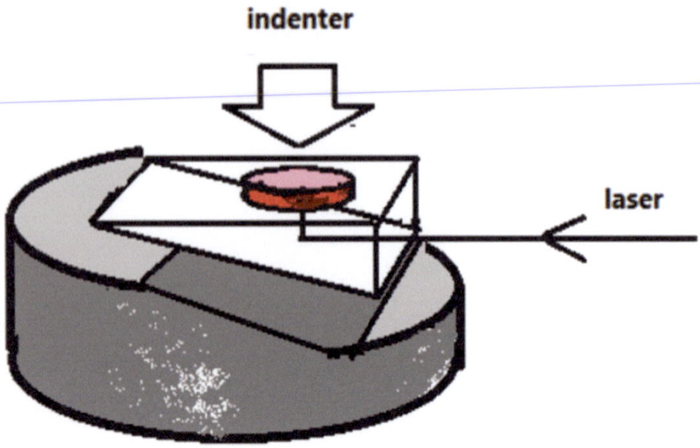

Fig. 10.7 Experimental set-up for investigation of the photoplastical effect in CGs

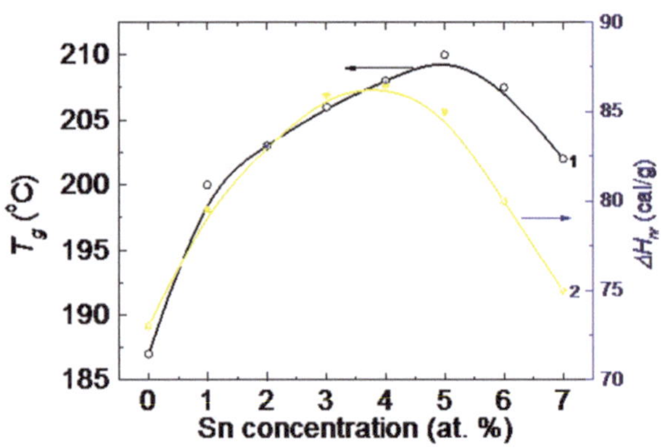

Fig. 10.8 Variations in T_g (1), $\Delta H_{nr}(x)$ for As_2Se_3:Sn_x glasses. The *smooth lines* are computer fitting

The maximum indentation load was 5 mN; therefore, the maximal penetration depth did not exceed 15 % of the film thickness. The hardness was calculated using the expression

$$H_B = (1570 \cdot P)/l^2, \quad (10.1)$$

where P is the applied load and $L$is the height of the triangle of the remaining imprints [34]. For both the bulk samples and the amorphous thin films, the applied load was 10 g and the depth of deposited imprints did not exceed 20 % of the films thickness.

Figure 10.8 shows variations in glass transition temperature $T_g(x)$ for the bulk As_2Se_3:Sn_x glasses and the nonreversing heat $\Delta H_{nr}(x)$. At low concentrations of the Sn additive, the glass transition temperature T_g of the base glass increases with x, suggesting that the base glass becomes more bound. However, as x approaches 5 at % Sn, the glass transition temperature T_g exhibits a threshold behavior [11].

The maximum in T_g suggests [11, 24] that the above threshold additives no longer form part of the base glass structure, e.g., they undergo nanoscale phase separation. These nanoscale phase separation effects have already been found in As_2Se_3 base glass ($x = 0$) and are apparently attributed to higher concentrations of Sn ($x > 5$ at % Sn). Mössbauer spectroscopy results show the presence of a single line (A) feature in the spectra at low x (<3 at %). The isomer-shift of this line has been previously assigned [35] to Sn that is tetrahedrally coordinated to 4 Se near-neighbors as in a $Sn(Se_{1/2})_4$ local structure [11]. Apparently, the introduction of a Sn additive in As_2Se_3 base glass contributes to the growth of $Sn(Se_{1/2})_4$ units and makes the base glass As-rich. The effect can occur in one of two modes: through the formation of either polymeric ethylene-like $As_2(Se_{1/2})_4$ units and/or monomeric As_4Se_4 cages based on the Realgar structure. It is found that, in the base glass, the polymeric species are first nucleated at$x$slightly below 0.4, and monomeric ones are formed afterwards ($x > 0.42$).

The experimental results of investigation of hardness for bulk and $As_2Se_3:Sn_x$ amorphous thin films are presented in Figs. 10.9 and 10.10 and are in good agreement with experimental results obtained earlier by Borisova [32, 33]. The hardness values of bulk samples depend on the Sn concentration in the $As_2Se_3:Sn_x$ glasses and vary between 1300 and 1700 MPa.

The hardness H of the amorphous $As_2Se_3:Sn_x$ thin films exhibits a nonmonotonic behavior in dependence on the Sn concentration and is higher than the hardness of bulk samples. An increase in hardness is registered at an impurity concentra-

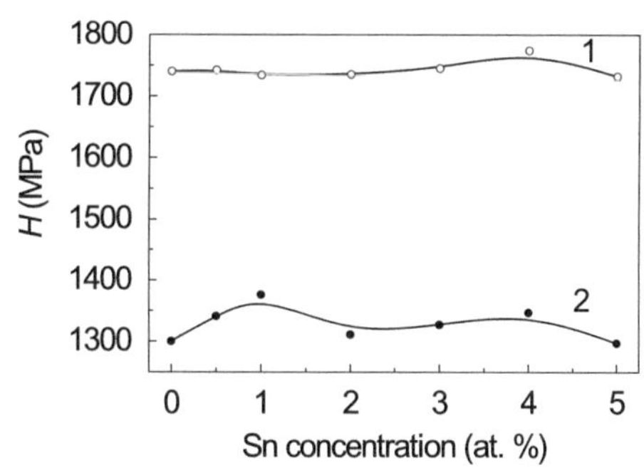

Fig. 10.9 Hardness H of $(As_2Se_3)_{1-x}:Sn_x$ samples, *thin films* (1) and *bulk* (2)

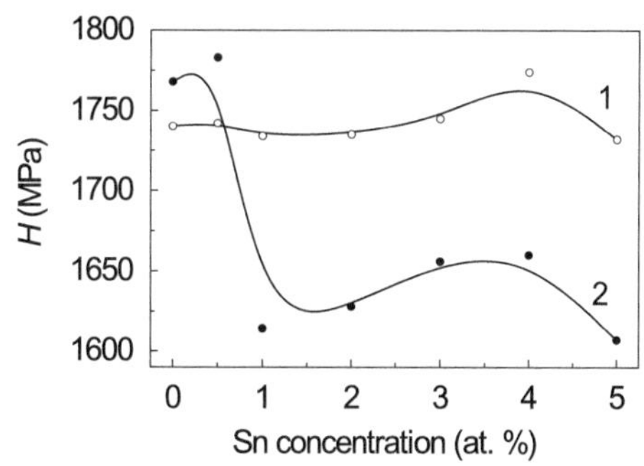

Fig. 10.10 Hardness H of $(As_2Se_3)_{1-x}:Sn_x$ as-deposited (1) and *illuminated* (2) *thin films*

tion of about 4 % Sn. The hardness H of $(As_2Se_3)_{1-x}$:Sn_x films varies between 1730÷1780 MPa. For both bulk samples and thin films, the hardness is higher according to the data obtained in [32, 33]. This fact can be attributed to some technological features in preparing the ChG. Under illumination, the hardness H of $(As_2Se_3)_{1-x}$:Sn_x decreases, and this decrease is specific especially of the doped samples (Fig. 10.10).

For investigation of microhardness and Young's modulus, a Nano-Indentation Tester (NIT) was used; it provides experimental data by indenting to depths at a nanometer-micron scale. This nanoindenter can be used to characterize organic, inorganic, soft or hard materials, and coatings. The mechanical characterization of the surface of bulk materials can also be performed, including metals, semiconductors, glasses, composites, and in vitro biomaterials.

The operating principle of the tester is as follows: an indenter diamond tip, normal to the sample surface, is driven into the sample by applying an increasing load ($P = 100$ mN) up to a certain present value. The load is then gradually decreased until partial or complete relaxation of the material occurs. After the removal of the load, the diagonal length is measured (imprint d_1 and d_2). The offset of diagonal tip was <0.25 μm and the load resolution was 0.001 N. To avoid overlapping of surface stresses developed around neighboring indentations, the separation between indentation diagonals was kept more than ten times the diagonal length of indentation impressions.

The dimensions of both diagonals $d$made at a particular load P were measured, and the average diagonal d was calculated [14]. The value of microhardness H_V was computed from the $P(d)$ data using the standard well-known relation:

$$H_V = \frac{kP}{d_2}, \tag{10.2}$$

where k is the geometrical conversion factor for the indenter used ($k = 0.1891$):

$$k = \frac{1}{g}\left(2\sin\frac{136^\circ}{2}\right), \tag{10.3}$$

where g is the acceleration of gravity, 136° is the angle between its opposite faces of pyramidal tip. The average values of indentation diagonal d and microhardness H_V for at least 10 indentations were used in the analysis of indentation size effect and hardness measurements.

Figure 10.11 shows the Young's modulus and microhardness dependences for the bulk samples of As_2S_3, As_2Se_3, and for $(As_4S_3Se_3)_{1-x}$:Sn_x($x = 0÷10$ at %) vs tin concentration. It was observed that the addition of tin in the host material initially leads to an increase in the microhardness up to 2.0 at % Sn. A further increase in the Sn concentration decreases the microharness.

In general, the hardness varies around one value $H_{\text{middle}} = 1282$ MPa. A similar behavior was observed for the system $(As_2S_3)_{1-x}$:$(Sb_2S_3)_x$, where a curve with the minimum at about 2–3 at % of (Sb_2S_3) was obtained [36]. At the same time,

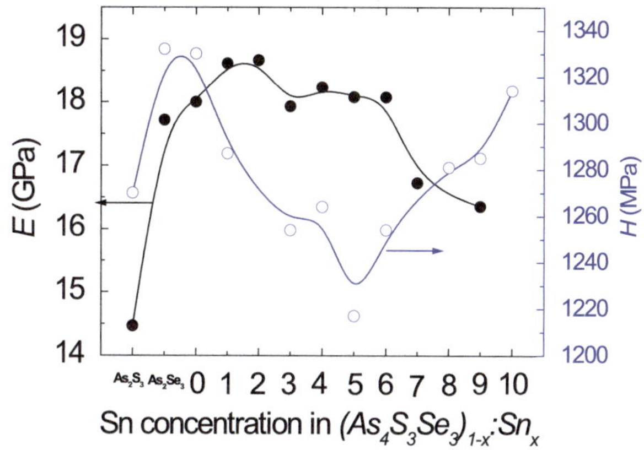

Fig. 10.11 Concentration dependence of Young's modulus E and microhardness H of the bulk samples As_2S_3, As_2Se_3, and $(As_4S_3Se_3)_{1-x}$:Sn_x

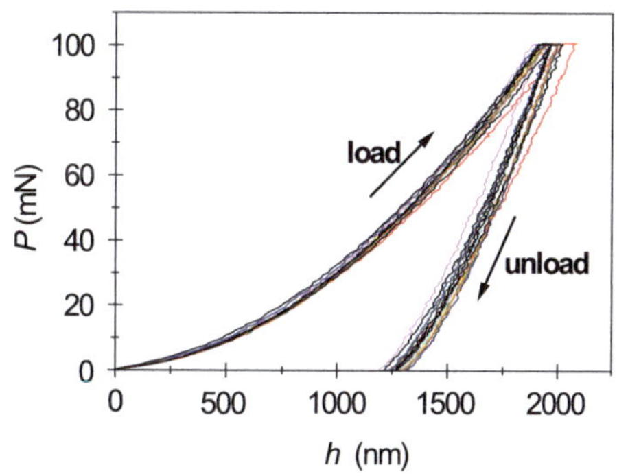

Fig. 10.12 Depth dependence of the load and unload values on chalcogenide samples of $(As_4S_3Se_3)_{1-x}$:Sn_x

the value of microhardness for vitreous As_2S_3 and As_2Se_3 better coincides with the values cited in [37], although the Young's modulus in our case is higher by an order. According to [36], the observed increase in the hardness with x in the $(As_2S_3)_{1-x}$:$(Sb_2S_3)_x$ system indicates structural changes, which lead to lower molecular or configuration mobility.

Figure 10.12 represents the depth dependence of the load and unload values for chalcogenide samples of $(As_4S_3Se_3)_{1-x}$:Sn_x. Similar dependences were obtained for amorphous As_2S_3 thin films prepared by thermal deposition in vacuum [38]. The author mentions this fact in these experiments because the hold segment is rather long and the unloading process is very fast, any further viscoplastic creep during unloading is negligible compared to the elastic recovery during the unloading procedure. Figure 10.13 represent the imprint on the surface of the diamond tip on a $(As_4S_3Se_3)_{1-x}$:Sn_x sample at a load of $P = 100\,mN$.

Figure 10.14 shows the surface morphology of the as-deposited amorphous $(As_2Se_3)_{0.90}$:$Sn_{0.10}$ thin film under the indentation load after (a) and before light illumination (b, upper patterns). For comparison, at the bottom of the figure, the pictures of the light-induced anisotropic plasticity in amorphous $As_{20}Se_{80}$ thin films are shown: (a) the image of the nonirradiated film surface and (b) the image of this surface by irradiated linearly polarized laser beam ($\lambda = 633\,nm$) [39]. In both cases,

Fig. 10.13 Example of imprint on the surface of the *diamond tip* on bulk $(As_4S_3Se_3)_{1-x}$:Sn_x samples at a load of $P = 100\,\text{mN}$

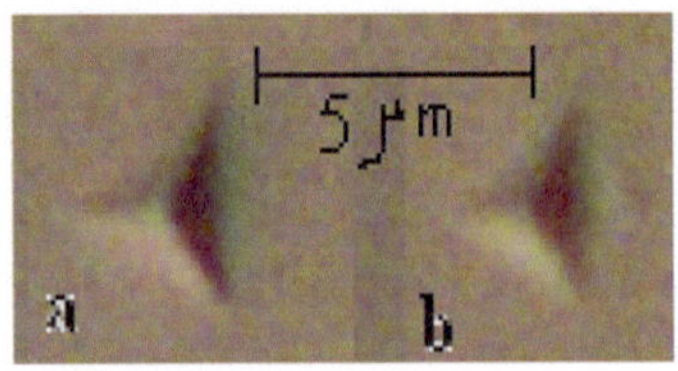

Fig. 10.14 Surface morphology of the as-deposited amorphous $(As_2Se_3)_{0.90}$:$Sn_{0.10}$ thin film under indentation load after (**a**) and before light illumination (**b**, upper patterns)

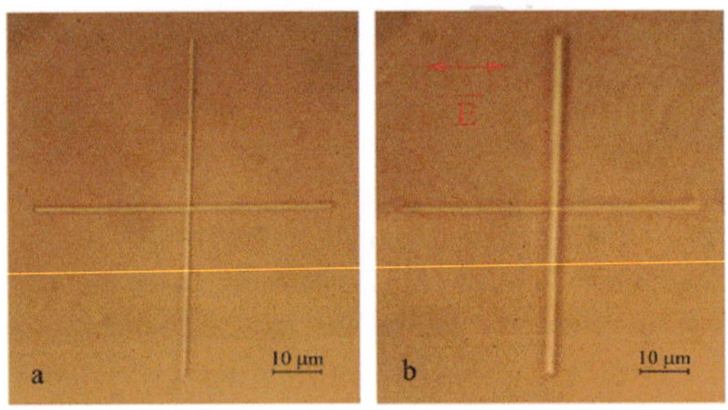

after light irradiation, the plasticity of the investigated amorphous films increases. For explanation of this phenomenon, the authors of [38] proposed a mechanical model of anisotropic plasticity in CGs, according to which the anisotropic softening consists in the weakening of mechanical compliance in the orthogonal direction to the light polarization.

10.4 Conclusions

The bulk of $(As_4S_3Se_3)_{1-x}$:Sn_x ($x = 0 \div 10$ at % Sn) CGs have been investigated by X-ray diffraction and nanoindentation methods. It has been found that the addition of these amounts of tin does not lead to significant changes in the glass physical properties, such as values of stress and Young's modulus related to the modification of the density and compactness. The additions of Sn atoms in the CG matrix considerably decrease the elasticity of this material or increase the hardness at some tin concentrations. The XRD measurements have shown that the Sn impurities in the $(As_4S_3Se_3)_{1-x}$:Sn_x do not significantly change the shape of the FSDP of the X-ray diffraction patterns either; the intensity and the position of the FSDP nonmonotonically depend on the Sn concentration. The addition of As_2Se_3 in As_2S_3 shifts the FSDP towards higher diffraction angles from $2\theta = 22.47°$ for As_2S_3 up to $2\theta = 24.60°$ for As_2Se_3. The FSDP intensity increases at about 5.0–7.0 at % Sn in the $(As_4S_3Se_3)_{1-x}$:Sn_x glasses.

It has been found that the addition of these amounts of tin in $(As_4S_3Se_3)_{1-x}$:Sn_x does not lead to significant changes in the glass physical properties, such as values of stress and Young's modulus related to the modification of the density and compactness. The study of the photoplastic effect has been performed in situ, with illumination of the bulk and thin film samples during indentation as well as their indentation after illumination with a green laser ($\lambda = 532$ nm) at a power of $P = 50$ mV/cm^2. The hardness is calculated from load-displacement curves by the Oliver–Pharr method. A sharp increase in hardness is registered if the tin concentration exceeds a value of 3–4 % Sn. The hardness H of $(As_2Se_3)_{1-x}$:Sn_x films varies between 115 and 130 kg/mm^2. It has been found that the hardness H of amorphous thin films is generally higher than the hardness of bulk samples with the same chemical composition.

References

1. Inoue A, Shen B, Nishiyama N, Miller M, Liaw P (eds) (2008) Microstructure and tribological behavior of spark plasma sintered iron-based amorphous coatings bulk metallic glasses. Springer, Dordrecht, The Netherlands, pp 1–25
2. Inoue A, Takeuchi A (2011) Recent development and application products of bulk glassy alloys. Acta Mater 59:2243–2267
3. Inoue A (2009) Bulk amorphous and nanocrystystaline alloys with high functional proertis. Mater Sci Eng A 1:304–306
4. Johnson WL (1999) Bulk glass forming metallic alloys: science and technology. MRS Bull 24:42
5. Zakery A, Elliot SR (2007) Optical nonlinearities in chalcogenide glasses and their applications. Springer, Berlin/Heidelberg/New York
6. Andriesh AM, Iovu MS, Shutov SD (2002) Optical phenomena in chalcogenide glasses and their application in optoelectronics. J Optoel Adv Mater 4(3):631
7. Nemec P, Jedelsky J, Frumar M, Stabl M, Vlcek M (2004) Photoinduced phenomena in amorphous and glassy chalcogenides. J Phys Chem Solids 65(7):1253
8. Gerbreders A , Teteris J (2007) Recording of surface-relief gratings on amorphous As-S-Se films. J Optoel Adv Mater 9(10):3164
9. Reinfelde M, Teteris J (2011) Surface relief and polarization holographic formation in amorphous As-S-Se films. J Optoel Adv Mater 13(11–12):1531
10. Iovu M, Shutov S, Popescu M (2002) Relaxation of photodarkening in amorphous As-Se films doped with metals. J NonCryst Solids 299–302:924–928
11. Boolchand P, Georgiev D, Iovu M (2005) Molecular structure and quenching of photodarkening in As_2Se_3:Sn amorphous films. Chalcogenide Lett 2(4):27
12. Iovu M, Boolchand P, Georgiev D, Optoel J (2005) Photoinduced phenomena in chalcogenide glass doped with memtals. Adv Mater 7(2):763

13. Iovu M, Harea D, Colomeico E (2008) Optical properties and photoinduced absorption in As-Se and As_2Se_3:Sn thin films. J Optoel Adv Mater 10(4):862
14. Yee LY, Chaudhri MM (2006) Indentation of elastic solids with a rigid Vickers pyramidal indenter. Mech Mater 38:1213
15. Trunov ML, Bilanich VS (2003) Photoplastic phenomena in chalcogenide glasse. J Optoel Adv Mater 5(5):1085
16. Yannopoulos SN, Trunov ML (2009) Photoplastic effects in chalcogenide glasses: a review. Phys Status Solidi B 246(8):1773
17. Osipyan YA, Savchenko IB, Zh P (1968) Experimental evidence for the influence on the plastic deformation of cadmium sulfide. Eksp Teor Fiz 7:130
18. Gerasimov AB, Chiradze GD, Kutivadze NG (2001) On the physical nature of a photomechanical effect. Semiconductors 35:72
19. Carlsson L, Svensson C (1970) Photoplastic effect in ZnO. J Appl Phys 41:1652
20. Matsuda A, Mizuno H, Takayama T, Saito M, Ki-kuchi M (1974) Photoinduced relaxation in amorphous As-S films. Appl Phys Lett 25:411
21. Igo T, Noguchi Y, Nagai H (1974) Photoexpansion and 'thermal contraction' of amorphous chalcogenide glasses. Appl Phys Lett 25:193
22. Koseki H, Odajima A (1982) Photo-induced stress relaxation in amorphous selenium films. Jpn J Appl Phys 21:424
23. Iovu MS, Shutov SD, Toth L (1996) Transient photocurrents under optical bias in time-of-flight experiment with amorphous films of As_2Se_3:Sn and As_2S3:Sb_2S_3. Phys Status Solidi (b) 195:149
24. Iovu MS, Syrbu NN, Shutov SD, Vasiliev IA, Rebeja S, Colomeico E, Popescu M, Sava F (1999) Spectroscopical study of amorphous AsSe:Sn films. Phys Status Solidi (a) 175(2):623
25. Iovu MS, Shutov SD, Arkhipov VI, Adriaenssens GJ (2002) Effect of Sn doping on photoconductivity in amorphous As_2Se_3 and AsSe films. J Noncryst Solids 299–302:1008
26. Boolchand P, Georgiev DG, Iovu MS (2005) Molecular structure and quenching of photodarkening in As_2Se_3:Sn amorphous films. Chalcogenide Lett 2(4):27
27. Iovu MS, Harea DV, Cojocaru IA, Colomeico EP, Prisacari A, Ciorba VG (2007) Optical properties and photoinduced absorption in As-Se and As_2Se_3:Sn thin films. J Optoel Adv Mater 9(10):3138
28. Popescu M, Andries A, Ciumas V, Iovu M, Sutov S, Tiuleanu D (1996) Fizica sticlelor calcogenice. Editura Stiintifica Bucuresti – I.E.P., Stiinta, Chisinau
29. Tatarinova L (1983) The structure of amorphous solids and liquids. Nauka, Ìoscow, p 151
30. Popescu M (1996) Medium range order in chalcogenide glasses. In: Andriesh A, Bertolotti M (eds) Physics and applications of noncrystalline semiconductors in optoelectronics. Springer, Dordrecht, The Netherlands, vol 36, p 215
31. Popescu M, Tudorica F, Andriesh A, Iovu M, Shutov S, Bulgaru M, Colomeyko E, Malkov S, Verlan V, Leonovici M, Mihai V, Steflea M (1995) Buletinul Academiei de Stiinte a Republicii Moldova. Fizica si tehnica 3:3
32. Borisova ZU (1981) Glassy semiconductors. Plenum Press, New Yok, pp 215–220 (and references therein)
33. Borisova ZU (1983) Chalcogenide semiconducting glasses. Khimiya, St. Petersburg (in Russian)
34. Boyarskaya YS (1986) Physics of microindentitation processes. Stiinta, Kishinev (in Russian)
35. Kumar P, Thangaraj R (2006) Glassy state and structure of Sn-Sb-Se chalcogenide alloy. J Noncryst Solids 352:2288
36. Sava F (2001) Structure and properties of chalcogenide glasses in the system (As_2Se_3)1-x(Sb_2S_3)x. J Optoel Adv Mater 3(2):425
37. Shchurova NN, Savchenko ND (2001) Correlation between mechanical parameters for amorphous chalcogenide films. J Optoel Adv Mater 3(2):491
38. Trunov ML (2008) Photoplastic effect in non-crystalline materials: a nanoindentation study. J Phys D Appl Phys 41:074011 (9pp)
39. Yannopoulos SN, Trunov ML (2009) Photoplastic effects in chalcogenide glasses: a review. Phys Status Solidi B-Basic Solid State Phys 246(8):1773–1785

Chapter 11
Thermal Memory and Thermal Induced Phase Transformation in Shape Memory Alloys

O. Adiguzel

Abstract Shape memory effect is a peculiar property exhibited by a series of alloy systems. This property is characterized by the recoverability of desired shape of the material at different conditions, and these alloys take place in a class of smart materials due to this property. Shape memory effect is based on martensitic transformation, which is a solid state phase transition and occurs on cooling the material from high temperature parent phase region. Martensitic transformations are first order diffusionless transitions and governed by changes in the crystal structure of the material at angstrom level. Copper based alloys exhibit this property in metastable β-phase field, which have bcc- structures at high temperature. These alloys undergo two ordered transitions on cooling, and bcc structures turn into B2 (CsCl) or DO3 (Fe3Al) -type ordered structures. These ordered structures martensiticaly undergo the non-conventional complex layered structures on further cooling. The product phase has the unusual complex structures called long period layered structures such as 3R, 9R or 18R depending on the stacking sequences on the close-packed planes of the ordered lattice.

11.1 Introduction

Shape memory alloys have a peculiar property to return to a previously defined shape with changing temperature. These alloys recover original shape on heating after deformation in low temperature product phase condition. This kind of behaviour in shape memory materials is evaluated by the structural changes caused by internal stresses in nanometer scale depending on the external conditions. These alloys involve the repeated recovery of macroscopic shape at certain temperature intervals. The origin of this phenomenon lies in the fact that the material changes its internal crystalline structure with changing temperature. The crystallographic basis

O. Adiguzel (✉)
Department of Physics, Firat University, 23169 Elazig, Turkey
e-mail: oadiguzel@firat.edu.tr

J. Bonča, S. Kruchinin (eds.), *Nanomaterials for Security*, NATO Science for Peace and Security Series A: Chemistry and Biology, DOI 10.1007/978-94-017-7593-9_11

of this phenomenon is martensitic transitions, which govern the remarkable changes in internal crystalline structure of materials [1–5]. Martensitic transformation is a shear-dominant solid-state phase transformation, by which the ordered parent phase structures turn into complex layered structures. Martensitic structures occur as martensite variants in a self-accommodating manner through twinning by cooperative movement of atoms on close packed planes of the parent austenite matrix [3, 4]. The movement of atoms is confined into inter-atomic distances called lattice parameters, due to the diffusionless character of martensitic transformation. Shape memory properties are intimately related to the microstructures of the material, especially orientation relationship between the various martensite variants. Twinning and detwinning processes can be considered as elementary processes activated during the transformation. In particular, the detwinning is essential as well as martensitic transformation in reversible shape memory effect [3, 4]. By applying external stress, the martensitic variants are forced to reorient into a single variant leading inelastic strains. Deformation of shape memory alloys in martensitic state proceeds through a martensite variant reorientation of twins. The twinning occurs with internal stresses, while detwinning occurs with the external stresses. The deformed material recovers the original shape in the bulk level, and the crystal structure turns into the parent phase structure on first heating over the austenite finish temperature in both reversible and irreversible memory cases. The parent phase structure returns to the multivariant martensite structures in the irreversible shape memory effect on cooling below the martensite finish temperature; in contrast, the material returns to the detwinned martensite structure in the reversible shape memory case. The basic mechanism of shape memory effect is schematically illustrated in Fig. 11.1 [3, 4]. As seen from this figure; the ordered parent phase turns into twinned martensite in thermal manner on cooling from high temperature, and the twinned martensites turn into the oriented martensites in stress-induced manner by applying external forces. The microstructural mechanisms responsible for the shape memory effect are reorientation and martensitic transformation, and shape memory alloys cycle between the original and deformed shapes by heating over the austenite finish temperature, A_f, and cooling down below the martensite finish temperature, M_f. The lattice orientation process in shape memory alloys is of primary importance for the occurrence of shape memory effect [3]. Shape memory alloys have the important ability to remember the original shape on heating after deformation, and they are used as shape memory elements in devices due to this property.

These alloys cycle between two particular shapes with changing temperature. As seen from Fig. 11.1, the twinned martensite structures turn into the oriented structures when an external stress is applied, and oriented structures turn into the ordered parent phase structure on heating over the reverse transformation temperature after releasing the stress. In the cycling processes; the martensite normally

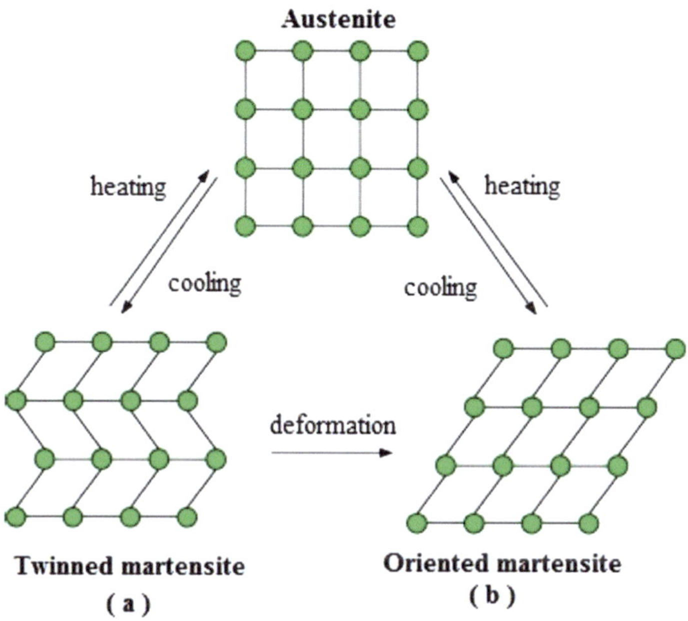

Fig. 11.1 Schematic illustration of structural changes with phase transformation; (**a**) formation of twinned martensite structure on cooling, (**b**) formation of oriented martensite structure with deformation in low temperature martensitic state

returns to a self-accommodated structure after cooling from austenite in irreversible shape memory effect, whereas the reversible shape memory effect causes the martensite to adopt an oriented-single variant configuration [3, 4]. Copper based alloys exhibit this property in metastable β-phase field. High temperature β-phase bcc-structures martensiticaly undergo the non-conventional structures following two ordered reactions on cooling, and structural changes in nanoscale level govern this transition cooling [1–3]. Martensitic transformations occur in a few steps with the cooperative movement of atoms less than interatomic distances by means of lattice invariant shears on a {110}-type plane of austenite matrix which is basal plane of martensite. These shears give raise the formation of unusual layered structures such as 3R, 9R or 18R depending on the stacking sequences on the close-packed planes of the ordered lattice. The periodicity and unit cell are completed through 18 layers in direction z in 18R case [1–5]. The complicated long-period stacking ordered structures mentioned above can be described by different unit cells. In case the parent phase has a B2-type superlattice, the stacking sequence is ABCBCACAB (9R) [2, 6]. The stacking of (110)β-planes in DO3-type structure and formation of layered structures are shown in Fig. 11.2.

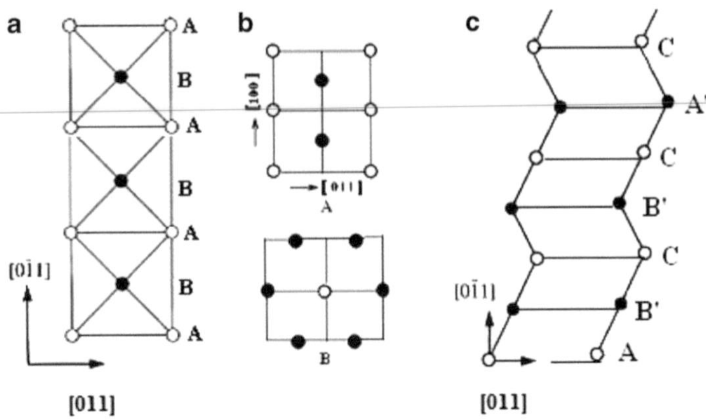

Fig. 11.2 (**a**) Stacking of (110)β planes viewed from [001]β direction, (**b**) Atomic configuration on first and second layers of (110)β plane in DO3-type structures, (**c**) inhomogeneous shear and formation of layered structures, stacking sequences of half 18R or M18R unit cell in direction z

11.2 Experimental

In the present contribution, two copper based Cu-26.1 %Zn-4 %Al and Cu-11 %Al-6 %Mn (in weight) alloys were selected for investigation. Powder specimens for X-ray examination were prepared by filling the alloys. Specimens for TEM examination were prepared from 3 mm diameter discs and thinned down mechanically to 0.3 mm thickness. These specimens were heated in evacuated quartz tubes in the β-phase field (15 min at 830 °C for CuZnAl and 20 min at 700 °C for CuAlMn) for homogenization and quenched in iced-brine. TEM and X-ray diffraction studies carried out on these specimens. TEM specimens were examined in a JEOL 200CX electron microscope, and X-ray diffraction profiles were taken from the quenched specimens using Cu-K_α radiation with wavelength 1.5418 Å.

11.3 Results and Discussion

An x-ray powder diffractogram taken from the quenched CuAlMn alloy samples is shown in Fig. 11.3, and an electron diffraction pattern taken from CuZnAl alloy samples is shown in Fig. 11.4. X-ray powder diffractogram and electron diffraction patterns reveal that these alloys exhibit superlattice reflections. A series of x-ray powder diffractograms and electron diffraction patterns were taken from the specimens in a large time interval and compared with each other. It has been observed that electron diffraction patterns exhibit similar characteristics, but some changes occur at the peak locations and intensities on the x-ray diffractograms with aging duration. In particular, some peak pairs satisfying a specific relation some peak pairs satisfying a specific relation between miller indices come close

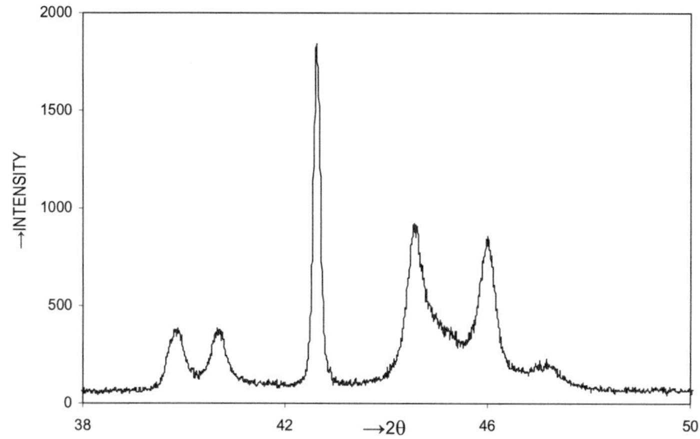

Fig. 11.3 An x-ray powder diffractogram taken from Cu-11 %Al-6 %Mn alloy sample

Fig. 11.4 An electron diffraction patterns taken from the CuZnAl alloy sample

each other with ageing [1]. These changes occur as redistribution of atoms in the material, and attribute to new transitions in diffusive manner [1, 7]. It means that some neighboring atoms change locations. This result can be attributed to a relation between interplane distances of these plane pairs and rearrangement of atoms on the basal plane. This result can be attributed to a relation between interplane distances of these plane pairs and rearrangement of atoms on the basal plane. In these changes, atom sizes play important role. The different sizes of atomic sites lead to a distortion of the close-packed plane from an exact hexagon and thus a more close-packed layered structure may be expected. In the disordered case, atom sizes can be taken nearly equal, and martensite basal plane becomes an ideal hexagon. Although martensitic transformations are displacive, the post-martensitic transitions have the diffusive character because this transition requires a structural change and this also gives rise to a change in the configurational order. In these changes, atom sizes play an important role. The ordered structure or super lattice structure is essential for the shape memory quality of the material. Crystallization is essential for shape memory quality as well as quenching process in suitable media. The quenching rate is also important for the formation of homogenous ordered structures and shape memory optimization. The martensitic transformation obtained on cooling is called thermally induced phase transformation. The obtained martensite consists of up to 24 variants, which are regions of the same structure but with different crystallographic orientations [7, 8]. The martensitic phase in copper-based β-phase alloys is based on one of the $\{110\}\beta$ planes of parent phase called basal plane for martensite. The (110) basal plane which has a rectangular shape in parent phase is subjected to hexagonal distortion and undergoes a hexagon. The different sizes of atomic sites lead to a distortion of the close-packed plane from an exact hexagon and thus a more close-packed layered structure may be expected. In the disordered case, atom sizes can be taken as nearly equal, and the martensite basal plane becomes an ideal hexagon. Although martensitic transformations are displacive, the post-martensitic transitions have a diffusive character because this transition requires a structural change and this also gives rise to a change in the configurational order.

On the other hand, post-quench ageing and service processes in devices affect the shape memory quality, and give rise shape memory losses. These kinds of results lead to the martensite stabilization in the reordering or disordering manner. In order to make the material satisfactorily ordered and to delay the martensite stabilization, copper-based shape memory alloys are usually treated by step-quenching after homogenization. In addition, the ageing treatments cause the formation of crystal imperfections, like dislocation and precipitations in the material [9, 10]. Thermo-mechanical treatment also contributes to the martensite stabilization as well as ageing [9, 10].

Stabilization is important factor and causes to memory losses, and changes in main characteristics of the material; such as, transformation temperatures, and x-ray diffraction peak location and peak intensities.

Although martensitic transformations are displacive, the post-martensitic transitions have a diffusive character because this transition requires a structural change and this also gives rise to a change in the configurational order.

11.4 Conclusions

It has been observed that electron diffraction patterns exhibit similar characteristics, but some changes have been observed in the locations and intensities of diffraction peaks on the x-ray diffractograms with aging duration. These changes imply new transitions which have diffusive character. It can be concluded from the above results that the copper-based shape memory alloys are very sensitive to the ageing treatments, and heat treatments can change the relative stability and the configurational order of atoms in the material. This result attributes to rearrangement of atoms in diffusive manner.

References

1. Adiguzel O (2007) Smart materials and the influence of atom sizes on martensite microstructures in copper-based shape memory alloys. J Mater Process Technol 185:120–124
2. Zhu JJ, Liew KM (2003) Description of deformation in shape memory alloys from DO3 Austenite to 18R Martensite by group theory. Acta Mater 51:2443–2456
3. Ma J, Karaman I, Noebe RD (2010) High temperature shape memory alloys. Int Mater Rev 55:257–315
4. Sun L et al (2012) Stimulus-responsive shape memory materials: a review. Mater Des 33:577–640
5. Sutou Y et al (2005) Effect of grain size and texture on pseudoelasticity in Cu-Al-Mn-based shape memory wire. Acta Mater 53:4121–4133
6. Pelegrina JL, Romero R (2000) Calorimetry in Cu–Zn–Al alloys under different structural and microstructural conditions. Mater Sci Eng A 282:16–22
7. Adiguzel O (2012) Martensitic transformation and microstructural characteristics in copper based shape memory alloys. Key Eng Mater 510–511:105–110
8. Li Z, Gong S, Wang MP (2008) Macroscopic shape change of Cu13Zn15Al shape memory alloy on successive heating. J Alloys Compd 452:307–311
9. Malarriaa J, Lovey FC, Sade M (2009) Two way shape memory effect in CuZnAl single crystals after pseudoelastic cycling at low temperatures. Mater Sci Eng A 517:118–124
10. de Castro F, Sade M, Lovey F (2012) Improvements in the mechanical properties of the 18R ↔ 6R high-hysteresis martensitic transformation by nanoprecipitates in CuZnAl alloys. Mater Sci Eng A 543:88–95

Chapter 12
Spectroscopic Properties of Nanoceria Allowing Visualization of Its Antioxidant Action

V. Seminko, A. Masalov, P. Maksimchuk, V. Klochkov, I. Bespalova, O. Viagin, and Yu. Malyukin

Abstract Two distinct luminescence centers were revealed in ceria nanocrystals: first one – Ce^{3+} ions with 5d-4f luminescence at 390 nm, and second one – Ce^{4+}–O^{2-} complexes showing charge transfer (CT) luminescence at 630 nm. Intensity of Ce^{3+} luminescence depends directly on the concentration of oxygen vacancies in nanoceria and can be varied by means of change of both heat treatment atmosphere from oxidizing to reducing and the size of nanocrystal. Ce^{3+} luminescence can be used for visualization of the processes of interaction between ceria nanoparticles and reactive oxygen species using relative intensity of Ce^{3+} band as a measure of Ce^{4+}/Ce^{3+} ratio during oxidation reaction.

12.1 Introduction

As was shown in the past decades, reactive oxygen species (such as hydroxyl radicals, superoxide anions and hydrogen peroxide) are not only one of the most widespread products of cell metabolism, but they also can serve as mediators at signal transmission inside the cell [13]. However, an increase of the ROS concentration beyond normal limits can lead to very dangerous consequences, such as peroxidation of lipid membranes leading in turn to strong diseases [4]. So, the problem of the control of ROS concentration is very important. This control can be realized as by cell's own systems, such as enzymes (superoxide dismutase and catalase), so by chemical antioxidant molecules (ascorbic acid, glutathione, alpha-lipoic acid etc.). Recently a number of new antioxidant materials based on the nanoparticles of oxides with variable valency were proposed as ROS scavengers [15]. One of the topmost materials of this kind is cerium dioxide (nanoceria) that has demonstrated strong protective effect on the different lines of cells [14, 17] increasing their lifespan up to four times.

V. Seminko (✉) • A. Masalov • P. Maksimchuk • V. Klochkov • I. Bespalova • O. Viagin
Yu. Malyukin
Institute for Scintillation Materials, 60 Lenina ave., 61001 Kharkiv, Ukraine
e-mail: seminko@isma.kharkov.ua

J. Bonča, S. Kruchinin (eds.), *Nanomaterials for Security*, NATO Science for Peace and Security Series A: Chemistry and Biology, DOI 10.1007/978-94-017-7593-9_12

Another problem closely connected with the problem of ROS control is the creation of effective ROS sensors (and, first of all, of hydrogen peroxide concentration as a most widespread type of ROS). In the last years a number of papers were published with different materials proposed as ROS sensors; these materials include fluorescent proteins [1], organic nanoparticles [9], rare-earth doped inorganic nanoparticles [2] and carbon nanotubes [6]. However, these materials do not possess any antioxidant activity only sensing the concentration of ROS. Instead the development of ROS sensor and ROS scavenger on the base of single material would have a number of promising applications allowing to get antioxidant material along with ability to control its antioxidant action.

Cerium dioxide has strongly pronounced oxygen buffering properties, so it can easily absorb and release oxygen depending on external conditions. So, it has found its application as ionic conductor, catalyst and antioxidant [18]. One of the main causes of such properties of nanoceria is the low energy of the formation of oxygen vacancies in this material. It was shown in the number of theoretical papers that formation of oxygen vacancy in ceria lattice is always accompanied by the transfer of two remaining electrons from oxygen to two Ce^{4+} ions which became Ce^{3+} ones. If these ions are situated close to oxygen vacancy, only 0.26 eV is required to create such a vacancy [16].

The concentration of oxygen vacancies in nanoceria (and so, Ce^{3+}/Ce^{4+} ratio) can vary in the broad range dependent on the external conditions (temperature, pH etc.), and can be used as a measure of oxidation/reduction properties of biological media. Using this fact, along with the strong dependence of the luminescence properties of nanoceria on the concentration of oxygen vacancies which was shown in our previous paper [12], we propose that the luminescence of ceria nanocrystals can be used for detection of their antioxidant properties and visualization of their antioxidant action in vivo.

12.2 Experimental

CeO_2 nanocrystals (50 nm) were obtained by Pechini method described in [12]. CeO_2 nanocrystals (3 nm, 10 nm) were obtained by the methods of colloidal synthesis described in [8].

CeO_2 nanocrystals were annealed during 2 h in different atmospheres – oxidative (air), neutral (argon) and reducing (hydrogen) at 1000 °C.

Luminescence spectra were obtained using spectrofluorimeter based on the grating monochromator, luminescence was excited by He–Cd laser with λ_{exc} = 325 nm. Excitation spectra in UV and VUV range were obtained at SUPERLUMI setup (DESY) in Hamburg, Germany using synchrotron radiation.

For experiments imploring hydrogen peroxide (HP) sensing water solutions of CeO_2 nanoparticles (10 nm) with ceria concentration of 1 g/l were taken. Luminescence intensity was measured in the maximum of Ce^{3+} band (390 nm) before and 5 min after HP addition. Concentration of hydrogen peroxide in the experiments was varied in the range from 0.1 to 0.8 mM.

12.3 Results and Discussion

12.3.1 *Identification of Luminescence Centers in CeO_2 Nanocrystals*

CeO_2 (ceria) is a dielectric with band gap of ~6 eV, which valence band is formed by 2p oxygen energy states and conduction band by 5d and 6s cerium energy states [11]. In the band gap of ceria a narrow band (from 0.5 to 1 eV, according to different authors) formed by $4f^0$ states of Ce^{4+} is present. The optical gap in CeO_2 is only about 3.2 eV and optical properties of ceria are determined by charge transfer from 2p oxygen states to the empty 4f shell of Ce^{4+} [5].

Electronic structure of both stoichiometric CeO_2 and non-stoichiometric CeO_{2-x} was investigated previously by methods of density functional theory (DFT). It was shown in [3] that for non-stoichiometric ceria formation of positively charged oxygen vacancies followed by capture of excess electrons by Ce^{4+} ions provides splitting of $4f^0$ band into two sub-bands: empty upper band (about 3–4 eV above the valence band) formed by unfilled $4f^0$ levels and lower one formed by filled $4f^1$ levels of Ce^{3+} ions (about 1.5–2 eV above the valence band). Filling of the lower band can occur also in stoichiometric ceria at electron transfer from the valence band (determined by O^{2-} ions) to empty $4f^0$ orbital of Ce^{4+} ion and further relaxation of "Ce^{3+}-hole" complex. Reverse charge transition can occur as in radiative (CT-luminescence), so in non-radiative way. So, in non-stoichiometric ceria besides CT-luminescence of Ce^{4+} ion also luminescence transitions from excited levels of Ce^{3+} ions (for instance, 5d→4f transition) can be expected.

CeO_2 nanocrystals with high degree of oxygen non-stoichiometry can be obtained by heat treatment in the reducing atmosphere (in our experiments – H_2 atmosphere). Luminescence spectra of CeO_2 nanocrystals after treatment at 1000 °C measured at 300 and 77 K are shown in the Fig. 12.1 ($\lambda_{exc} = 325$ nm). At 77 K luminescence spectrum consists of two bands with maxima at 390 and 630 nm, while at room temperature 630 nm band is absent. Also at 77 K 390 nm band demonstrate splitting to two components ($\Delta = 1900\ cm^{-1}$).

For determination of the nature of 390 and 630 nm luminescence bands (Fig. 12.1), luminescence excitation spectra at different registration wavelengths were investigated.

Excitation spectrum of 630 nm band is shown in the Fig. 12.2. It consists of single wide band (~4000 cm^{-1}) with maximum at 340 nm. This band was observed previously in the excitation spectra of CeO_2 nanocrystals doped by Eu^{3+} and Sm^{3+} ions and was ascribed to Ce^{4+}-O^{2-} charge transfer [10]. Position of the charge transfer band in the excitation spectrum can be estimated by $\sigma\ (cm^{-1}) \approx 30{,}000 \cdot [\chi_{opt}(X) - \chi_{opt}(M)]$ [7] where $\chi_{opt}(X)$ and $\chi_{opt}(M)$ are optical electronegativities of ligand and metal ion, respectively. For oxygen and cerium ions $\chi_{opt}\ (O^{2-}) = 3.2$ eV and $\chi_{opt}\ (Ce^{4+}) = 2.1$ eV, so, $\sigma \approx 33{,}000\ cm^{-1}$ and $\lambda \approx 310$ nm that is close to experimentally observed value (340 nm). Strong temperature dependence of 630 nm luminescence band also confirms the CT-nature of this band.

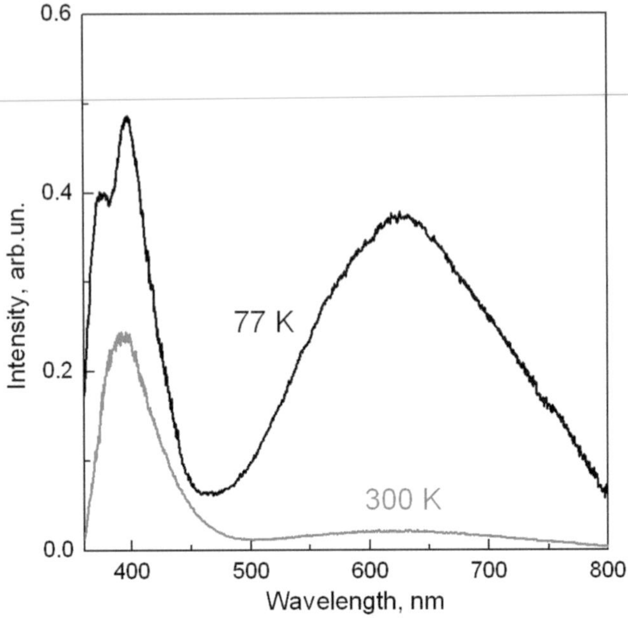

Fig. 12.1 Luminescence spectra of CeO_2 nanocrystals ($d \sim 50$ nm) after treatment in reducing atmosphere ($\lambda_{exc} = 325$ nm) at 77 K (1) and 300 K (2)

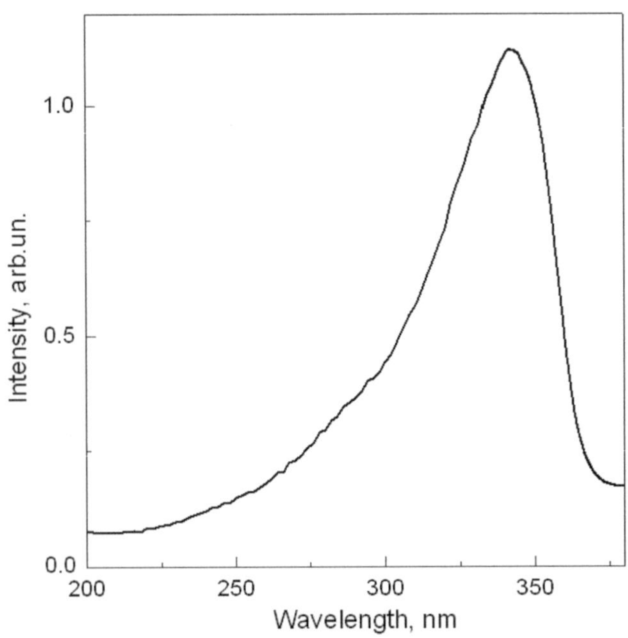

Fig. 12.2 Excitation spectra of 630 nm luminescence band of CeO_2 nanocrystals treated in reducing atmosphere

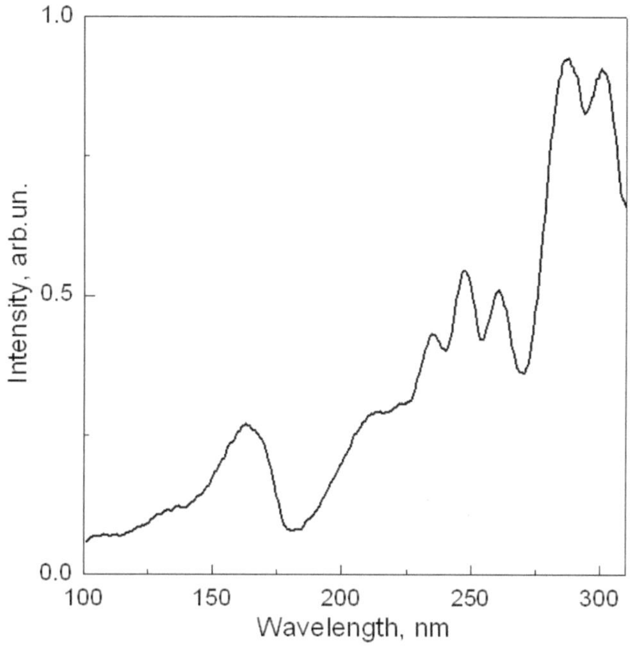

Fig. 12.3 Excitation spectra of 390 nm luminescence band of CeO_2 nanocrystals treated in reducing atmosphere

Excitation spectrum of 390 nm band is shown in the Fig. 12.3. This spectrum consists of five spectral lines that can be divided into two groups – two low energy ones at 300 and 285 nm (33,350 and 35,000 cm^{-1}) and three high energy ones at 260, 246 and 234 nm (38,500, 40,650 and 42,750 cm^{-1}). Energy gap between two groups of spectral lines was about 3500 cm^{-1} that is much more than the distance between lines in the group (1000–2000 cm^{-1}). Such spectra are typical for 4f→5d transitions of rare earth ions at full removal of 5d level degeneracy.

Character of splitting allows to determine the symmetry of ligand surrounding of rare-earth ion. If the surrounding of the ion has cubic symmetry (O_h point group), splitting of 5d level to three low energy and two high energy components takes place. Such splitting could be observed in nonstoichiometric CeO_{2-x} if the compensation of excess negative charge due to replacement of Ce^{4+} to Ce^{3+} ion was non-local. Oxygen vacancies in this case are formed preferentially outside the first coordination sphere of cerium ion and symmetry of cerium surrounding remains intact. However experimentally observed splitting to two low energy and three high energy components corresponds to lower symmetry than cubic – namely, to C_{3v} symmetry. Decrease of symmetry of local center to C_{3v} can be expected if charge-compensating defect (oxygen vacancy) is formed in the first coordination sphere of cerium ion. So, it can be suggested that 390 nm luminescence band is determined by the optical center connected with Ce^{3+} ion with oxygen vacancy in the first coordination sphere. Splitting of 390 nm band observed at 77 K can be unambiguously ascribed to splitting of ^2F electronic term of Ce^{3+} ion to $^2F_{5/2}$ and $^2F_{7/2}$ terms due to spin-orbital interaction. For oxide crystals the value of such splitting is about 2000 cm^{-1} that agree well with experimental data obtained (1900 cm^{-1}).

12.3.2 Dependence of Luminescence Properties of Ceria Nanocrystals on Atmosphere of Treatment and Size

As was mentioned in Introduction, formation of oxygen vacancy in ceria lattice is always accompanied by the transfer of two electrons from oxygen to two Ce^{4+} ions which became Ce^{3+} ones. So, the concentrations of oxygen vacancies and Ce^{3+} ions in ceria nanocrystals are interdependent and change of the content of oxygen vacancies must lead to the change of Ce^{3+} content as well. While Ce^{3+} luminescence intensity is directly proportional to the concentration of Ce^{3+} ions, change of intensity of 390 nm band should be ascribed to increase or decrease of the degree of oxygen stoichiometry of nanoceria. CeO_2 nanocrystals with different oxygen stoichiometry can be obtained by variation of treatment atmosphere. At transition from oxidation to neutral and from neutral to reducing atmosphere increase of Ce^{3+} luminescence band intensity as compared to CT-band is observed (Fig. 12.4a), so the number of Ce^{3+} ions with increase of reducing properties of treatment atmosphere increases as it would be expected if one supposes the Ce^{3+} nature of this band.

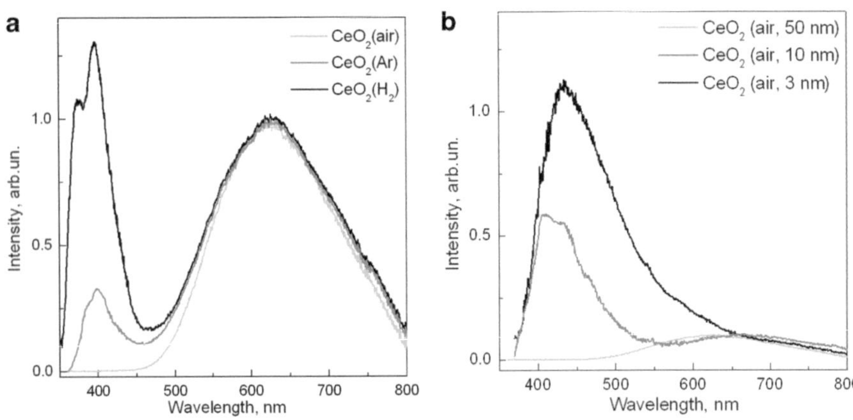

Fig. 12.4 Luminescence spectra of CeO_2 nanocrystals at variation of: (**a**) atmosphere of treatment ($d \sim 50$ nm), (**b**) size of nanocrystals

Variation of the size of ceria nanocrystals allow to obtain the particles with different antioxidant activity. The luminescence spectra of CeO_2 nanocrystals ($\lambda_{exc} = 325$ nm) with sizes of about 3, 10 and 50 nm are shown in the Fig. 12.4b. Decrease of the size from 10 to 3 nm leads to increase in 2 times of intensity of 390 nm band associated with 5d→4f luminescence of Ce^{3+} ions. Intensity of this band is directly proportional to Ce^{3+} concentration, which, in turn, is closely connected to antioxidant activity of nanocrystals which thereby is more pronounced for smaller nanoparticles.

12.3.3 Visualization of Interaction of Hydrogen Peroxide with Ceria Nanoparticles

As was shown in the number of papers [16, 17], antioxidant action of ceria is based on the ability to switch easily between Ce^{3+} and Ce^{4+} states of cerium ion. Interaction of ceria nanoparticle with oxidant leads to $Ce^{3+} \rightarrow Ce^{4+}$ transition, so the number of Ce^{3+} ions decreases. As an oxidant in our experiments hydrogen peroxide was chosen. The most pronounced antioxidant properties are observed for small ceria nanoparticles (less than 15 nm) which are connected with high content of Ce^{3+}. For the following experiment ceria nanoparticles with the size of 10 nm were chosen. As it was expected an addition of hydrogen peroxide to the solution of ceria nanoparticles led to quenching of Ce^{3+} luminescence band due to transition of the part of Ce^{3+} ions to Ce^{4+} ones. The fall of intensity took place in the range of 1–2 min after hydrogen peroxide addition and after that intensity did not change.

In Fig. 12.5 the ratio between intensity of Ce^{3+} luminescence band 5 min after hydrogen peroxide addition and the one before hydrogen peroxide addition is shown. The portion of Ce^{3+} ions underwent $Ce^{3+} \rightarrow Ce^{4+}$ transition increased with increase of the concentration of the oxidant – from 42 % at 0.1 mM to 70 % at 0.8 mM. Even at higher concentrations of H_2O_2 some portion of Ce^{3+} luminescence always remained due to inaccessibility of some portion of Ce^{3+} ions to hydrogen peroxide.

So, our investigations have shown that in ceria nanocrystals two optical centers are observed: Ce^{3+} ions with 5d-4f luminescence, and Ce^{4+}–O^{2-} complexes with charge transfer (CT) luminescence. Intensity of Ce^{3+} luminescence band depends on the atmosphere of heat treatment and the size of nanoparticle. Addition of hydrogen peroxide leads to fall of Ce^{3+} band intensity due to transfer of the part of Ce^{3+} ions to Ce^{4+} ones. So, using intensity of Ce^{3+} luminescence band as a measure, the antioxidant action of ceria can be easily controlled by conventional spectroscopic technique. ROS scavenging properties of nanoceria along with absence of toxicity for nanoparticles with sizes in the range of 5–15 nm shown by [15] allow using this material as an effective antioxidant material able both to detect and to regulate ROS concentration in living cells.

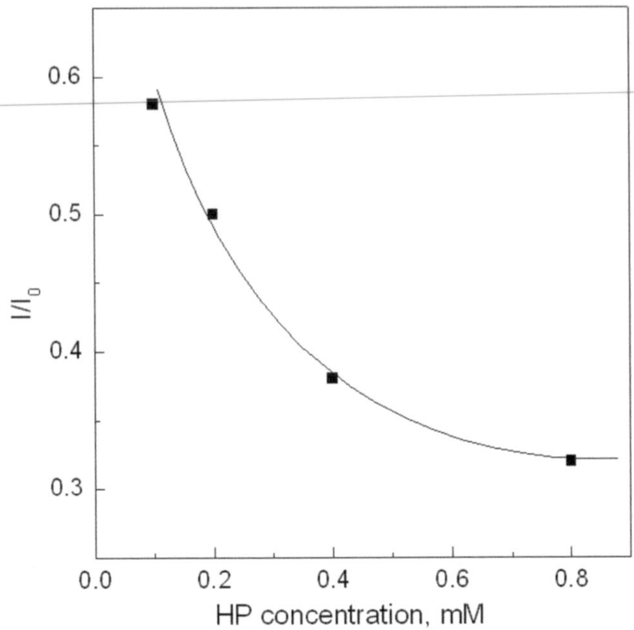

Fig. 12.5 Dependence of the ratio of intensity of Ce^{3+} luminescence band 5 min after hydrogen peroxide addition (I) to the one before hydrogen peroxide addition (I_0) on the concentration of hydrogen peroxide

References

1. Belousov VV, Fradkov AF, Lukyanov KA, Staroverov DB, Shakhbazov KS, Terskikh AV, Lukyanov S (2006) Genetically encoded fluorescent indicator for intracellular hydrogen peroxide. Nat Methods 3:281–286
2. Casanova D, Bouzigues C, Nguyen T, Rivo O, Bouzhir-Sima RL, Gacoin T, Boilot J, Tharaux P, Alexandrou A (2009) Single europium-doped nanoparticles measure temporal pattern of reactive oxygen species production inside cells. Nat Nanotechnol 4:581–585
3. Castleton CWM, Kullgren J, Hermansson K (2007) Tuning LDA + U for electron localization and structure at oxygen vacancies in ceria. J Chem Phys 127:244704–244704
4. Finkel T, Holbrook NJ (2000) Oxidants, oxidative stress and the biology of ageing. Nature 408:239–247
5. Goubin F, Rocquefelte X, Whangbo MH, Montardi Y, Brec R, Jobic S (2004) Experimental and theoretical characterization of the optical properties of CeO_2, $SrCeO_3$, and Sr_2CeO_4 containing Ce^{4+} (f^0) ions. Chem Mater 16:662–669
6. Hong J, Heller DA, Kalbacova M, Kim J, Zhang J, Boghossian AA, Maheshri N, Strano MS (2010) Detection of single-molecule H_2O_2 signalling from epidermal growth factor receptor using fluorescent single-walled carbon nanotubes. Nat Nanotechnol 5:302–309
7. Jorgensen K (1970) Electron transfer spectra. Wiley, New York
8. Klochkov VK, Grigorova AV, Sedyh OO, Malyukin YV (2012) The influence of agglomeration of nanoparticles on their superoxide dismutase-mimetic activity. Colloids Surf A: Physicochem Eng Asp 409:176–182

9. Lee D, Khaja S, Velasquez-Castano JC, Dasari M, Sun C, Petros J, Taylor WR, Murthy N (2007) In vivo imaging of hydrogen peroxide with chemiluminescent nanoparticles. Nat Mater 6:765–769
10. Li L, Yang HK, Moon BK, Fu Z, Guo C, Jeong JH, Lee HS (2008) Photoluminescence properties of CeO_2: Eu^{3+} nanoparticles synthesized by a sol-gel method. J Phys Chem C 113:610–617
11. Marabelli F, Wachter P (1987) Covalent insulator CeO_2: optical reflectivity measurements. Phys Rev B 36:1238–1242
12. Masalov A, Viagin O, Maksimchuk P, Seminko V, Bespalova I, Aslanov A, Malyukin Yu, Zorenko Yu (2014) Formation of luminescent centers in CeO_2 nanocrystals. J Lumines 145:61–64
13. Rhee SG (2006) H_2O_2, a necessary evil for cell signaling. Science 312:1882–1883
14. Rzigalinski BA (2005) Nanoparticles and cell longevity. Technol Cancer Res Treat 4:651–659
15. Schubert D, Dargusch R, Raitano J, Chan SW (2006) Cerium and yttrium oxide nanoparticles are neuroprotective. Biochem Biophys Res Commun 342:86–91
16. Skorodumova NV, Simak SI, Lundqvist BI, Abrikosov IA, Johansson B (2002) Quantum origin of the oxygen storage capability of ceria. Phys Rev Lett 89(16660):1–8
17. Tarnuzzer RW, Colon J, Patil S, Seal S (2005) Vacancy engineered ceria nanostructures for protection from radiation-induced cellular damage. Nano Lett 5:2573–2577
18. Trovarelli A, Fornasiero P (2002) Catalysis by ceria and related materials. Imperial College Press, London

Chapter 13
Vibration Based Microstructure Replication and Analysis

R. Šakalys, G. Janušas, A. Palevičius, V. Grigaliūnas, and P. Palevičius

Abstract The main focus of the paper is the development of technological process for the production of complex 3D microstructure, from designing it by using computer generated holography to its physical 3D patterning by exploiting the process of electron beam lithography. The logo image was chosen for numerical generation, which was performed by using Gerchberg–Saxton algorithm for computer generated hologram design. Physical implementation of microstructure was performed by using a single layer polymethyl methacrylate (PMMA) as a basis for 3D microstructure, which was exposed by using e-beam lithography system e-LINEplus. After production, verification of 3D microstructure is performed by exposing it under the laser beam and qualitative analysis is performed by using atomic force microscope. Finally, improvement of mass production of complex microstructures, designed by using computer generated holography, is presented.

13.1 Introduction

It is hard to imagine everyday life without products, created by using micro/nano-technologies. Such technologies allow to overcome many problems, related to medicine, optics, ecology etc. The trend of miniaturization and incorporation is becoming very important for the health sector because it reduces costs and improves performance. Various micro devices are used in medical diagnosis, therapy

---

R. Šakalys • G. Janušas • A. Palevičius (✉)
Department of Mechanical Engineering, Kaunas University of Technology, Studentų 56 – 338, 51424 Kaunas, Lithuania
e-mail: rokas.sakalys@ktu.lt

V. Grigaliūnas
Institute of Materials Science, Kaunas University of Technology, K. Baršausko 59, 51423 Kaunas, Lithuania
e-mail: viktoras.grigaliunas@ktu.lt

P. Palevičius
Department of Mathematical Modelling, Kaunas University of Technology, Studentų 50 – 147, Kaunas, 51368, Lithuania
e-mail: paulius.palevicius@ktu.lt

J. Bonča, S. Kruchinin (eds.), *Nanomaterials for Security*, NATO Science for Peace and Security Series A: Chemistry and Biology, DOI 10.1007/978-94-017-7593-9_13

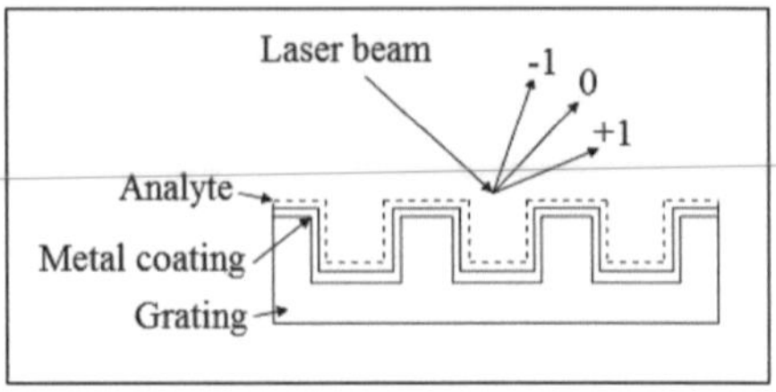

Fig. 13.1 The system of microperiodic biosensor

etc. Novel microdevices work more effectively than former macroscopic devices, manufactured by standard production means.

The usage of Micro-Opto-Electro-Mechanical Systems (MOEMS) sensors for gathering and processing real time information about cardiovascular health status is one of the most important application fields in the medical diagnostic. The most important components of such systems are periodical microstructures (Fig. 13.1).

This system allows evaluate the environment (analyte), which is being investigated, since different analytes differently affect the diffraction efficiency. Thus it is very important to create universal microstructure, which would enable precise identification of material, and to process or evaluate materials and dynamic processes. In order to create MOEMS optical element, computer generated holography (CGH) is employed in this article.

CGH is defined mathematically by calculating the phase and amplitude information of the wave propagation created by an object. There exist lots of applications which employ CGH method such as:

- diffractive-optical elements for storage of digital data and images [1],
- interferometric measurements [2],
- recognition of the patterns [3],
- data encryption [4],
- 3D displays [5].

The major advantage of CGH is that the object used for recording CGH holograms does not need to exist physically, as it is recorded mathematically. The Fourier transformation and Gerchberg–Saxton algorithms are used to create the CGH in this article.

13.2 Generation of CGH

CGH was generated by using Gerchberg-Saxton algorithm, which employs Fast Fourier Transform (FFT). In our case, as we have discrete data, the discrete Fourier transform is used, and it is the primary tool of digital signal processing. The DFT transforms time or space-based data into frequency-based data. The 2D discrete transform is defined as follows:

$$F(u, v) = \frac{1}{NM} \sum_{x=0}^{M-1} \sum_{y=0}^{N-1} f(x, y) e^{-i2\pi \left(\frac{xu}{M} + \frac{yn}{N} \right)},$$

where u and v are the discrete spacial frequencies, M and N are the number of samples in x and y directions in spacial and frequency domains, $F(u, v)$ is the 2D discrete spectrum of $f(x, y)$.

In the beginning, a random number generator is used to generate a phase distribution $\varphi[-\pi, \pi]$. The Gerchberg–Saxton algorithm steps are given as follows:

(i) Calculate initial field in the hologram plane.
(ii) This field is propagated from the image-plane to the object-plane. The amplitude information is discarded leaving only the phase information (for the phase mask).
(iii) The amplitude and phase of the illumination field are added to the phase information in order to obtain the resulting object field.
(iv) This field is propagated from the object plane to the image plane.
(v) The resulting reconstructed image (the square of the field amplitude) is compared with the expected one. The decision to terminate the process (or continue to iterate) is made by computing the correlation between both images.
(vi) The phase from the reconstructed image is combined with the field amplitude obtained from the expected irradiance. The process is repeated from step (i).

Gerchberg–Saxton algorithm for generation of CGH is implemented by using Matlab software. As direct calculation of N-point DFT requires N^2 complex multiplications and $N(N-1)$ complex additions, Fast Fourier Transform (FFT) was used to reduce computation speed. During the generation of CGH iterative quality evaluation was introduced. Calculations are considered as finished only then, when the quality of hologram reaches desired level, i.e. new iterations of the algorithm do not produce better quality, defined by some specific threshold. For verification of the algorithm and establishment of technologic process, Kaunas University of Technology (KTU) logo (Fig. 13.2) was used.

The generated hologram is 2×2 mm size and contains 106 pixels. It is generated for exposition on (polymethyl methacrylate) PMMA, which is 35 K in molecular

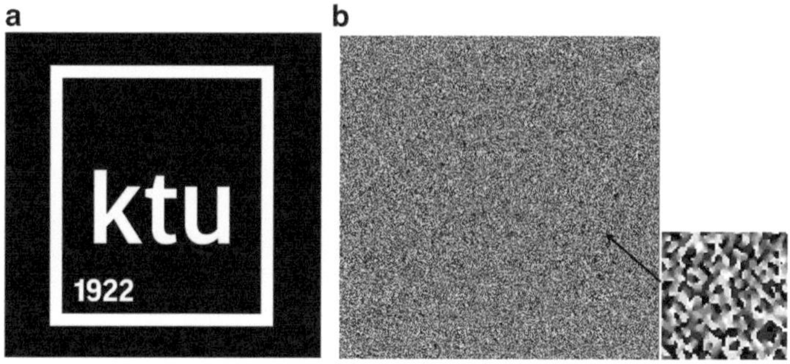

Fig. 13.2 The logo of Kaunas University of Technology (**a**) and its CGH (**b**)

mass with refractive index of 1.49. Laser with wavelength of 632.8 nm is used for the reconstruction of the hologram. Exposed pixels are of eight levels. In the map of hologram exposition, black colour identifies the location, which is not exposed and white location identifies areas with maximum exposition.

13.3 E-Beam Lithography

Raith e-LiNEplus high resolution electron beam lithography system (Fig. 13.3) equipped with Schottky thermionic field emission gun was used for EBL process.

Silicon wafer, with the size of 10 × 10 mm, was exposed to oxygen plasma for 5 min and heated for 30 min at 150 °C on a hotplate. Then PMMA resist of 35 K molecular weight was spin coated at 2000 rpm and dried at 200 °C on a hotplate. E-beam patterning was performed with acceleration voltage of 10 kV, using a 30 μm aperture. PMMA resist was developed in 1:3 methyl isobutyl ketone (MIBK) and isopropyl alcohol (IPA) solvent. Development was finished in IPA and the holograms were washed with water and dried with a compressed air flow. The simplified scheme of the process is presented in the Fig. 13.4.

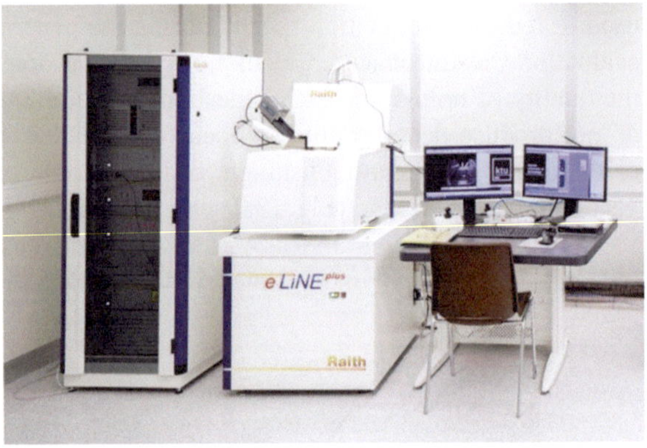

Fig. 13.3 Raith e-LiNEplus electron beam lithography system

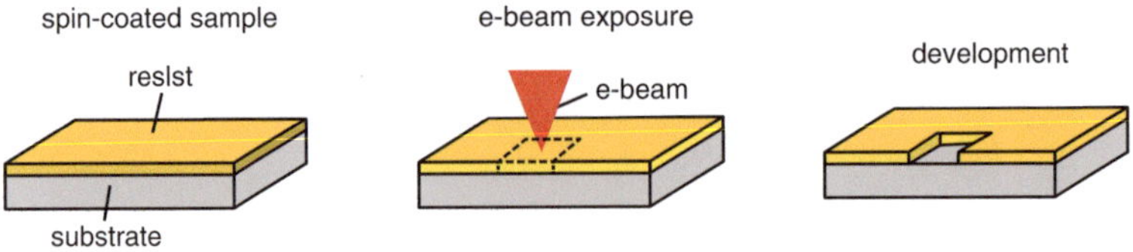

Fig. 13.4 E-beam lithography process

13.4 Investigation of CGHs

Usually design patterns are divided into writing fields. Within a writing field the electron beam is deflected in order to expose one pixel after another. The write fields must be stitched together. The "stitching" means dividing your design pattern into smaller pieces in order to expose the bigger areas. Optical microscope images of the exposed holograms are shown in Fig. 13.5. We used a writing field size of 100 μm square with 2 μm pixel size. The writing field "stitching error" and overlapping pixels for the both (*x* and *y* coordinates) can be well seen in Fig. 13.5a, b. The exposed area without "stitching" error is shown in Fig.13.5c.

Atomic force microscope (AFM) 3D view of CGH at different scales is shown in Fig. 13.6. The step-wise structure within the range of PMMA height from 0 μm to 1.64 μm is well visible in AFM pictures. The height of steps corresponds to the 8 greyscale levels of the individual pixels (200, 400, 600, 800, 1000, 1200, 1400 and 1600 nm).

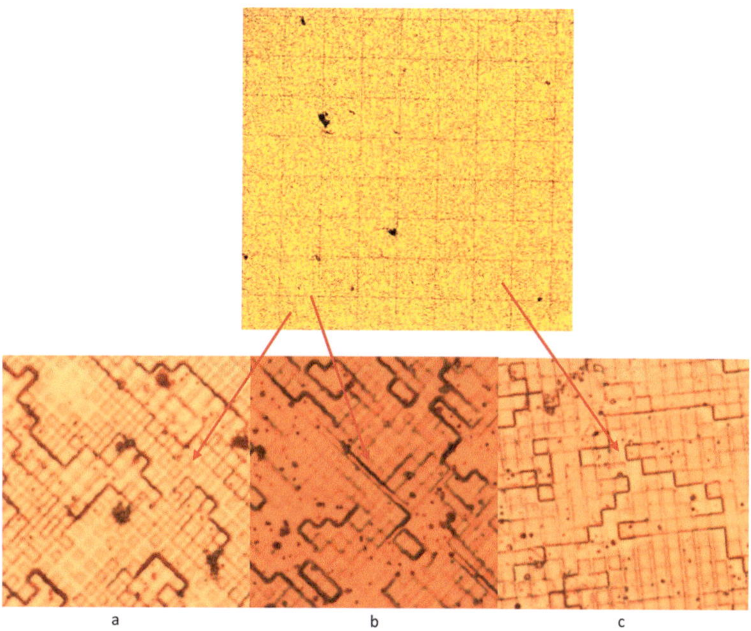

a b c

Fig. 13.5 Images of CGHs, generated on the PMMA (the size of one pixel is 2 × 2 μm)

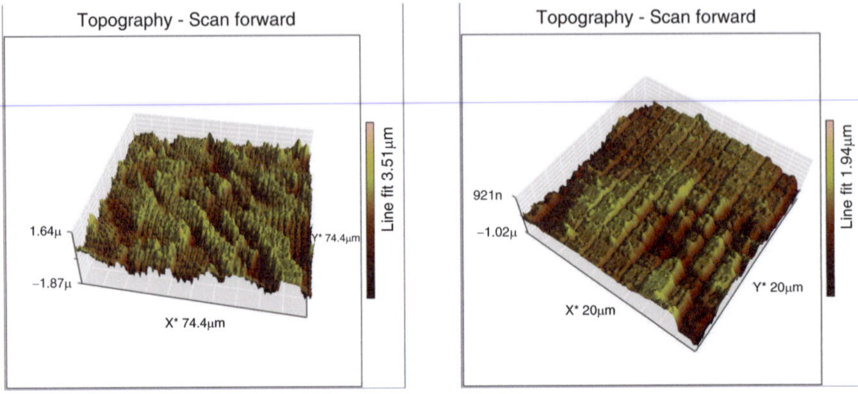

Fig. 13.6 3D view of CGH, exposed on PMMA

Fig. 13.7 CGHs generated on the PMMA

13.5 Results

During the improvement of technological process six holograms were produced (Fig. 13.7). The latest specimen has the biggest diffraction efficiency – 58 %, which is 29 % more than in early trials. Diffraction view of KTU logo is presented in Fig. 13.8.

13.6 Mass Replication of Microstructures on the Polymer

Precise microstructures can be replicated by using different techniques: microinjection moulding, ultrasonic embossing, and hot imprint [6]. Due to advantages over other microstructure replication methods, hot imprint is chosen for microstructure mass fabrication. Advantages of this process are: comparatively simple equipment, low-cost production, and high resolution of final microstructure [7].

Fig. 13.8 Diffraction view of KTU logo CGH

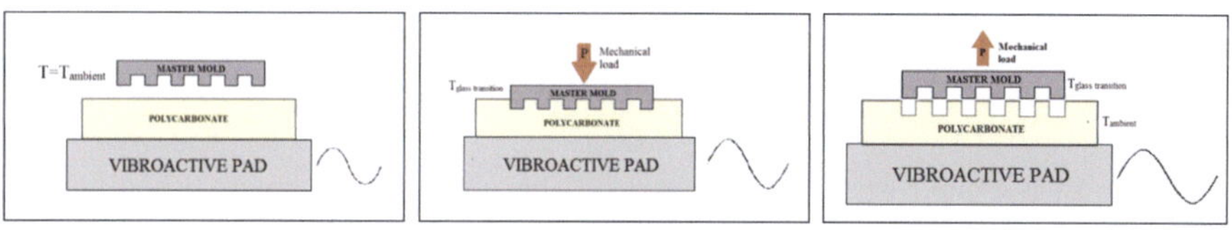

a b c

Fig. 13.9 Steps of the process of hot imprint with usage of ultrasonic excitation: heating (**a**), imprint (**b**), and demoulding (**c**)

Process of mechanical hot imprint (Fig. 13.9) comprises the following three steps:

1. Heating. Initial temperature of master microstructure and polycarbonate is 20 °C, which corresponds to ambient temperature. As the master microstructure reaches the surface of polycarbonate, the heating is applied till glass transition temperature of the polymer is achieved. The polymer starts to experience elastic and plastic deformations under the action of mechanical stress. Heat transfer during the heating phase is described by the following equation:

$$\rho(T)c_p(T)\frac{\partial T}{\partial t} + \nabla(-k\nabla T) = q,$$

   where k is thermal conductivity; ρ is density; c_p is heat capacity; T is temperature; q is rate of the heat generation.
2. Imprint. The master microstructure imprints preheated polymer till nominal pressure (from 101.325 to 506.625 kN/m^2) is reached. The pressure is applied for 10 s. As the master moves into the slave, its deformations from purely elastic becomes plastic.

3. Demoulding. The master microstructure in this step is withdrawn backwards and polycarbonate is cooled down until initial process temperature is reached [8].

On the other hand this process has some disadvantages, such as: material shrinkage, which occurs due to adhesion between master microstructure and polymer; low filling ratio of microstructure; uneven master imprint; high surface roughness; and long fabrication time [9–12]. Appropriate choice of hot imprint process parameters (temperature, pressure, and imprinting duration) is one of the techniques to improve quality of the final product [13–17]. However, to cope with these problems, it is not enough to regulate the parameter – ultrasonic excitation during the process of mechanical hot imprint is recommended for further improvement of the quality. It has twofold effect in the processes of precise microstructure fabrication: forces preheated polymer to flow to the master microstructure under the action of ultrasonic excitation, thus better filling empty cavities between master microstructure and polymer. It also helps during demoulding step, by diminishing adhesion between master microstructure and polymer, this helps to avoid cracks and distortions.

13.7 Master Microstructure for Replication on Polymer

In order to use microstructure, created by the method of e-beam lithography it is necessary to metalize the microstructure (Fig. 13.10), which was produced by the method of e-beam lithography with the layer of metal. This starts with metal deposition, then resist lift off takes place, and finally, master microstructure is obtained in this way.

13.8 Ultrasonic Excitator for Mass Production

The piezoelectric low voltage stack type piezoactuator (PSt 150/4/20 VS9) is used as a source of high frequency excitation during the process. Piezoactuator is 9 mm in diameter and 31 mm in height.

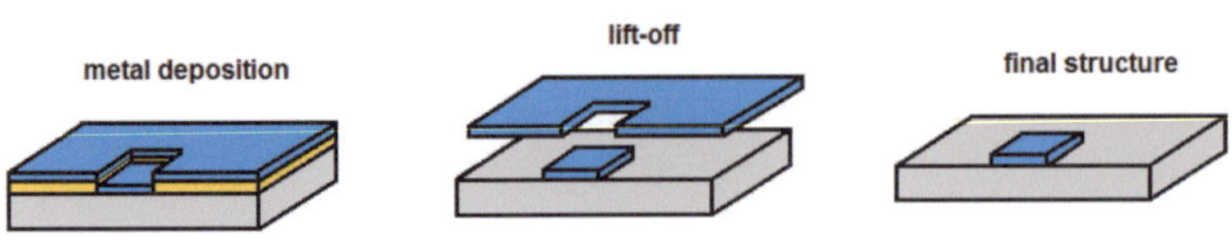

Fig. 13.10 Process of metallization of master microstructure

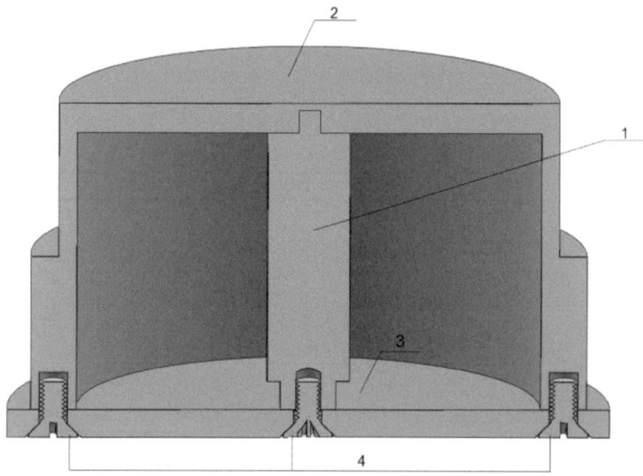

Fig. 13.11 Cross section of vibroactive pad: *1* – multilayer piezoactuator PSt 150/4/20 VS9; *2* – *upper side* of the frame; *3* – *bottom side* of the frame; *4* – M3 bolts

Vibroactive pad (Fig. 13.11) is based on the multilayer actuator in the center of construction and aluminium frame, to increase operating area, protect the pad against mechanical load, and possible damage of actuator. Material and geometrical parameters of the construction are chosen according to the application of the device: construction should sustain the pressure of 506,625 N/m^2, which corresponds to the pressure of mechanical hot imprint. On the other hand, construction should be flexible enough to transmit vibrations to the polymer. In order to do this in a more efficient way, the upper side of wall was turned more than the lower side. This decreases the stiffness of the upper side of the frame and allows get bigger displacements and lower resonant frequencies.

13.9 Mechanical Hot Imprint Experiment

In order to prove the positive effect of ultrasonic excitation during the process of mechanical hot imprint, experiments were performed with and without high frequency excitation, while other process parameters, such as temperature, pressure and duration of imprint were the same [18, 19]. Experimental matrix is presented in Table 13.1.

During the experiments periodical diffraction grating (Fig. 13.12), with a period of 4 μm was imprinted.

Table 13.1 Experimental matrix of mechanical hot imprint

Vibroactive pad	With ultrasonic excitation	Without ultrasonic excitation
Impressing time t, s	10	10
Pressure P, N/m^2	303,900	303,900
Temperature T, °C	148	148

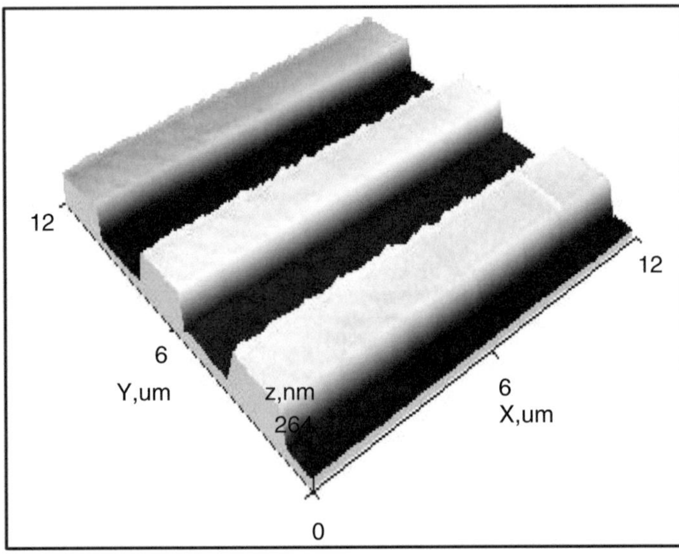

Fig. 13.12 Atomic force microscope image of the master mold

13.10 Examination Techniques, Used to Define the Quality of Fabricated Microstructures

Measurement of diffraction efficiency was performed by using laser ($\lambda = 632.8$ nm) and photodiode BPW-34 (Fig. 13.13). The scheme is connected with ammeter. The diffraction efficiency is calculated by using following formula:

$$RE_{i,j} = \frac{I_{i,j}}{I_j}, \tag{13.1}$$

$$I_j = \sum_i I_{i,j}, \tag{13.2}$$

where $RE_{i,j}$ is relative diffraction efficiency; $I_{i,j}$ is the light intensity at particular maxima; and I_j is the sum of all light intensities in all maxima. The current, registered with ammeter, is proportional to the illumination, which reaches the

Fig. 13.13 Diffractometer and measuring scheme: sample (*1*); photodiode (*2*); tester (*3*); and distribution of maximas (*4*)

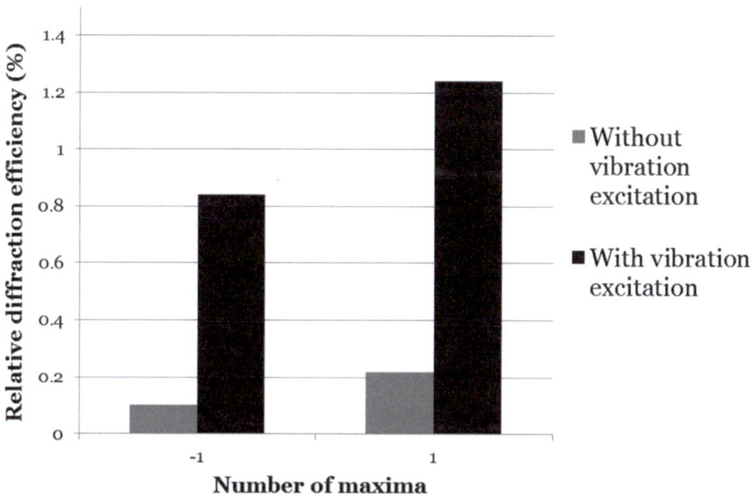

Fig. 13.14 Distribution of diffraction efficiency

photodiode. The relative diffraction efficiency is then multiplied by 100, thus values in percent are obtained. Usually +1 and −1 maxima are the most essential, because they are desirable property in many application areas [20].

13.11 Results of Experiments

In this section results, obtained after performing experiments of mechanical hot imprint are presented. Diffraction efficiency measurement results (Fig. 13.14) revealed, that usage of ultrasonic excitation helps to increase diffraction efficiency. First (1.24 %) and minus first (0.84 %) maxima in grating, made with vibration excitation, are respectively 5.63 and 8.4 times higher than in grating, made without vibration excitation (0.22 % first maxima and 0.1 % minus first maxima).

13.12 Conclusions

In this article computer generated hologram was designed for human health diagnostic purposes by using Fourier and Gerchberg–Saxton methods. The physical implementation of CGH on the silicon was performed by using e-beam lithography, where PMMA was used as a microresist. Production revealed, that exposing of pixels in rows or columns causes their overlapping, thus division of the specimen into parts was proposed to overcome this problem. Finally, pixel division into four parts was proposed, in order to avoid rounded corners of the pixels. All these measures allowed increasing the diffraction efficiency by 29 %. Usage of ultrasonic excitation increased relative diffraction efficiency in plus and minus first maxima 5.63 and 8.4 times respectively.

Acknowledgements This research was funded by a grant (No. MIP-026/2014) from the Research Council of Lithuania.

References

1. Wilson LW et al (1998) High density, high performance optical data storage via volume holography: viability at last? Opt Quantum Electron 32:393–404
2. Gren P (2003) Four-pulse interferometric recordings of transient events by pulsed TV holography. Opt Lasers Eng 40:517–528
3. Saari P, Kaarli R, Ratsep M (1992) Temporally multiplexed fourier holography and pattern recognition of femtosecond-duration images. J Lumin 56:175–180
4. Nishchal KN, Joseph J, Singh K (2004) Fully phase encryption using digital holography. Opt Eng 43:2959–2966
5. Son YJ, Javidi B, Kwack DK (2006) Methods for displaying three-dimensional images. Proc IEEE 94:502–523
6. Krauss PR, Chou YS (1997) Nano-compact disks with 400 Gbit/in2 storage density fabricated using nanoimprint lithography and read with proximal probe. Appl Phys Lett 71:3174–3176
7. Lee JL et al (2001) Design and fabrication of CD-like microfluidic platforms for diagnostics: polymer based microfabrication. Biomed Microdevice 3(4):339–351
8. Jaszewski WR et al (1998) Hot embossing in polymers as a direct way to pattern resist. Microelectron Eng 41–42:575–578
9. Mekaru H, Goto H, Takahashi M (2007) Development of ultrasonic micro hot embossing technology. Microelectron Eng 84:1282–1287. doi:10.5772/8191
10. Hirai Y, Yoshida S, Takagi N (2003) Defect analysis in thermal nanoimprint lithography. J Vac Sci Technol B 21(6):2765–2770
11. Narijauskaitė B et al (2011) High-frequency excitation for thermal imprint of microstructures into a polymer. Experimental techniques. Published online 11 Apr 2011. http://onlinelibrary.wiley.com/enhanced/doi/10.1111/j.1747-1567.2011.00724.x/
12. Liu C et al (2010) Deformation behavior of solid polymer during hot embossing process. Microelectron Eng 87:200–207. doi:10.1016/j.mee.2009.07.014
13. Li MJ, Liu C, Peng J (2008) Effect of hot embossing process parameters on polymer flow and microchannel accuracy produced without vacuum. J Mater Process Technol 207:163–171
14. Yao GD, Vinayshankar VL, Byung K (2005) Study on sequeezing Flow during nonisothermal embossing of polymer microstructure. J Polym Eng Sci 45:652–660

15. Lin RC, Chen HR, Chen HC (2003) Preventing non-uniform shrinkage in ope-die hot embossing of PMMA microstructures. J Mater Process Technol 140:173–178
16. Ermakov V, Kruchinin S, Fujiwara A (2008) Electronic nanosensors based on nanotransistor with bistability behaviour. In: Bonca J, Kruchinin S (eds) Proceedings NATO ARW "Electron transport in nanosystems". Springer, Berlin, p 341–349
17. Kruchinin S, Pruschke T (2012) Thermopower for a molecule with vibrational degrees of freedom. Phys Lett A 378:157–161 (2014)
18. Narijauskaitė B, Palevičius A, Narmontas P, Ragulskis M, Janušas G. High-frequency excitation for thermal imprint of microstructures into a polymer. Exp Tech X(X):1–7 (2013)
19. Šakalys R, Janušas G, Palevičius A, Bendikienė R, Palevičius R (2015) Microstructure replication using high frequency vibroactive pad. Mechanika 21(2):5–12
20. Popov E (2012) Gratings: theory and numeric applications. Institut Fresnel, Université d'Aix Marseille, Institut Fresnel, Université d'Aix-Marseille, Marseille Cedex, 1st edn. Presses universitaires de Provence (PUP). ISBN:978-2-8539-9860-4

Part II
Nanosensors

Chapter 14
Nanotechnology and Microfluidics Based Biosensing

H. van Heeren and P. Salomon

Abstract Microfluidics based biosensing is increasingly seen as a major enabler for medical diagnostics, especially Point of Care applications. Although often seen as one application, Point of Care is divided into three area, each with its own specific demands and constraints. As with all new products, the diversity in technologies is bewildering. However the demands and constraints forces the industry in certain directions, of which the need for very sensitive sensor concepts not needing complicated pre-processing steps is the most dominant one. Another important trend is integration of functionalities in the biosensor enabler device. That integration is needed to create faster and cheaper devices. The last trend is towards plug and play microfluidics. As a big barrier to the integration and plug and play devices are universal microfluidic interconnections, there is a need for microfluidic standards. A roadmap for such microfluidic standards is discussed.

14.1 Introduction

Microfluidics, nanotechnology and biosensing are strongly linked items. They are increasingly being used in medical diagnostics where this combination enables the detection of biomarkers in very low concentrations of fluids: nanogram per litre down to picogram per litre or even femtogram per litre. At the same time, samples volumes are often limited – not only because of the availability (for instance in the case of blood samples) but also while smaller sample volume is linked to device miniaturisation and speed of detection. The fluidic technology used differs from the more traditional industrial segments like the semiconductor industry or factory automation that deal with liquids in the ml/minute area.

H. van Heeren (✉)
enablingMNT The Netherlands, Drakensteynlaan 34, 3319 RG Dordrecht, The Netherlands
e-mail: henne@enablingmnt.com

P. Salomon
enablingMNT, Berlin, Germany

J. Bonča, S. Kruchinin (eds.), *Nanomaterials for Security*, NATO Science for Peace and Security Series A: Chemistry and Biology, DOI 10.1007/978-94-017-7593-9_14

14.2 Application Area for Microfluidics and Biosensing

There are several application areas where the combination of microfluidics, biosensing and nanotechnology lead to new paradigms and finally and are a source of, sometimes disruptive, new products and applications.

14.2.1 Dispensing

The first important application for microfluidics, even before it was called microfluidics, was ink jetting. Originally developed for paper printing, it is now increasingly being used for printed electronic circuits or the functionalisation of biosensors by depositing small amounts of biomarker sensitive material in a pattern onto a sensor surface to create multiplexed sensors.

14.2.2 High Speed Sequencing

An important and fast growing market is high speed sequencing of DNA, workhorse for the quest for a better understanding of biological processes in the human body and understanding the relation between a person's DNA and a specific disease. Over the last 15 years the cost of sequencing decreased from 100 M$ per gnome in the beginning of this century to a few thousands of dollars this year [1]. Microfluidics played a key role in this impressive cost reduction.

14.2.3 Pharma

The pharma industry has a broad interest in microfluidics. Initial application was to couple a certain drug to a dosing method: e.g. think of inhalers with accurate miniature nozzles, microneedles for painless injection or drugs that are encapsulated by microfluidic technologies. Recently we have seen a growing interest in "organ on a chip" in the pharma industry; i.e. to mimic cell behaviour on a miniature scale. The holy grail is to improve the accuracy of preclinical trials and overcome the need for animal testing. But that's only one aspect of organ on a chip; another interesting opportunity is that of personalized medicine or individualized therapy to match specific drugs to the appropriate patients or, in some cases, design a targeted treatment for each individual patient. To estimate how effective a specific medicine will be on a patient, testing out the medicine on a lab scale before applying it to the patient is envisioned. Some even go further by predicting a future where DNA analyses will be the basis not only for diagnostics but also for therapy. A success in any of these steps will open significant market opportunities.

14.2.4 *Medical Diagnostics Outside the Lab*

The medical market is expected to become the biggest market for microfluidics and biosensing. Biomarkers in blood, urine, tears, etc. can help doctors identify health problems or can help the patient to control its medication. Microfluidics provides the toolbox to detect small amounts of these biomarkers faster and at lower cost compared to conventional methods. Generally, the microfluidic part is in a disposable. Ease-of-use paradigms demand that such a disposable integrates all the steps needed form the sample taking to the sensing. As the technology is designed for miniaturised lab instruments, it also offers a route towards portable testing equipment, making testing at the point of care possible. There is consensus about the usefulness of such tests, although there is strong discussion (and some disagreement) about the exact place in the care supply chain, where such tests will be needed most [2] and generate interesting revenues. Some of the technologies developed for this market found their way into other markets like food safety, home defence, etc.

In the literature, Point of Care (PoC) is often treated as one area. In practice there are three rather distinct application area. Firstly, in generic medical diagnostics: to determine what disease is causing a certain problem. Secondly, in emergency care: confirmation of a certain expected cause of the problem or determining what specific variation of the disease is responsible for the illness. Thirdly, applying PoC when the patient is already diagnosed with a certain disease and there is a need for regular testing to check the progress of the disease or the effectiveness of a treatment. For the sake of clarity we will use the term Point of Care in this article only for this last application.

The differences between these applications have their consequences for the acceptable cost levels, the needed time to result and the multiplicity of the test, i.e. how many biomarkers must be tested simultaneously. The Table 14.1 gives an overview of these three applications.

Other variations between the three are in terms of data handling, (e.g. communication with hospital information systems or patient databases) and safety/security issues.

14.3 Design Choices

A particular challenge for the companies that are aiming at microfluidics driven market segments, as for all new industries, is facing the wide range of technologies available. For every microfluidic/biosensing function in the devices to be developed there are a range of possibilities. For instance, to set the flow in motion: will we use capillary-, electrokinetic- or centrifugal forces, use overpressure or an external pump like gear-, peristaltic-, piston- or syringe pumps? For the detection of the biomarker, will we use electrical detection (SAW, conductivity, redox reaction, etc.)

Table 14.1 Main application area for medical diagnostic outside the lab

Application	Medical diagnostics	Emergency care	Point of care
Goal of testing	Estimate what disease causes the complaint	Confirm the disease or determine the specific variant	Progress of the disease, effectiveness of treatment or check of recurrence after treatment
Frequency	Incidental	Incidental	On a regular base
Biomarker	Often test for a range of diseases, i.e. several markers simultaneously	Optimized for one specific (set of) marker(s)	Optimized for one specific (or a very limited number of) marker(s)
Point of use	In a GP's office or in a hospital	In or near an ambulance, in a hospital	In a GP's office or at the patient's home
Indication of allowable cost/disposable	<10 $	<100 $	<1 $
Indication of allowable time to result	<5 min for GP; <20 min other settings	<20 min	<5 min in the GP's office; in a home setting up to 1 h is acceptable

or optical detection (fluorescence, chemiluminescence, wavelength changes etc.) or measure a mechanical effect (membrane or cantilever deformation)? Can we use direct detection or do we need labelling? Or, probably the biggest challenge, sample preparation: is there a need to lyse the cells, do we need pre-concentration, filtering, amplification etc. Each of the choices has its consequences for the reliability, sensitivity and specificity of the device to be designed, and has a strong influence on the, usually, most important factor: cost.

14.4 Point of Care

Point of Care (PoC) is perhaps the most challenging area for microfluidic based biosensing. Not only there are, as shown earlier, severe restrictions to cost and time to result, the test should still have the same accuracy as established lab tests, which are regarded as the gold standard. Most of the currently used testing methods necessary for PoC systems, like those using target amplification, labelling or complex sample concentration technologies (for instance using magnetic beads) will have difficulty meeting that standard. That doesn't say that there isn't a market for such tests, they will find ample use in other areas, but unlikely in Point of Care. Still, the state of the art in microfluidics/biosensing enabled PoC is complex pre-processing and labelling with fluorescence markers. No wonder there are many developments into diagnostic devices either omitting or limiting pre-processing as much as possible or integrating the steps in the disposable.

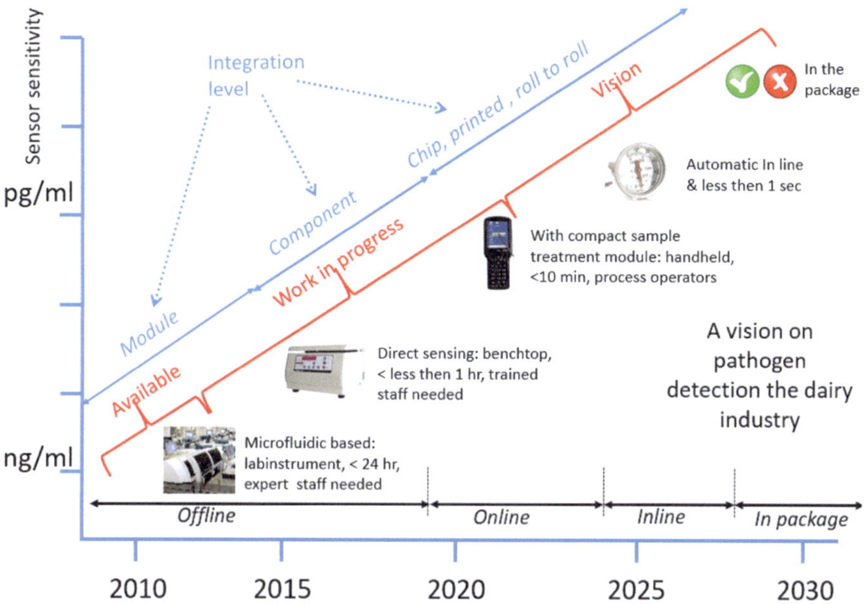

Fig. 14.1 Roadmap for testing dairy products (FoodMicroSystems project)

Currently the state of the art for biosensing is labelling of the markers and detecting the fluorescence signal using a benchtop device, often needing some pre-processing of the sample by the operator. The next generation is still likely to be also a benchtop instrument, but the time to result is expected to be much shorter. Portable instruments will follow after that, ultimately leading to handheld devices and fully integrated disposables that reach the wanted price and speed specifications.

Not only in medical diagnostics this trend towards miniaturisation is visible, we see the same trend in testing agricultural products as was discussed recently in a report of the FoodMicroSystems project. The summary roadmap for the implementation of microsystems in the dairy sector [3] is shown in the Fig. 14.1.

14.5 Who Will Be the Future Leader in Microfluidics?

No wonder there is a broad interest in microfluidics, and there is hardly a university or a large medical or pharma company not actively investigating the opportunities of the combination of microfluidics with other technologies; not only for medical diagnostics but also for pharma or for chemical processing. At this moment there are more than 700 companies worldwide active in microfluidics (see Fig. 14.2) with hundreds of microfluidic components, instruments and equipment on the market and this number continues to rise as researchers are publishing new ideas nearly every day, and one of these may have the potential for the large market breakthrough with microfluidics.

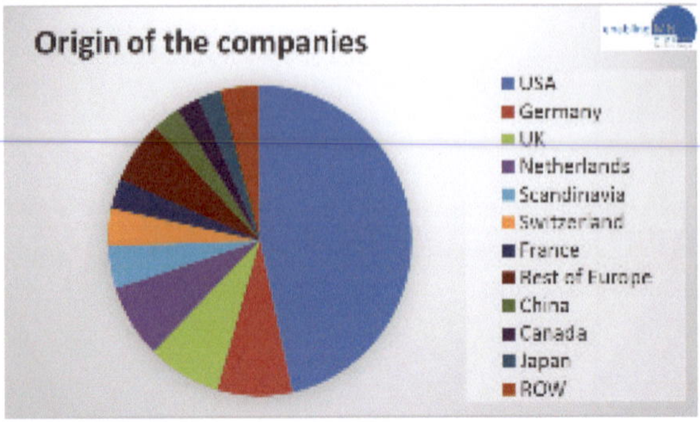

Fig. 14.2 Microfluidic companies per country

As can be seen by the figure above, the interest is not restricted to one country or region, but the USA is clearly leading. At least in the number of companies pursuing (potential) high volume applications like PoC, high throughput sequencing or organ on a chip. Europe has a strong focus on the supply chain. The leading companies providing the "nuts and bolts" for this industry are nearly all located in Europe.

14.5.1 Time to Market, Financing, Successes and Failures

As shown above the interest of the market in microfluidics/biosensing is genuine, but there is a barrier: the timeline to the market is annoyingly long. Taking, for instance, Atlas Genetics that is developing a Point of Care device for infectious diseases. They started in 2005 as a university spin off. After 10 years and over 40 million dollars investment, they are still not on the market, although very near. Their disposable cassette is a good demonstration of the state of the art technology described earlier. It is a stack of injection-moulded parts with three quarters of the cassette volume needed for the microfluidic processes to prepare the blood sample for the electrochemical detection technology. This complicated pre-processing is also the main reason the testing takes about 20 min.

A real success story is Abaxis with an annual turnover over 140 million dollar, but their investors had to spent $100 million and had to wait almost 20 years to see the product entering the market. The 7.9 cm disk made of moulded 2 cm PMMD runs a full chemistry analysis (several panels available) from just five drops of blood in about 10 min. The disposable is a circular disk with is spun fast around after applying the sample. The blood is forced by the centrifugal forces through a series of interlinked internal chambers and passages. The movement of fluid is controlled by a series of stop junctions, capillaries and siphons managed by the use of centrifugal forces. The chemicals needed for the processing are delivered with the disposable and don't need to be added manually. The complexity of ensuring the testing quality is illustrated by the fact that only 15 of the 25 cuvettes are used to analyse the patient's sample; the other are used for internal quality control. The read out is optical. The time to measure is 12 min and the company offers now 31 tests, conveniently configured into 16 completely self-contained reagent discs, 11 of which are CLIA approved.

14.6 The Need for Direct Sensing

Both examples demonstrate the difficulty to get below 10 min time to measure when the sample needs pre-processing before detection of the biomarker becomes possible. It seems to be only realistic when the pre-processing of the sample to be tested is limited as much as possible. Take for instance the case of Blue4Green. This company is selling a device that measures the amount of calcium in cow's blood. The positive ions in the blood sample are separated by capillary electrophoreses, the only treatment needed of the sample. The conductivity change over time gives then a measure of the presence of the different positive ions. All of the steps are integrated in the device and the measurement time is only 3 min. The simplicity of the fully integrated disposable, just a glass chip in a plastic housing, opens the door to the fabrication of very low cost disposables. After applying a droplet of blood, the ions are driving through a small channel to a crossing with another channel. From the sample volume at the crossing, the positive ions are forced through a long small channel. The different ions become separated and by measuring the conductivity changes over time, they different ions are detected (see Fig. 14.3).

Not in all cases the biomarkers can be separated that easily. Sometimes the biomarkers need to be released from a cell; disturbing other biomarkers need to be removed or the concentration of the marker needs to be increased. No wonder many research and development groups are looking into new technologies enabling very sensitive sensing technologies to avoid the troublesome concentration steps. Those sensitive sensors have an additional advantage. Most of the ones that are developing biosensing instruments decide using fluorescence detection. For that purpose a fluorescence marker is attached to the biomarker. This combination is

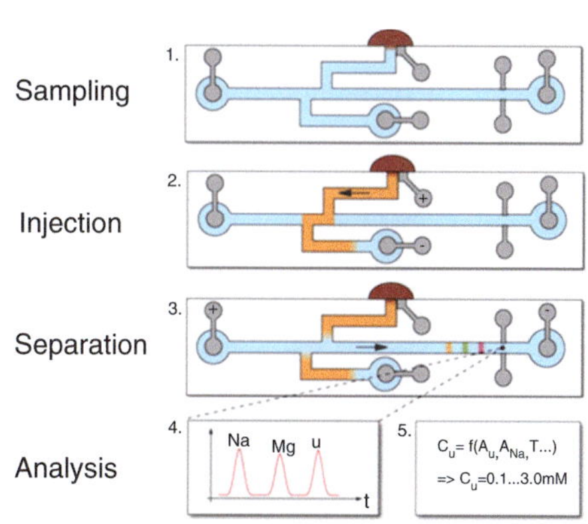

Fig. 14.3 Working principle of capillary electrophoresis device

then fixated on a small surface and the measured optical signal is a measure of the original amount of biomarker. A very effective method, but it leads again to an increase of complexity while chemicals have to be added and extra process are needed.

14.6.1 Direct Sensing Technologies

There are several approaches for direct sensing, for instance:

- Cantilever: measuring stress induced bending caused by the adherence of a biomarker to the surface.
- Planar waveguide: measuring the difference in optical properties of two parallel channels, one exposed to the liquid to be tested.
- SAW sensors: the sensor generates a mechanical wave. Due to adherence of the biomarker to be detected to the functionalized surface, the properties (phase, frequency etc.) of the wave are changed. By comparing the difference in properties between generated and measured waves, the concentration of biomarker is determined.

Cantilevers have been studied extensively, but most efforts to commercialize this technology have failed up till now.

Planar waveguides tend to be relative expensive; they also need accurate optical coupling, making them less suitable for low cost disposables, although there are some interesting opportunities for very sensitive multiplexed sensors.

Electrical sensing has some interesting advantages: They are generally made with semiconductor technologies and can therefore be made at low costs. It is relatively easy to multiplex them on one chip. Reliable electrical connections can be made easily at low cost.

An example of such an electrical sensor is the one in development in the recently started PoC-ID project [4], see Fig. 14.4.

PoC-ID will combine the detection of both pathogens and host responses leading to more accurate diagnosis as compared to the current standard which is focused on detection of pathogens only. This novel approach will support prevention and control of pathogen spread and enable faster and more personalized patient treatment. Improved performance in terms of robustness, sensitivity and selectivity will be reached by a combination of innovative nanomembrane technology, capture molecules and two novel sensing concepts. Further advances will be realized in terms of usability and speed of data-analysis arising from the integration of sensors, read-out electronics and microfluidics into one user friendly point-of-care (PoC) platform. Costs of the new disposable sensors will be low at high volumes, thanks to designing into microelectronics production flows.

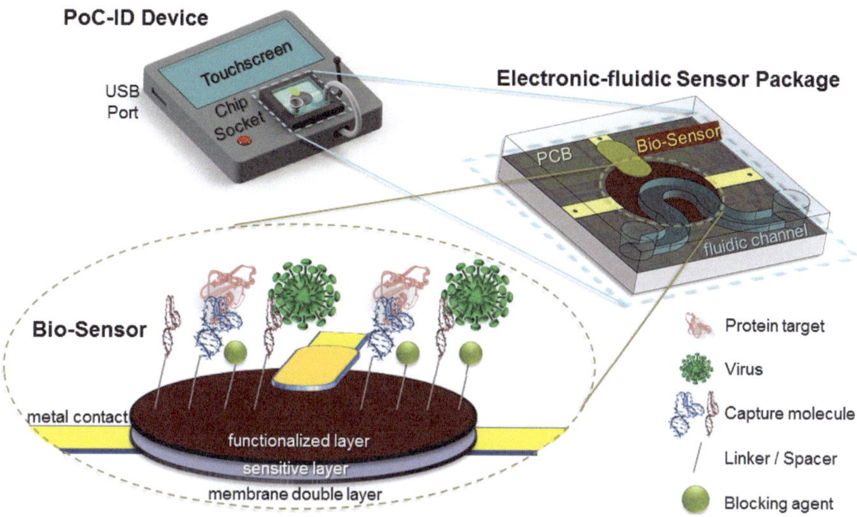

Fig. 14.4 Concept of the point-of-care device from macro to nano scale; device setup is applicable for graphene-FET- and MEMS-based biosensing as well as other electronic sensing concepts with similar chip size (PoC-ID project)

14.7 Integration

Besides direct sensing, integration is seen as the way forward. Integration offers smaller handheld devices and, by limiting the distance for the liquid to flow, it faster operation. Integration also opens a door to efficient high volume manufacturing. There are however substantial barriers. In the first place the difficulty of combining electronic, mechanical, fluidic and optical components or structures. Secondly integration means making microfluidic connections. Where for electrical connections there are number of reliable, low cost and manufacturing friendly solutions, microfluidic connections are often unreliable, expensive and not (yet) suitable for high volume manufacturing. Lastly, but perhaps most importantly, microfluidics doesn't scale as easy as electronics. There are two main reasons for that: in the first place in microfluidics the wall of the tube or channel has a profound influence on the flow performance: not only varies the flowrate over the channel, biological material tend to stick to surfaces. With smaller structures, the tube or channel surface starts to dominate or even prohibit the flow. Another reason is that in microfluidics often a certain minimal amount of liquid is needed to make the measurement statistically valid or even technically possible.

14.8 Plug and Play Microfluidics

The development of integrated devices will be stimulated by the appearance of industry wide standards. In general standards promote the uptake of new technologies. Standardized products and services are seen as safer, more secure, and having high quality. Furthermore the industry expects them to fit better in existing infrastructure, to enable multi-market access and to create more active markets. There are also specific benefits from a microfluidic perspective. Researchers don't want to spend much time on side issues like correct connection of tooling. Providers of analytical services don't want their limited lab space cluttered with a multitude of incompatible instruments. Chemical engineers want easy interconnection between pumps, sensors end reactors [5].

The discussion about standards for microfluidics has been a long one; nearly as old as the industry itself. As developing standards must be a bottom up process with broad industrial support the discussion started to get some momentum when a group of companies from several parts of the supply chain got involved in this discussion during a session of the Microfluidic Consortium [6] in 2009 and later followed up in the MFManufacturing project [7]. Already in the beginning of the discussion it became clear that microfluidic interconnections was to become the major point of discussion. Such interconnections should enable users to make a number of fluidic connections easily, fast and reliable. As a first step the discussion focused on the pitches between microfluidic ports. It was decided to use a 1.5 mm grid, on the grid points position of a port is allowed. This pitch was chosen while it will cover 3 mm, which is regarded as the current state of the art and 4.5 mm which is an often used port pitch for the well-known miniLuer connectors (see Fig. 14.5). It also offers a roadmap to smaller pitches needed for more integrated products.

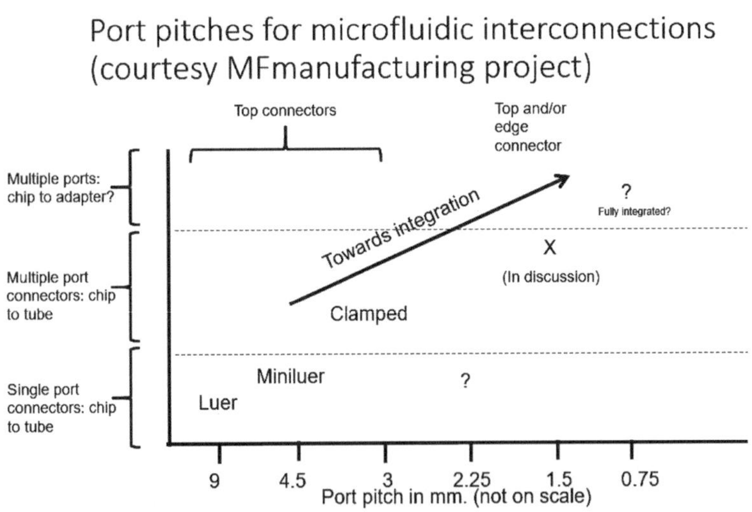

Fig. 14.5 Microfluidic interconnection roadmap

Although the roadmap above takes the first step towards smaller port pitches, it might be that in the future smaller port pitches are needed, for instance for edge connectors or fully integrated devices; in that case a finer grid might be needed.

Worth mentioning that this industry agreement, found its way into ISO; a first step towards an official standard will be a workshop organised buy ISO in the beginning of 2016. The discussion continues however, and there are now also proposals in the making for chip formats, sensor interfaces, standard tests etc. Much of that work is published in a white paper (www.mf-manufacturing.eu/downloads-and-links).

14.9 Conclusions

The diversity in microfluidics, even in the segment that deals with microfluidics enabled biosensing, is huge. The market constraints and needs are guiding the developments and there are clearly trends in industry and science toward direct sensing, higher levels of integration and more plug and play devices.

Acknowledgements This work was supported by the MFManufacturing project, co-funded by grants from the UK, France and the Netherlands and the ENIAC/ECSEL Joint Undertaking and the PoC-ID project, which is funded by the EU H2020 programme.

References

1. Wetterstrand K (2015) DNA sequencing costs: data from the NHGRI genome sequencing program (GSP). National Human Genome Research Institute. Retrieved 24 Oct 2015
2. Kaman WE et al (2013) Perceptions of point-of-care infectious disease testing among European medical personnel, point-of care test kit manufacturers, and the general public. Patient Prefer Adherence 7:559
3. Salomon P, van Heeren H, Roadmap: implementation of microsystems in the dairy sector, FoodMicroSystems project. http://www.foodmicrosystems.eu/?page_id=3063
4. PoC-ID Project Platform for ultra-sensitive point-of-care diagnostics for infectious diseases. www.PoC-ID.eu
5. van Heeren H (2012) Standards for connecting microfluidic devices? Lab Chip 12:1022
6. The Microfluidic Consortium. www.cfbi.com/index_files/microfluidics.htm
7. Van Heeren H. (2015) Interoperability of microfluidic components. http://www.mf-manufacturing.eu

Chapter 15
Resistivity Sensors of Metal Oxides with Metal Nanoparticles as Catalysts

G.A. Mousdis, M. Kompitsas, D. Tsamakis, M. Stamataki, G. Petropoulou, and P. Koralli

Abstract The metal oxide resistivity sensors are the most common sensors for reducing and oxidizing gases, due to their low cost, easiness of preparation and signal processing. The sensing properties of a metal oxide film depend on the surface roughness, porosity, crystallinity and some other factors, which differentiate for each preparation method. Moreover, the addition of noble nanoparticles on the surface of the films, improves the sensing properties of the films. In this work two different techniques were used to prepare thin films of metal oxides doped with noble metals and their sensing properties for different reducing gases were investigated. Namely we used the (a) PLD technique to prepare thin films of Cu_xO ($1 < x < 2$) doped with Au and testing them as CO and CH_4 sensors. (b) Sol-gel technique to prepare ZnO thin films doped with Au as H_2 and acetone sensors. All the films are characterized. A significant response to several concentrations of the analytes was demonstrated for all the films at temperatures lower than 200 °C.

15.1 Introduction

In past decades, improvements in gas sensor manufacturing technologies have occurred, driven by the development of low-power and low-cost microelectronic circuits [1], chemometrics [2], and microcomputing [3]. Nowadays, the development of solid state gas sensors for the detection of inflammable and toxic gases, such as hydrogen, carbon monoxide, etc. is a major concern. The world market for gas sensors is increasing and due to commercial competiveness there is a need for enhance quality and product reliability. There are many types of gas sensors such as electrochemical, optical (including IR and holographic sensors), semiconducting

G.A. Mousdis (✉) • M. Kompitsas • G. Petropoulou • P. Koralli
NHRF-National Hellenic Research Foundation, Theoretical and Physical Chemistry Institute-TPCI, 48 Vass. Constantinou Avenue, Athens 11635, Greece
e-mail: gmousdis@eie.gr

D. Tsamakis • M. Stamataki
Electrical and Computer Engineering Department, National Technical University of Athens, Zografou Campus, Athens 15700, Greece

J. Bonča, S. Kruchinin (eds.), *Nanomaterials for Security*, NATO Science for Peace and Security Series A: Chemistry and Biology, DOI 10.1007/978-94-017-7593-9_15

(or metal oxide), capacitance, calorimetric, ultrasonic etc. The type of sensor to be used depends on the application demands, such as: gas & concentration range, environment (humidity, temperature, pressure, gas velocity, chemical poisons and/or interfering species), power consumption, response time, maintenance interval, fixed or portable, point or open path etc.

The most common category of gas sensors are the resistivity sensors (also known as chemiresistors). In this type of sensors the resistivity is changing on the presence of the analyte. In most cases, the sensing material is an organic polymer or a metal oxide. Although the number of commercial polymer, based on chemiresistors, is increasing the majority of those are based on metal oxides.

The popularity of the metal oxides sensors (MOSs) is due to: their low cost, small size, simple construction, large number of detectable gases, high compatibility with microelectronic processing and many possible application fields. Their major disadvantages are their pure selectivity and the fact that they work at elevated temperature. A method to increase the selectivity is the use of catalyst that increases the sensitivity to a specific analyte.

The MOSs are used to detect various types of gases, reductive (e.g., H_2, CO, Hydrocarbons), oxidative (e.g., O_2, O_3) or neutral (e.g., CO_2, H_2O). They employed in environmental monitoring mainly for the detection of toxic or explosive gases [4], for chemical and industrial processes for the prevention of hazardous gas leaks [5] and in many other activities such as human health [6], food and wine quality monitoring [7], traffic safety [8] etc.

Recently the increasing demand for reliable sensors has triggered a large amount of research worldwide for metal oxide to improve the well-known "3S" criterion: sensitivity, selectivity, and stability. A plethora of devices have been developed using different materials and methods mainly by an empirical approach, and a lot of basic research has been carried out [9].

For the preparation of metal oxides sensors different metals can be used with most common some transition or post transition metals because their cations have a d^n configuration with a small energy difference with the d^{n+1} or d^{n-1} configurations [10]. The most effective metal oxide sensors are those with cation configurations d^0 (e.g. TiO_2) or d^{10} (e.g. SnO_2, ZnO).

The metal oxides used as sensors, are *n* or *p*-type semiconductors. Their sensing mechanism based on the interaction of analyte with surface groups such as O_2^-, O_2^{-2}, O^-, OH. This interaction alters the resistivity of the metal oxide sensor. The change of the sensor sensitivity, depends on the type of semiconductor and whether the analyte is an oxidizing or reducing gas (Table 15.1) [11].

The fundamental sensing mechanism of metal oxide based, gas sensors relies on a change in electrical conductivity due to the interaction process between the

Table 15.1 Dependence of resistance change from the type of semiconductor and the gas

Classification	Oxidizing gases	Reducing gases
n-type	Resistance increase	Resistance decrease
p-type	Resistance decrease	Resistance increase

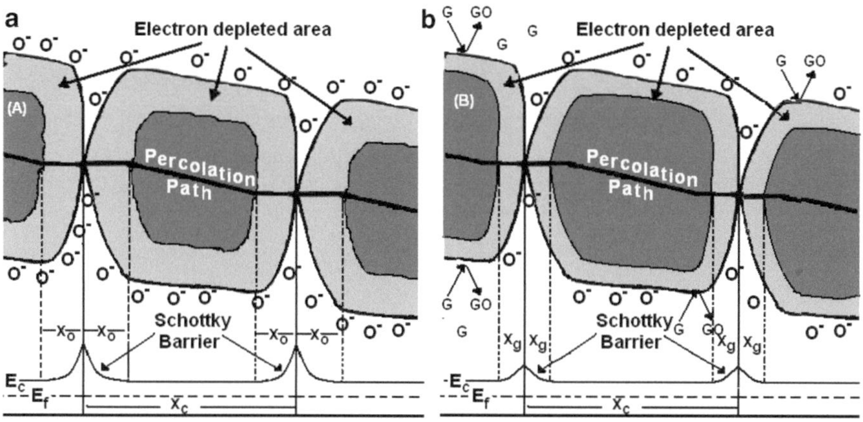

Fig. 15.1 Schematic representation of some crystallites of a n type metal oxide with and without the presence of analyte gas (G). x_c is the grain size and x_o and x_g is the depth of the depletion layer in the absence and presence of the analyte, respectively

surface complexes such as O_2^-, O_2^{-2}, O^-, and OH reactive chemical species and the gas molecules to be detected.

The surface of the metal oxide microcrystals in the presence of O_2 adsorb molecules of O_2 that depending on the temperature [11] attract 1 or 2 electrons to give O_2^-, O_2^{-2}, O^-. The reactions are:

$$O_2\ (\text{gas}) + e^- \leftrightarrow O_2^-\ (\text{adsorb})\ (T \leqslant 200\ ^\circ\text{C})$$

$$1/2O_2\ (\text{gas}) + e^- \leftrightarrow O^-\ (\text{adsorb})\ (T \geqslant 200\ ^\circ\text{C})$$

$$1/2O_2\ (\text{gas}) + 2e^- \leftrightarrow O^{-2}(\text{adsorb})\ (T \geqslant 200\ ^\circ\text{C})$$

Due to the reduction of e^- the mobility carriers of the semiconductor decreased (n-type) or increased (p-type) resulting to increase or decrease of the resistivity respectively. In most cases the sensing layer consists of single crystalline grains. So the free charge carriers have to overcome the surface barriers appearing at the surface of the grains (for a n-type sensor see Fig. 15.1) the so called Shcottky barrier. But in the presence of the gas there is a reaction with the surface complexes and the Shottky barrier decrease (or increase for the p-type) and the resistivity altered.

The degree of resistivity change depends on the number of the surface complexes reacted with the gas, which is better proportional to the gas concentration. But in general, it depends also on the number of surface complexes and their accessibility to the gas, so the morphology of the sensing layer [12] (e.g. porosity, crystallinity, surface smoothness etc.) is very important for the properties of the sensor. As a consequence the preparation method and the processing of the sensor affect significantly its properties.

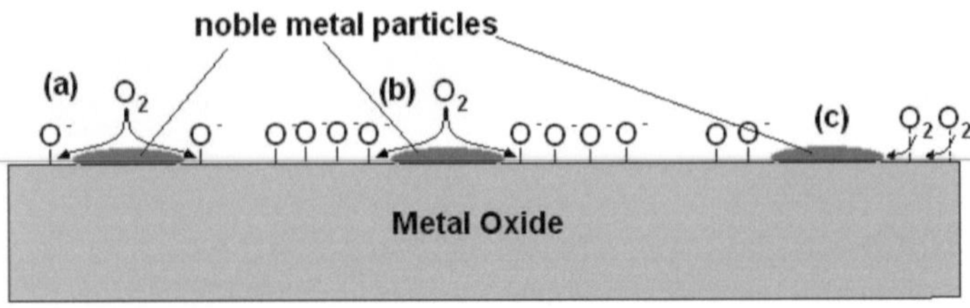

Fig. 15.2 Schematic depiction of the major processes of molecular O_2 dissociation at a noble metal nanoparticle on a metal oxide surface

A method to improve the sensing properties is the impregnation with catalysts (mostly noble metals) that will enhance the reactions of gas on the sensor surfaces. The impregnation with noble metals can be done with many methods either after the preparation of the film or simultaneously with the preparation of the film. There are many methods (e.g. sol-gel, PLD, sputtering, thermal evaporation etc.) for introducing noble metal additives into metal oxide films. The results do not only depend on the concentration of the noble metal but also on the methodology used.

The explanation of the noble metal role in the enhancement of the sensing properties is based on their highly effective dissociation catalytic ability [13]. The noble metals are far better oxygen dissociation catalysts than metal oxides and catallytically dissociate the molecular oxygen to ions (Fig. 15.2a). These ions are diffused to the surface of the metal oxide (Fig. 15.2b).

Moreover the O_2 molecule can reside on the metal oxide surface and diffuse to a catalyst and dissociate before desorption (Fig. 15.2b). There is an "effective area" [14] around the noble particle where the O_2 molecules dissociate before desorption. The optimum will be the noble metals to be dispersed so that these "effective areas" cover all the surface of the film (Fig. 15.2c). The latest process is called "back-spillover effect" [15].

Furthermore, in many cases the noble metal catalyses and the reaction of gas with the O species e.g. in the case of H_2 the following reactions occur:

$$\text{(a) } H_2(g) \rightarrow 2H_{ads}, \quad \text{(b) } H_{ads} \rightarrow H^+_{ads} + e^-,$$

$$\text{(c) } H^+_{ads} + O^-_{ads} \rightarrow OH, \quad \text{(d) } 2OH \rightarrow H_2O + O_{ads},$$

where the first 2 reactions (a and b) were catalyzed by noble metals.

As a part of research to understand and find the parameters that influence the sensing properties of MOSs, we prepared a series of MOSs doped with Au by 2 different methods. Namely we prepared ZnO thin films doped with Au by the sol gel method and tested them as sensors for acetone and H_2, and Cu_xO ($1 < x < 2$), doped with Au, by Pulse Laser Deposition technique (PLD) as CO and CH_4 sensors.

15.2 Experimental

15.2.1 Synthesis

The substrates, microscope glass slides, were cleaned by soaking for 24 h in a sulforochromic bath, then washed copiously with distilled water and kept in isopropanol until used.

Au Precursor

$HAuCl_4$ ethanolic solution was prepared by dissolution of Au in HNO_3/ 3HCl, dried under vacuum and then dissolved in a specific volume of methoxyethanol to obtain a 0.3 M solution.

ZnO Precursor

Zinc acetate dihydrate was dissolved in 2-methoxyethanol and 2-aminoethanol (DEA) was added slowly under magnetic stirring. The molar ratio of DEA to zinc acetate was maintained to 1.0, while the concentration of zinc acetate was 0.5 M. A volume of the 0.3 M $HAuCl_4$ methoxy-ethanolic solution was added to obtain a 5 % Au/Zn atoms. The resultant solution was stirred at 80 °C for 2 h to yield a clear and homogeneous solution and then maintained at room temperature to aging for 24 h.

Sol-Gel Film Growth and Thermal Annealing

The ZnO films were prepared by spin coating at 3000 rpm for 30 s and were heat treated at 300 °C for 5 min in air to evaporate the solvent and organic residues (Fig. 15.3). This procedure was repeated for a number of times varying from 1 to 10 in order to obtain films with different thicknesses. Finally these films were annealed in air at 500 °C for 1 h.

Films Prepared by PLD

The Cu_xO thin films were deposited under a dynamic oxygen pressure of 40 Pa for 90 min deposition time on heated SiO_2 substrates at 200 °C by pulsed laser deposition. The oxygen pressure of 40 Pa was chosen because previous results showed that at lower pressures we obtain films of Cu_xO with less sensitivity [16].

An Nd:YAG laser (λ = 355 nm (THG)) source at 10 Hz repetition rate was focused on a metallic Cu target (Fig. 15.4). The laser fluence incident on the target surface was set at ~2 J/cm^2. After the Cu_xO film deposition, the chamber was

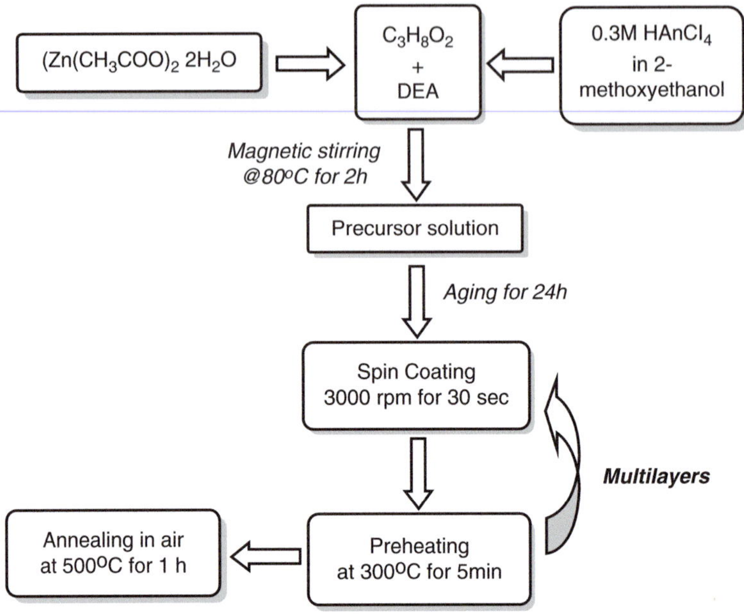

Fig. 15.3 Flow chart of sol-gel spin coating deposition process for ZnO-Au films

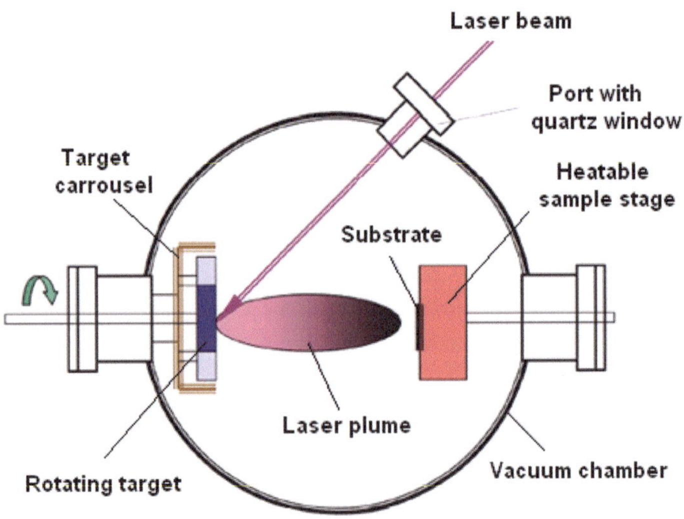

Fig. 15.4 Schematic diagram of the PLD set-up for the preparation of Cu_xO and Cu_xO-Au doped films

evacuated down to 10^{-4} Pa pressure and an Au target (99.998 % purity) was ablated by the focused laser beam for about 2 min.

Films Characterization and Sensing Testing

The films were characterized by Thermogravimetric Analysis (TGA), Atomic Force Microscopy (AFM), and X-Ray Diffraction (XRD). Additionally, the absorbance and the reflectivity were measured.

Hydrogen sensing tests were performed in an aluminum vacuum chamber. The chamber was evacuated down to 1 Pa, and filled with dry air at atmospheric pressure. The samples were mounted on a resistively heated surface at the operating temperature. Hydrogen concentration was calculated based on the partial pressures of the sensing gas and air inside the chamber. A bias of 1 V was applied, and the current through the film was measured with a Keithley Mo. 485 Picoammeter. Hydrogen sensing was monitored by recording the current changes in real time on a lab computer.

15.3 Results and Discussion

15.3.1 Cu_xO films

The surface structure of the Cu_xO thin films was studied by AFM. Figure 15.5 shows the film surface consisting of nanoparticles of various sizes of some tens of nanometers on a rather smooth background. This is typical for PLD grown thin films where the nanoparticles consist of pure target material.

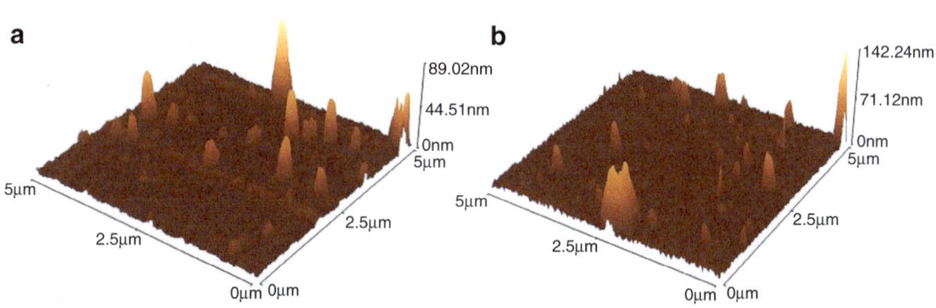

Fig. 15.5 Typical AFM images showing the surface structure of the samples (**a**) without Au nanoparticles, (**b**) with Au nanoparticles

In Table 15.2 we observe that maximum height of the nanoparticles on sample (b) is much larger than the one of sample (a), despite the fact that the average roughness is almost the same. This is due to the fact that Au nanoparticles have been deposited during a post-growth process on the film surface.

Cu_xO thin films have been tested as gas sensors towards CO and CH_4 for 0.3 % and 0.5 % in air. The operating temperature was in the 120–220 °C range.

We observe that the films with nanoparticles show a slower time response than the pure ones, but the signal is ~3 times larger, see Fig. 15.6.

The reaction leading to CH_4 sensing could be:

$$CH_4 + 2O^-_{2-ads} \rightarrow CO_{2-gas} + 2H_2O + 2e^-_{bulk}.$$

As a result, two electrons are released back to the bulk of the film, recombine with two charge carriers (holes) of the *p*-type Cu_xO film and the film resistivity is increased.

Similarly, Fig. 15.7 shows the sensor response towards CO. We observe that the films with nanoparticles show both a faster time response and a much stronger signal by a factor of ~8 than the pure ones.

Table 15.2 Quantitative results of the AFM recordings

Sample	(a)	(b)
Au nanoparticles	No	Yes
Roughness (av.), nm	2.91	2.83
Max. height of particles, nm	89	142

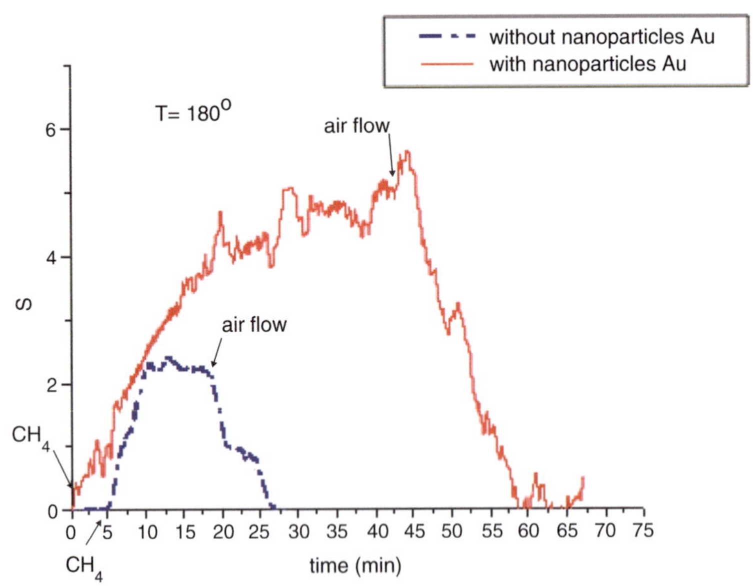

Fig. 15.6 Response of the Cu_xO films towards 0.5 % CH_4/air at 180 °C operating temperature

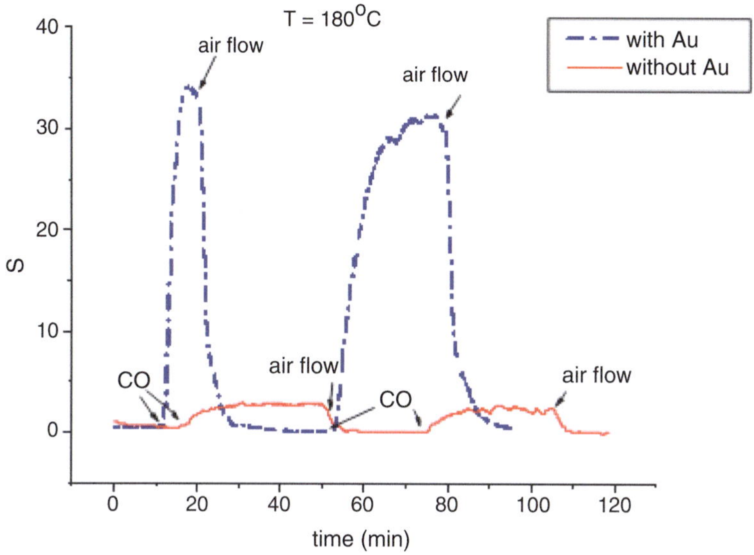

Fig. 15.7 Response of the Cu_xO films towards 0.5 % CO/air at 180 °C operating temperature

The chemical reaction that leads to the increase of the film resistivity is as follows:

$$CO + O^-_{2-ads} \rightarrow CO_{2-gas} + e^-_{bulk}.$$

In the following, we compare the time response and the signal strength of the films with and without nanoparticles towards CO and CH_4. In Fig. 15.8a we show the sensor response for pure Cu_xO films under the same experimental conditions. We notice that the signal strength is comparable although the time response is shorter for CH_4. In Fig. 15.8b, the presence of the Au nanoparticles, shows that both signal strengths are larger than without nanoparticles, as expected, when the vertical scales are compared. Moreover, the responses towards the two gases CO and CH_4 differ dramatically. The response towards CO is both faster and shows a signal increase by a factor of 12 in respect to the pure Cu_xO.

This later fact indicates that Cu_xO with Au nanoparticles at the 180 °C operating temperature shows efficient CO selectivity in a CH_4/CO gas mixture.

15.3.2 *ZnO Films*

Thermogravimetric Analysis (TGA) of the Au doped ZnO thin films showed that: the solvents evaporate at temperatures lower of 150 °C and the remaining organics

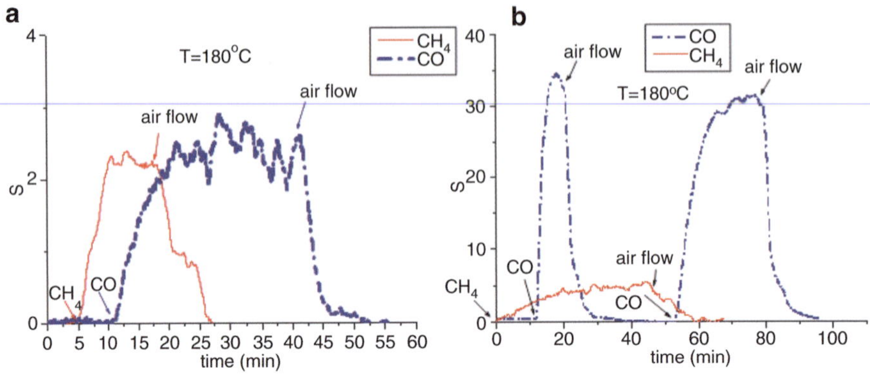

Fig. 15.8 Response of the Cu_xO films towards 0.5 % CO/air and 0.5 % CH_4/air at 180 °C operating temperature: (**a**) without nanoparticles, (**b**) with nanoparticles

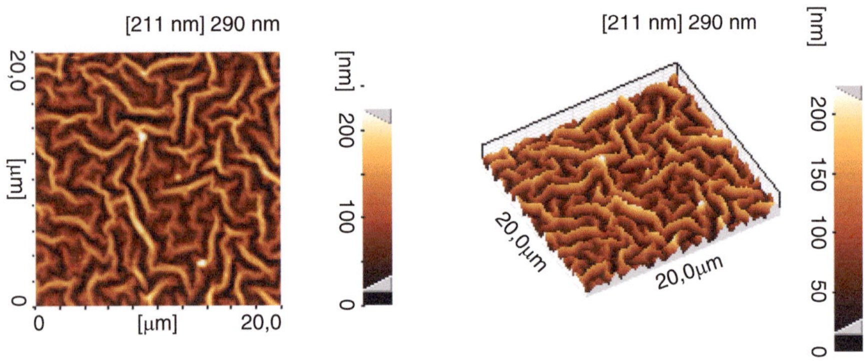

Fig. 15.9 AFM Image of a ZnO film doped 5 % with Au

are burned up to 400 °C, at temperatures higher than 400 °C the $HAuCl_4$ is reduced to metallic Au. After 500 °C there is no loss of mass.

The AFM images (Fig. 15.9) show a "worm" like surface that we expect to improve the sensing due to increased surface.

The XRD diagram shows that the ZnO is mainly amorphous with some microcrystallites (Fig. 15.10a). Anealing at 500 °C for 2 h does not improved the crystallinity. Also from the transmission spectra (Fig. 15.10b) we can calculate the optical band gap for the 5 % Au (3.52 eV) and for the 2.5 % Au (3.44 eV) that are higher than that of undoped ZnO (3.3). Also in the spectra we can see clearly the plasmonic band of Au nanoparticles at 552 nm.

The prepared films doped of ZnO doped with 5 % Au were tested as acetone sensors at 200 °C working temperature.

The film shows a fast and linear response to acetone concentration in the range 100–1600 ppm (Fig. 15.11), at higher concentrations the signal is reducing probably due to saturation.

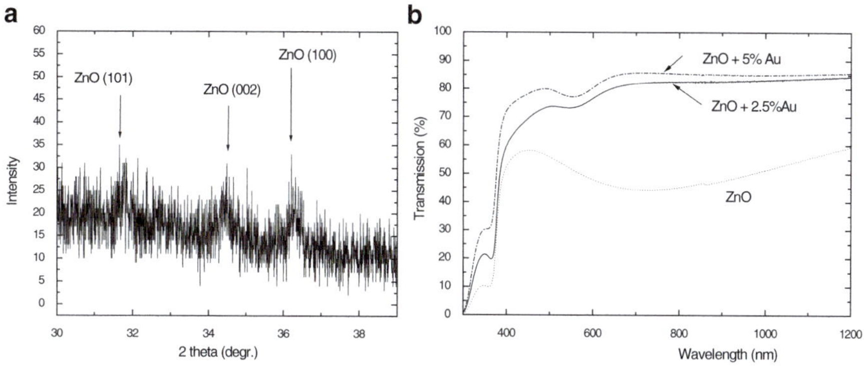

Fig. 15.10 XRD and transmition spectra of ZnO doped with Au

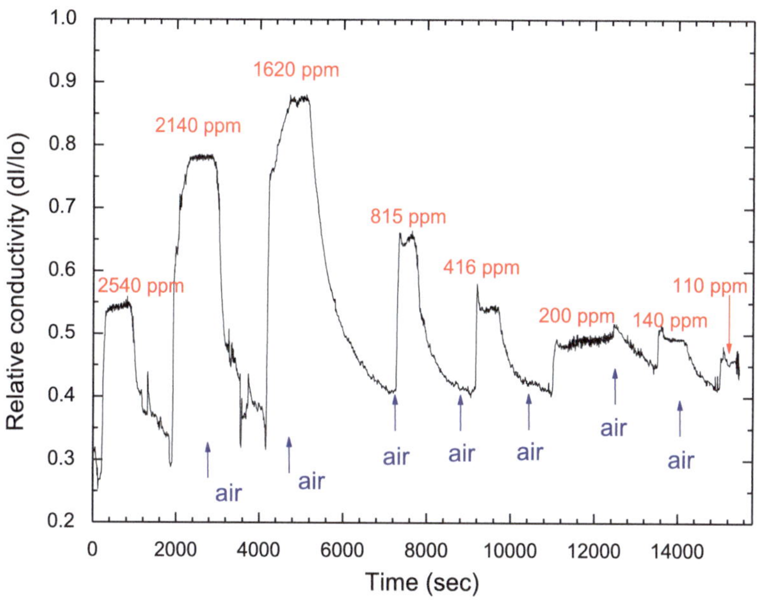

Fig. 15.11 The response of ZnO-5 % Au to different acetone concentrations

We prepare also ZnO films doped with 3 % Au and tested them as H_2 sensors for concentrations of H_2 in air above 40,000 ppm which is the low explosion limit (LEL)[5]. For this application is very important for the sensor to give a clear signal at low temperature for security reason (to avoid explosion by its hot surface).

The films show a strong response to concentrations range from 47,000 ppm up to 286,000 ppm and at temperatures down to 137 °C (Fig. 15.12). Although the signal is very strong the response time is getting bigger by lowering the temperature and at 137 °C is up to 30 min. Experiments with ZnO films prepared by the same

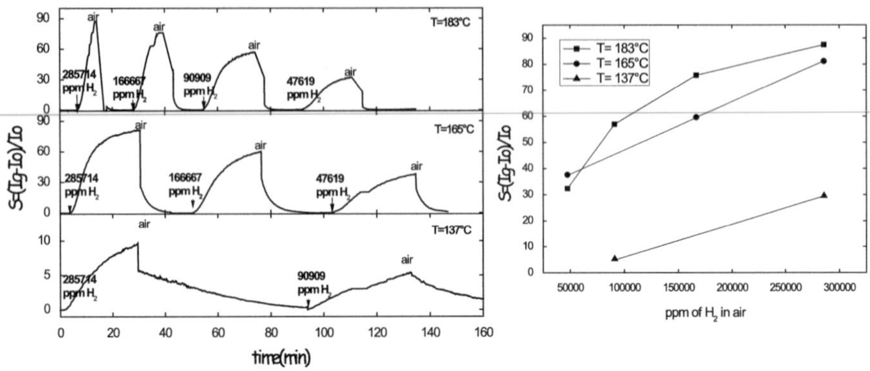

Fig. 15.12 The response of ZnO-3 % Au to different H_2 concentrations

method but without the addition of Au gave no signal or small signal with the same concentrations of acetone and H_2 at the same temperature.

15.4 Conclusions

The experimental results showed that in all cases the presence of Au nanoparticles improved the sensitivity of the sensor significantly at low temperatures. In the case of Cu_xO there is no selectivity between CH_4 and CO, but the addition of Au nanoparticles although improves the sensitivity for both gases the sensitivity for CO is seven times bigger than that of CH_4. The experiments will be continued with lower concentrations of the analytes, investigating parameters such as the concentration and the type of noble metals.

Acknowledgements The authors would like to acknowledge the financial support from the Greek General Secretariat for Research and Technology and the European Commission, through the European Social Fund for Regional Development, NHRF 2007–2013 action 'Development of Research Centres – ΚΡΗΠΙΣ', project 447963 'New Multifunctional Nanostructured Materials and Devices–POLYNANO'.

References

1. Dai C-L, Liu M-C, Chen F-S, Wu C-C, Chang M-W (2007) A Nanowire WO3 humidity sensor integrated with micro-heater and inverting amplifier circuit on chip manufactured using CMOS-MEMS technique. Sens Actuators B 123:96
2. Pravdova V, Pravda M, Guilbault GG (2002) Role of chemometrics for electrochemical sensors. Anal Lett 35:2389
3. Oyabu T, Ohta Y, Kurobe T (1988) Intelligent' alarm detector of gas leaks. Sens Actuators 13:323

4. Fasaki I, Suchea M, Mousdis G, Kiriakidis G, Kompitsas M (2009) The effect of Au and Pt nanoclusters on the structural and hydrogen sensing properties of SnO_2 thin films. Thin Solid Films 518:1109
5. Oyabu T (1991) A simple type of fire and gas leak prevention system using tin oxide gas sensors. Sens Actuators B 5:221
6. Xing R, Xu L, Song J, Zhou C, Li Q, Liu D, Wei Song H (2015) Preparation and gas sensing properties of In_2O_3/Au nanorods for detection of volatile organic compounds in exhaled breath. Sci Rep 5:10717
7. Ghasemi-Varnamkhasti M, Mohtasebi SS, Siadat M, Lozano J, Ahmadi H, Razavi SH, Dicko A (2011) Aging fingerprint characterization of beer using electronic nose. Sens Actuators B 159:51
8. Seetha M, Mangalaraj D (2012) Nano-porous indium oxide transistor sensor for the detection of ethanol vapours at room temperature. Appl Phys A 106:137
9. Comini E, Faglia G, Sberveglieri G (2009) Solid state gas sensing. Springer, New York
10. Henrich VE, Cox PA (1994) The surface science of metal oxides. Cambridge University Press, Cambridge
11. Williams DE (1999) Semiconducting oxides as gas-sensitive resistors. Sens Actuators B **57**, 1
12. Barsan N, Weimar U (2001) Conduction model of metal oxide gas sensors. J Electroceram 7:143
13. Kolmakov A, Klenov DO, Lilach Y, Stemmer S, Moskovits M (2005) Enhanced gas sensing by individual SnO_2 nanowires and nanobelts functionalized with Pd catalyst particles. Nano Lett 5:667
14. Tsu K, Boudart M (1961) Recombination of atoms at the surface of thermocouple probes. Can J Chem 39:1239
15. Henry CR, Chapon C, Duriez C (1991) Precursor state in the chemisorption of CO on supported palladium clusters. J Chem Phys 95:700
16. Stamataki M, Mylonas D, Tsamakis D, Kompitsas M, Tsakiridis P,Christoforou E (2013) CO-sensing properties of CuxO-based nanostructured thin films grown by reactive pulsed laser deposition. Sens Lett 11:1964

Chapter 16
Iono-Electronic Interface Based on Innovative Low Temperature Zeolite Coated NMOS (Circuits) for Bio-nanosensor Manufacture

A.S. Fiorillo, S.A. Pullano, R. Tiriolo, and J.D. Vinko

Abstract Nanoporous thin layer of Zeolite was deposited on Silicon wafers by using a technique fully compatible with standard integrated circuit manufacture. Differently from previous attempts of Zeolite layer growth or sieve micro-membrane fabrication in situ by calcination, the proposed technique is performed at low temperature by mixing the nanoporous material with selected vegetable oils, having high iodine value, and spin coating on different substrates, including also NMOS circuits. Various tests have been carried out on urea molecules and miRNA trapping, confirming the suitability of using Zeolite as iono-electronic interface in biochemical or biophysical to electrical energy transduction processes.

16.1 Introduction

The purpose of our work is to directly transduce biological signals into electrical signals by using a layer of nanoporous material, such as synthetic zeolite deposited on a MOSFET gate, in order to interface bio-systems to microelectronic circuits integrated on Si wafers. In this article, we describe methods, measurements, and results of synthetic zeolite spin-coating onto silicon wafers which include integrated field-effect transistors (FET), with the aim of investigating the feasibility of fabricating low noise iono-electronic microsystems in order to not only acquire and process in situ biochemical signals, but also to measure physico-chemical parameters involved in matter conversion.

One of the most important properties of zeolite is the capability to exchange ions and trap molecules, compatibly with dimensions of porosity [1, 3], what means in other words, to change its electrical state modifying the electric field,

A.S. Fiorillo (✉) • S.A. Pullano • R. Tiriolo
Department of Health Sciences, School of Biomedical Engineering, University Magna Grœcia of Catanzaro, Italy
e-mail: Nino@unicz.it

J.D. Vinko
HERA Physics, Velletri, Italy

J. Bonča, S. Kruchinin (eds.), *Nanomaterials for Security*, NATO Science for Peace and Security Series A: Chemistry and Biology, DOI 10.1007/978-94-017-7593-9_16

which modulates the FET channel, when deposited on the gate of an N channel metal–oxide–semiconductor (NMOS) circuit. On the basis of this assumption we investigated a simple and effective low temperature technique to deposit a thin layer of zeolite, compatible with standard Integrated Circuits (IC) technology. Zeolite powders, with different degree of porosity, were first spun at relatively low temperature on silicon wafer in order to fabricate the iono-electronic interface. Afterwards the thin nanoporous layer was activated both in O_2-plasma reactor [5] and under UV rays, so its chemical composition was verified with microanalysis. A total activation of the zeolite layer was shown by ATR-IR investigation after Plasma and UV activation of the samples. Finally, various attempt where performed in order to test its trapping capabilities with respect to water, urea, miRNA, and Rhodamine B molecules [14].

16.2 Materials and Methods

Zeolite (from Greek: "zeo" means to boil and "lithos" means stone) is a nanoporous crystalline material with excellent mechanical properties and a modulus of elasticity equal to 35 GPa. It has a low dielectric constant ($k = 2.7$), good heat conductivity, variable hydrophilicity/hydrophobicity, protein and enzyme supporting capacities [13, 17]. This nanoporous crystalline material contains exchangeable cations of group IA and IIA such as sodium, potassium, magnesium and calcium, expressed as

$$M_{2/n}O \cdot Al_2O_3 \cdot xSiO_2 \cdot yH_2O. \qquad (16.1)$$

In the basic formula (16.1), M is a metal cation of n valence, which compensates the excess of negative charge from the AlO_4 tetrahedra. Nanoporous are related to SiO_4 and AlO_4 tetrahedra, joined by shared oxygen atoms [18]. Thanks to its molecular structure it has been largely used as excellent molecular sieve [16] and catalyst in chemical reactions [6]. Porous dimensions range from 3 Å, in zeolite type 3A, up to more than 20 Å, in aluminophosphate as zeolite type AlPO.

16.2.1 Zeolite 3A

In our first experiment we used zeolite 3A, arising by LTA framework zeolite type that is able to absorb molecules smaller than 3 Å. This is a potassium form of alumino-silicate which structure is shown in Fig. 16.1.

Its chemical formula is

$$K_{12}[(AlO_2)_{12}(SiO_2)_{12}] \cdot 27H_2O. \qquad (16.2)$$

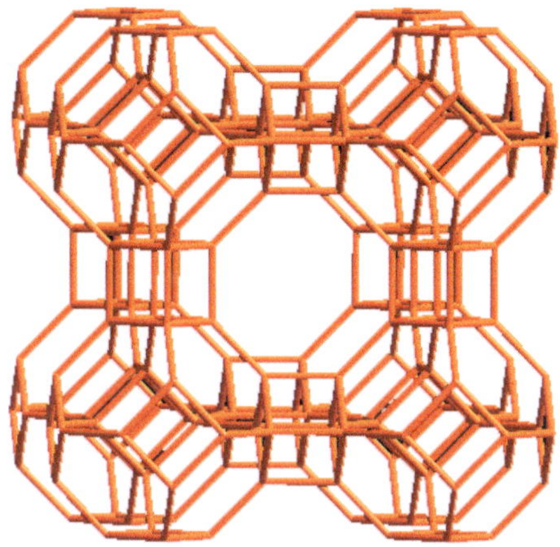

Fig. 16.1 Zeolite 3A LTA framework type view along [100] axis

Table 16.1 Physico-chemical parameters of vegetable oils

Oil	Density (kg/m^3)	Viscosity 35 °C (cSt)	Smoke point (°C)	Flash point (°C)	Iodine value	References
Sunflower	914	31	232	274	145	[15]
Soybean	914	33	232	254	143	[15]
Corn	910	35	232	277	128	[15]
Peanut	903	40	232	271	100	[15]
Castor	955	251	200	–	86	[15]

The zeolite powder was deposited on a silicon wafer by spin coating process. Differently from usual zeolite suspension [20], it was mixed with different vegetable oil in different percentages w/w (grain dimensions ranged from several hundred nm to 3 μm approximately, with a porosity of 2/3 Å). Being formed before mixing with oils, no calcination process is necessary (standard temperature for calcination process exceeds 450 °C, which is not compatible with integrated circuits). Zeolite 3A was mixed respectively with sunflower, soybean, corn, peanut, and castor oil which physico-chemical parameters as density, viscosity, smoke point, iodine value etc. are reported in Table 16.1. Best mixture in terms of spinning process complexity and repeatability was 50 % w/w, 10 g of powder plus 10 g of vegetable oil.

Experiments were carried out on 2-in (1 1 1) p-type silicon wafers, with a resistivity of 20 Ω·cm. In order to eliminate superficial impurities and electrical charges, which could be trapped in the nanoporous layer modifying the electrical state at the interface, the silicon substrates were shaken with isopropyl alcohol for 15 s, then dried under a stream of N_2 gas, rinsed in deionized water, and, again, dried under N_2 gas.

After cleaning, the mixtures of zeolite 3A and the various oils (a mole ratio of 40–60 zeolite 3A and 60–40 oil, respectively) were spun onto the silicon substrate for different time (30–60 s) and at different speeds (3000–5000 rpm/min), and then heated, for 16 h, in a furnace at 120 °C, a temperature which is normally well supported by integrated circuits with medium-high performances. The samples were then left in the furnace cooling to room temperature. The so cooked oil exhibits a supporting matrix for the nanoporous grains in contact with the silicon substrate. A lower oil viscosity allows the deposition of more uniform and thinner layers, while higher iodine values allow the best crosslinking and favor the curing process of the zeolite mixture at lower temperatures. Table 16.1 shows the most important physico-chemical parameters of tested vegetable oils.

Best results in creating a thinner layer at the lowest curing temperature were obtained by mixing zeolite 3A with sunflower oil or soybean oil, which have the higher iodine value and the lower viscosity.

16.2.2 Zeolite 13X

In the second experiment we used zeolite 13X (courteously furnished, as well as zeolite 3A, by the UOP-Honeywell plant in Reggio Calabria, Italy) which porosity dimension is more than 4 Å, more than that of zeolite 3A, and the ratio Si/Al ranges between 1 and 1.5 (while is equal to 1 in zeolite 3A). Its chemical formula is given in (16.3)

$$(Na_{86}[(AlO_2)_{86}(SiO_2)_{106}] \cdot 264H_2O). \tag{16.3}$$

and its molecular structure, FAU framework type class, is shown in Fig. 16.2.

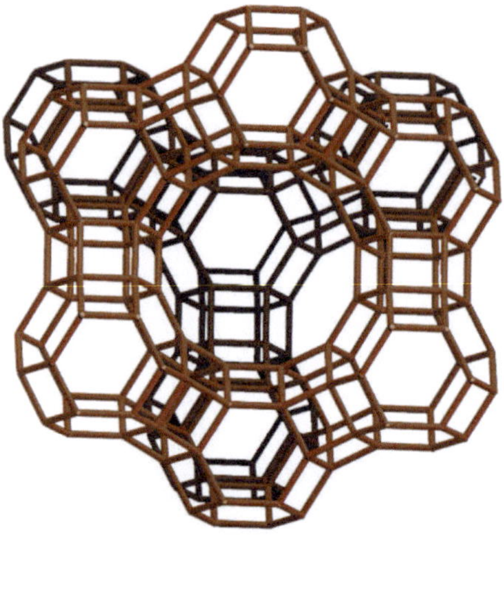

Fig. 16.2 Zeolite 13X FAU framework type view along [111] axis

Nanoporous layers on silicon wafers were deposited according to the same procedure as for zeolite 3A previously described, and best results were related to oils with higher iodine value and the lower viscosity, sunflower oil and soybean oil also in this experiment as well.

The final layer of zeolite, both 13X or 3A, on the silicon substrate exhibits excellent mechanical strength, good degree of adhesion, and good resistance to scratches. Reducing the curing temperature relatively to the degree of cross-linking and the number of double bonds of the selected oil, makes the process fully compatible with integrated circuit technology. Microanalysis performed shown that the spectrum of the chemical composition, was in accordance with that of standard zeolite 3A or 13X except for the presence of carbon atoms produced during the heating process of oils. Thicknesses, measured with a profilometer VEECO model "Dektak 6M", are in the order of 5–8 μm with a mean roughness of about or less than 1 μm [7].

16.3 The Iono-Electronic Interface

The same procedure was repeated on top of the gate of a silicon integrated NMOS-FET circuits. Although we don't have yet an evidence of the direct acquisition of signals originated by biochemical species entrapped in the zeolite framework, through on silicon microcircuits coupled with the iono-electronic interface, we believe that the drain current of MOSFET amplifiers can be modulated by the change of the electrical state of the nanoporous interface as a consequence of the entrapped species. In other words, the electrical field $\bar{\mathscr{E}}$ acting in the direction orthogonal to the gate of the MOS is modified according to bioenergy transduction phenomena occurring in the zeolite interface, as shown in Fig. 16.3. After spin-deposition the zeolite layer was activated (removal of organic templates and

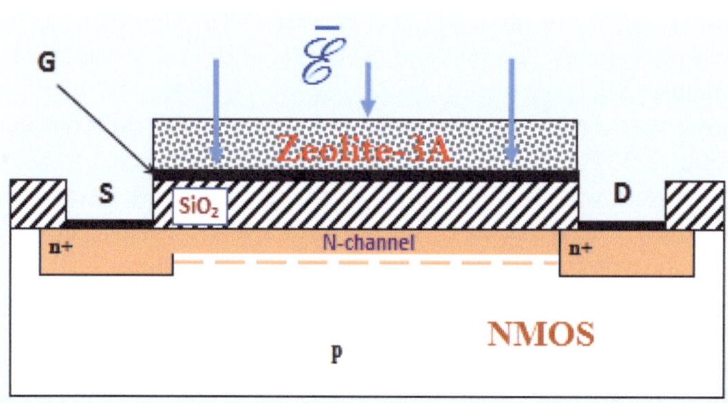

Fig. 16.3 Cross section of the proposed NMOS circuit coupled with the zeolite layer on the gate electrode

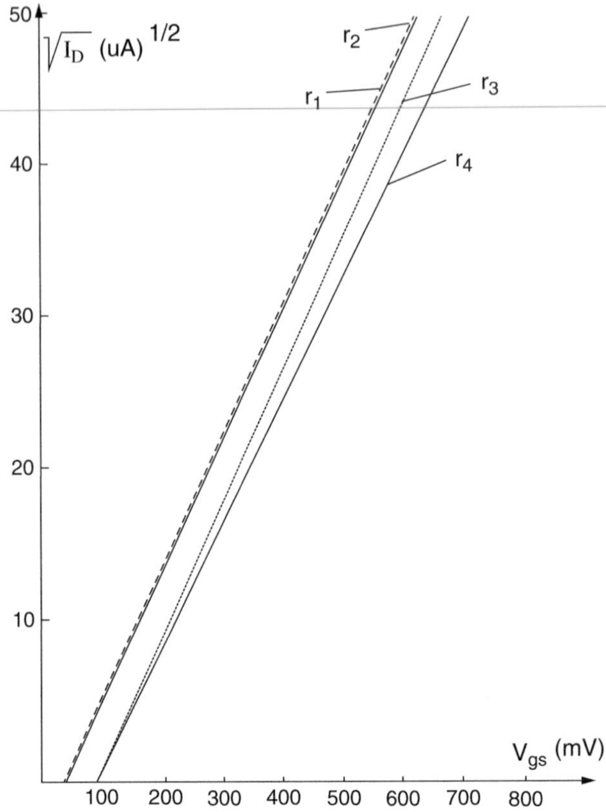

Fig. 16.4 I-V characteristics of the NMOS experiment vs Gate-source voltage after annealing and activation steps of the zeolite layer at 120 °C for two different samples; continuous line before and *dotted line* after low temperature curing process

molecules) at 120 °C in an O_2-plasma reactor (STS Multiplex, an inductively coupled plasma system), at a pressure of 350 Pa, after evacuation in a vacuum for several minutes [20].

Activation was verified by hygrometric characterization (in a dry environment at a pressure of 13 Pa for 10 h, first, and subsequently exposed to air with 75 % humidity to absorb water molecules). It results in an increase in the weight of the zeolite layer (16 % total zeolite measured with a micro balance PCEABZ 200C) according to the material specification provided by the producer [5].

As shown in Fig. 16.4, the annealing and activation steps of the nanoporous layer deposited on top of the gate electrode, carried out at 120 °C, doesn't modify the electrical characteristic of the NMOS, particularly the threshold voltage V_t. In order to increase the surface of the iono-elctronic interface, an extended gatewas

Fig. 16.5 Design of the extended gate MOSFET device with thin zeolite layer deposited on the top and its cross section

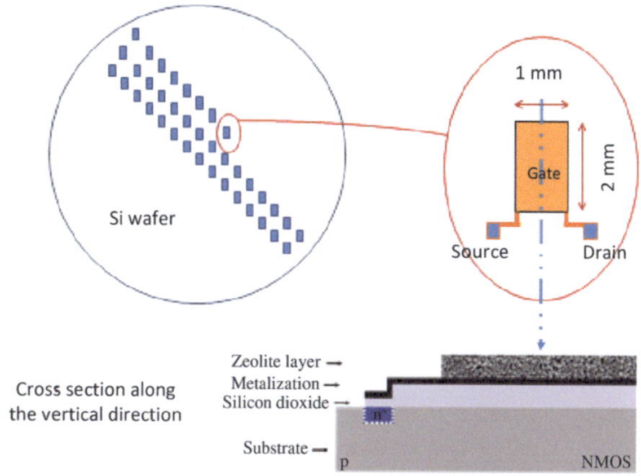

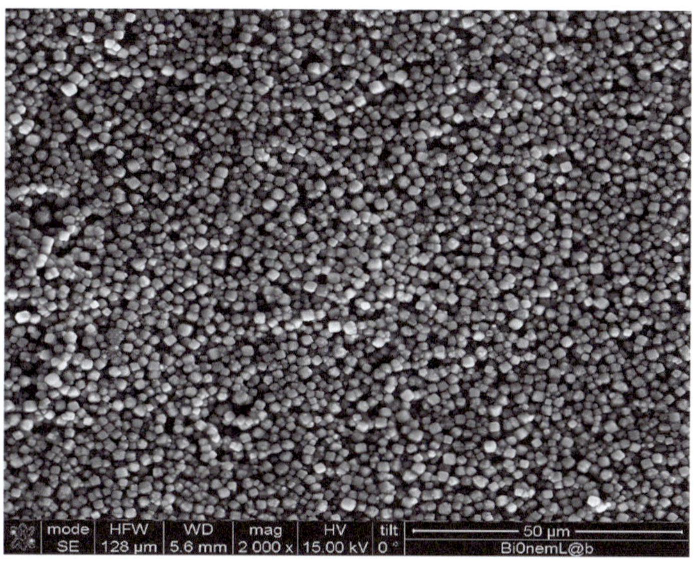

Fig. 16.6 Zeolite 3A grains spun onto the surface of the p-silicon wafer

used [10]. The cross section of the circuit along the w (channel width) is shown in Fig. 16.5.

In Fig. 16.5 a picture of the interface is shown including the top view of the zeolite layer spun onto the extended gate (1 × 2 mm) of the NMOS transistor. Low temperature deposition of thin zeolite 3A and 13X layers carried out for the characterization of biomolecules absorption are shown in Figs. 16.6, 16.7, 16.8 and 16.9.

Fig. 16.7 Magnification of the Zeolite grains, spun onto the surface of the p-silicon wafer, obtained at 15 kV, with a current of 0.15 nA

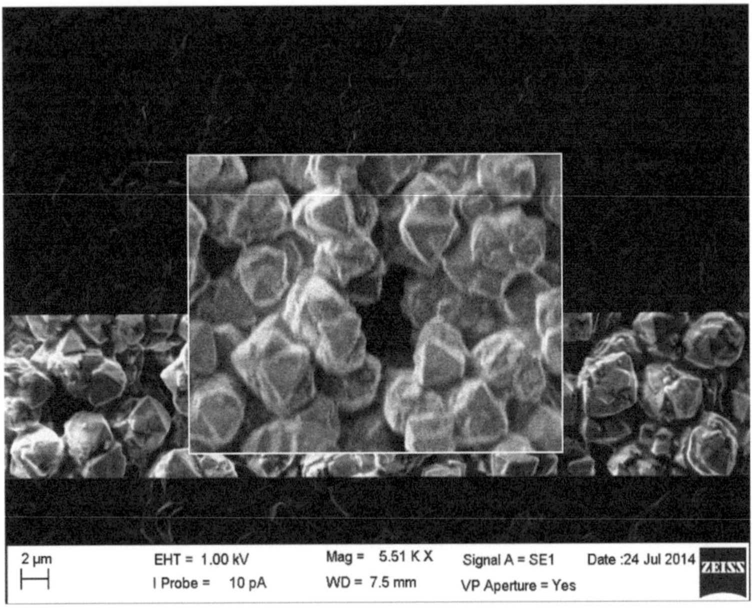

Fig. 16.8 Zeolite 13X grains spun onto the surface of the p-silicon wafer

Fig. 16.9 Magnification of the Zeolite grains, spun onto the surface of the p-silicon wafer, obtained at 10 kV, with a current of 0.10 pA

Table 16.2 Urea molecule characteristics

Form	Powder
Chemical formula	CH_4N_2O
Molecular weight	60.06 g/mol
Solubility	1 g per 10 ml of water

16.4 Biomolecules Absorption and Interface Characterization

The absorption of different biomolecules as urea and miRNA, and their interaction with the zeolite framework was investigated using different techniques, based mainly on FT-IR [2, 8, 12].

16.4.1 Urea Molecules Trapping and Characterization

Urea FT-IR spectroscopy was performed with an infrared spectrometer NICOLET 6700, Thermo Scientific. Urea (purchased from Sigma Aldrich) main characteristics are reported in Table 16.2.

The urea solution (2 g powder plus 20 ml of deionized water) was placed on the surface of the 11 μm zeolite layer. After the urea deposition, the sample was washed

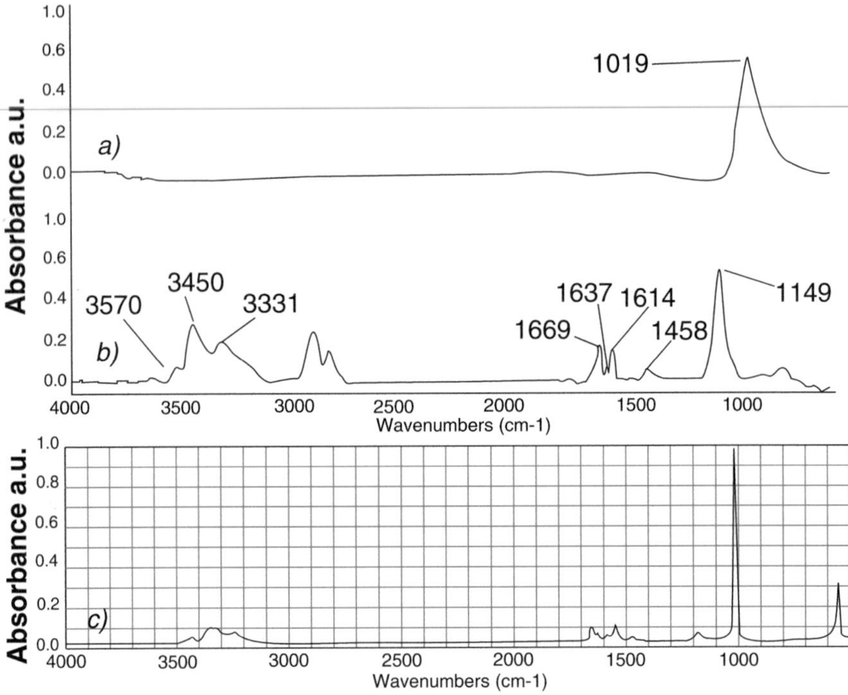

Fig. 16.10 Comparison between IR spectra of pure zeolite (**a**), deposited urea on zeolite layer (**b**) and Sigma Aldrich urea (**c**)

with isopropyl alcohol and then dried with N_2, in order to remove the excess of urea from the sample. In order to verify the entrapment of Urea molecules into zeolite pores FT-IR analysis were performed on the zeolite coating the silicon wafer. FT-IR spectrum shows the characteristic functional groups and vibration frequencies of urea molecules and their interaction with the zeolite framework. In Fig. 16.10 that spectrum was compared with the standard urea spectrum [15] and the pure zeolite [9]. The zeolite spectrum shows a strong absorption centered at 1019 cm^{-1} that represents at least five vibrational modes of asymmetric stretching of the orthosilicate group in accordance with a previous paper already reported in literature [11]. A comparison between the two urea spectra highlights a significant difference in two peaks around 3000 and 2880 cm^{-1}, which correspond to the soybean oil of the deposited mixture. In particular they refer to the symmetric CH2 stretching and the asymmetric CH_3 and CH_2 stretching [4]. The Urea spectra show a band characterized by three peaks in the higher frequency regions, a main peak centered at 3450 cm^{-1} with a shoulder component at 3570 cm^{-1} and a sharp component at 3331 cm^{-1}.

The higher frequency peaks correspond respectively to the h-bonded and free asymmetric stretching modes of the NH_2 group, while the lower frequency peak corresponds to its symmetric stretching mode. Two bands of medium intensity are observable, the first at 1669 cm^{-1} which represents the C=O stretching mode and the second with two peaks, at 1637 and 1614 cm^{-1}, representing respectively the NH_2 in-plane and out-plane bending mode. Another band of medium intensity appears at 1459 cm^{-1} which is attributed to the asymmetric stretching mode of N–C–N group of urea. The peak at 1149 cm^{-1} represents the NH_2 rocking mode while the low intensity shoulders at lower frequencies correspond to the symmetric stretching of the N–C–N group [4].

16.4.2 Small Nucleic Acids Trapping and Characterization

The interaction between small nucleic acids and synthetic nanoporous zeolite with the aim of fabricating biosensors for cancer detection has been investigated. Several studies underline the fact that circulating small nucleic acids, such as miRNAs, are directly involved in many physiological and pathological processes, particularly in processes of cancer initiation and progression [19].

The absorption of single-strand DNA molecules was studied through FT-IR spectroscopy. In the experiment, 5 μl of solution of single-strand DNA, characterized by 22 nucleotides (100 pmol/μl)

$$\text{TCTCACTCCCTTCCCAGATA} \tag{16.4}$$

was deposited through a tube onto the zeolite layer, washed in deionized water and then examined by IR spectroscopy. Infrared analysis reveals that characteristic infrared oligonucleotides peaks such as 1160 and 832 cm^{-1} appear in the zeolite infrared spectrum after their deposition. The absorption site peaks of the zeolite framework are observed to be shifted in frequency of about 10 cm^{-1} due to the electrostatic- and Van Der Waals-interaction between the zeolite pores and the absorbed small nucleic acids as shown in the IR spectrum of Fig. 16.11.

The nanoporous characteristics of zeolite coupled with silicon wafer the deposition process of a thin-layer compatible with IC technology stand out as a promising, non-invasive device integrated onto a silicon wafer for use in small-scale biomolecule detection and could be implemented in biomedical engineering as an early diagnostic tool.

16.4.3 Zeolite 13X for Rhodamine B Entrapping

During the development of new devices such as silicon integrated sensors but even more specifically, drug-delivery systems, it was demonstrated that when zeolite structures are used as carriers encapsulating different anticancer agents, their basal potency is increased due to the cumulative effect of drug compound and carrier.

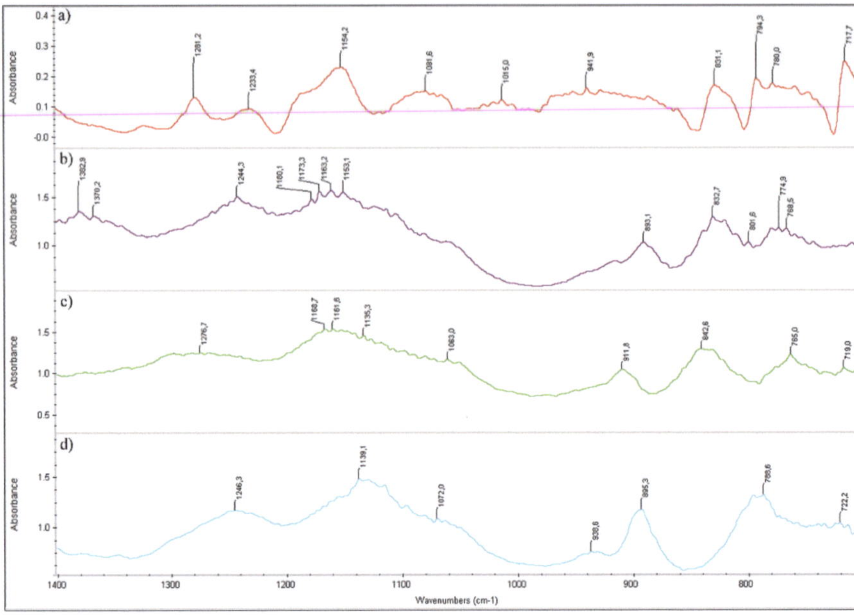

Fig. 16.11 Comparison between IR spectra of zeolite layer deposited on silicon wafer (**a**), and small nucleic acids after deposition on zeolite layer (**b**, **c**). Pure oligonucleotides spectra deposited on silicon wafer (**d**)

Rhodamine molecules were loaded into zeolite pores, following an experimental procedure present in literature [19]. It is a fluorescent material that emits electromagnetic waves in the wavelength of 544 nm. Referring to our integrated device we believe that fluorescence can also be detected by NMOS circuits The conjugation and absorption of Rhodamine B molecules into zeolite pores were investigated by FT-IR spectroscopy. The zeolite and Rhodamine B powder were individually investigated to compare their spectra with that of conjugated samples. The results showed that some Rhodamine characteristic peak appears in the zeolite absorption region infrared spectra, as showed in Fig. 16.12.

Referring to previous experiments on zeolite carriers for anticancer agents, we demonstrated the existence of a peculiar bimodal effect of low-dosages of zeolite nanoparticles that could be useful in designing drug-delivery systems. In fact, taking advantage of the nuclear localization of the zeolite nanoparticles and the fact that at low dosages there is relatively little genotoxic damage, the nanoparticles can be considered good carriers for the improvement of pharmacodynamics and the kinetic of cytotoxic agents which act at the nuclear level.

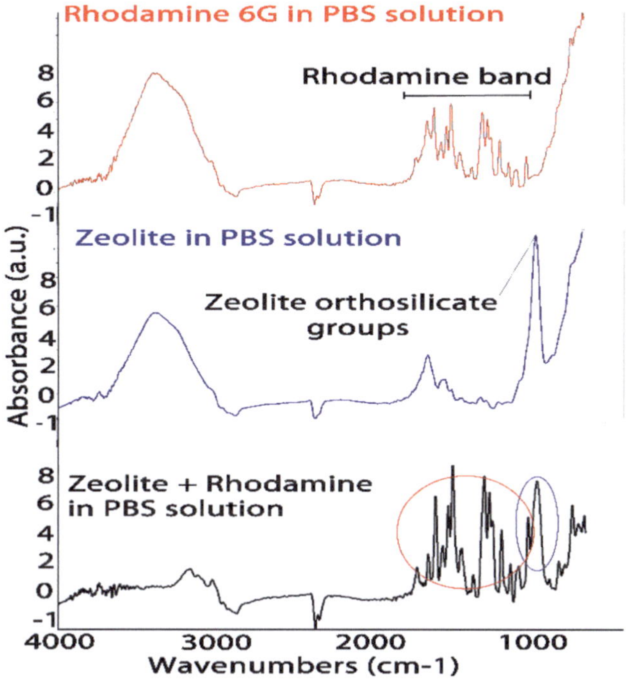

Fig. 16.12 Infrared spectra of Rhodamine (*red*), of PZNs (*blue*) and of a solution of Rho-PZN conjugation (*black*)

16.5 Conclusion

In this paper a method for the deposition of thin zeolite layer has been presented with emphasis on zeolite 3A and 13X. Different mixtures were chosen among those having the necessarily higher number of double bonds of the selected oil in order to reduce the curing temperature. Best results were obtained by mixing zeolite with soybean oil in a repeatable process performed at 120 °C. Absorption analysis were performed on the thin layer of zeolite following urea, miRNA and Rhodamine solution deposition by means of by the FT-IR spectroscopy. In all cases the nanoporous layer showed the capacity to easily entrap biological molecules in a way similar to pure zeolite powder. NMOS characterization allow us to establish a process, fully compatible with IC technology aimed at fabricating integrated sensors on silicon wafers capable of interacting with biomolecules.

References

1. Barrer RM (1982) Hydrothermal chemistry of zeolites. Academic, London, p 152
2. Bottoni U, Tiriolo R, Pullano SA, Dastoli S, Amoruso GF, Nistico SP, Fiorillo AS (2016) Infrared saliva analysis of psoriatic and diabetic patients: similarities in protein components. IEEE Trans Biom Eng 63(2):379–384

3. Breck DW (1984) Zeolite molecular sieve. Krieger, Malabar
4. Cortez MA, Bueso-Ramos C, Ferdin J, Lopez-Berestein G, Sood AK, Calin GA (2011) MicroRNAs in body fluids-the mix of hormones and biomarkers. Nat Rev Clin Oncol 8:467–477
5. El Roz M, Lakiss L, Valtchev V, Mintova S, Thibault-Starzyk F (2012) Cold plasma as environmentally benign approach for activation of zeolite nanocrystals. Microporous Mesoporous Mater 158:148–154
6. El Roz M, Lakiss L, Valtchev V, Mintova S, Thibault-Starzyk F (2012) Cold plasma as environmentally benign approach for activation of zeolite nanocrystals. Microporous Mesoporous Mater 158:148–154
7. Fiorillo AS (2009) Deposition of layers of porous materials, layers thus obtained and devices containing them, EP2132768 A2, 16/12/2009
8. Fiorillo AS (2012) Deposition of zeolite thin layers onto silicon wafers for biomedical use. IEEE Trans Nanotechnol **11**, 654–656
9. Fiorillo AS, Tiriolo R, Pullano SA (2015) Absorption of urea into zeolite layer integrated with microelectronic circuits. IEEE Trans Nanotechnol 14(2):214–217
10. http://www.sigmaaldrich.com/spectra/rair/RAIR014477.PDF
11. Kuligowski J, Quintás G, Garrigues S, de la Guardia M (2008) Determination of lecithin and soybean oil in dietary supplements using partial least squares-Fourier transform infrared spectroscopy. Talanta 77(1):229–234
12. Larrubia MA, Ramis G, Busca G (2000) An FT-IR study of the absorption of urea and ammonia over V2O5-MoO3-TiO2 SCR catalysts. Appl Catal B: Environ 27:145–151
13. Lee GS, Lee YJ, Choi SY, Park YS, Yoon KB (2000) Self-assembly of beta-glucosidase and D-glucose-tethering zeolite crystals into fibrous aggregates. J Am Chem Soc 122:12151–12157
14. Malara NM, Tiriolo R, Trunzo V, Pullano SA, Dattola E, Failla P, di Fabrizio E, Mollace V, Fiorillo AS (2014) Low-dosages of zeolite 13x produce a bimodal effect on breast cancer cells: geno-oxidative damage and reduction of lipid peroxidation. In: IEEE EMBS – micro and nanotechnology in medicine conference, Hawaii
15. Montanari T, Busca G (2008) On the mechanism of absorption and separation of CO2 on LTA zeolites: an IR investigation. Vib Spectrosc 46:45–51
16. Rhodes CJ (2010) Properties and applications of zeolites. Sci Prog 93:223–284
17. Um SH, Lee GS, Lee YJ, Koo KK, Lee C, Yoon KB (2002) Self-assembly of avidin and D-biotin-tethering zeolite microcrystals into fibrous aggregates. Langmuir 18:4455–4459
18. Universal Oil Products (2005) An introduction to zeolite molecular sieves. UOP, Des Plaines, [Online]. Available: www.eltrex.pl/pdf/karty/adsorbenty/ENG-Introduction%20to%20Zeolite%20Molecular%20Sieves.pdf
19. Vilaca N, Amorim R, Machado AF, Parpot P, Pereira MF, Sardo M, Rocha J, Fonseca AM, Neves IC, Baltazar F (2013) Potentiation of 5-fluorouracil encapsulated in zeolites as drug delivery systems for in vitro models of colorectal carcinoma. Colloids Surf B: Biointerfaces 112:337–244
20. Wang Z, Mitra A, Wang H, Huang L, Yan Y (2001) Pure silica zeolite films as low-k dielectrics by spin-on of nanoparticle suspension. Adv Mater 13:1463–1466

Chapter 17
SQUID Detectors for Non-destructive Evaluation in Industry

W. Nawrocki

Abstract This review paper describes DC-SQUID detectors and SQUID measurement systems used for non-destructive evaluation of materials. Extremely sensitive detectors of magnetic flux, SQUIDs can be used as preamplifiers in setups for measuring many physical quantities. DC-SQUIDs have better energy resolution ε and lower noise level Φ_r than RF-SQUIDs. Both parameters are discussed in this paper. The best DC-SQUIDs have an energy resolution as good as $0.5h$ (where h is Planck's constant). SQUID measurement systems are used not only for material testing and inspection in industry but also in biomagnetic studies (especially in medicine), in geology, military tasks, thermometry and many other fields.

17.1 Introduction

Superconducting quantum interference devices (SQUIDs) are extremely sensitive detectors of magnetic flux. They can be used as preamplifiers in setups for measuring a variety of physical quantities that can be converted to magnetic flux, e.g., magnetic induction, magnetic field gradient, magnetic susceptibility, voltage, electric current or temperature. The operation of a SQUID is based on two physical phenomena: the Josephson effect (Josephson tunneling) and magnetic flux quantization in a superconducting magnetic loop. There are two kinds of SQUIDs: the direct current-biased DC-SQUID, and the RF-SQUID biased by a radio-frequency signal of a frequency between 100 kHz and 3 GHz. The DC-SQUID has two Josephson junctions (JJ) connected in a superconducting ring. The RF-SQUID contains one JJ in a ring.

SQUIDs are made of superconducting materials with a low critical temperature (LTc), typically niobium, or of high critical temperature (HTc) superconductors, usually CuO-based ceramics (e.g. $YBaCu_3O_7$). LTc SQUIDs are cooled by liquid helium at 4.2 K, and HTc SQUIDs by liquid nitrogen at 77 K.

W. Nawrocki (✉)
Faculty of Electronics and Telecommunications, Poznan University of Technology, ul. Piotrowo 3, 60-965 Poznan, Poland
e-mail: nawrocki@et.put.poznan.pl

J. Bonča, S. Kruchinin (eds.), *Nanomaterials for Security*, NATO Science for Peace and Security Series A: Chemistry and Biology, DOI 10.1007/978-94-017-7593-9_17

DC-SQUIDs have better energy resolution ε and lower noise level Φ_r than RF-SQUIDs. Therefore only DC-SQUIDs and measurement systems based on them are discussed in this paper.

17.2 Operation of DC-SQUID Detector and Measurement Setup

Let us consider a DC-SQUID (Fig. 17.1, right) with two identical Josephson junctions with a critical current I_{crJ}. An input current I flows though the detector. The critical current I_c of the DC-SQUID, or the maximum current that ensures the superconductivity of the junctions, is described by the formula (17.1) [1].

$$I_c = 2I_{crJ}\left|\cos\pi\left(n - \frac{\Phi_e}{\Phi_0}\right)\right|. \tag{17.1}$$

where I_c is the critical current of the DC-SQUID, I_{crJ} is the critical current of each Josephson junction, Φ_0 is the magnetic flux quantum: $\Phi_0 = e/2h \cong 2.07\times10^{-15}$ Vs (e denotes the charge of the electron and h is Planck's constant), Φ_e is the external magnetic flux applied to the detector, and n is a natural number: $n = 0, 1, 2, \ldots$.

The general dependence that does not require the fulfillment of this condition is plotted in Fig. 17.2a. Figure 17.2b shows the critical current of a DC-SQUID versus the applied external magnetic flux for $LI_{crJ} \ll \Phi_0$, where L is the self-inductance of the DC-SQUID.

Figure 17.3 shows the current-voltage (I–V) characteristic of a DC-SQUID, contained between the curve starting from $I = I_{c\,\max}$, for the flux $\Phi_e = n\Phi_0$, and the curve starting from $I = I_{c\,\min}$, corresponding to $\Phi_e = (n + 1/2)\Phi_0$.

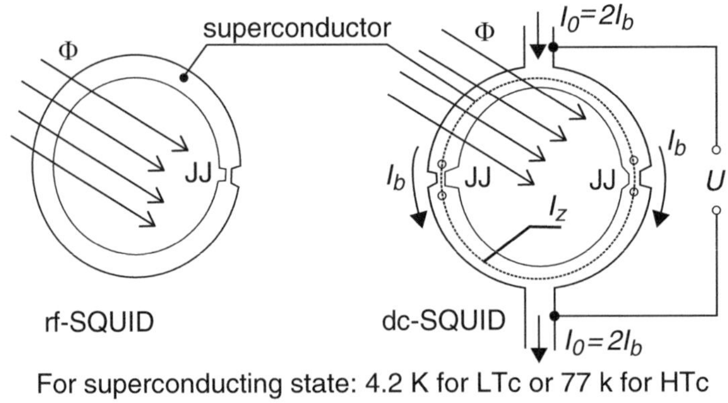

Fig. 17.1 Schematic of an RF-SQUID and a DC-SQUID

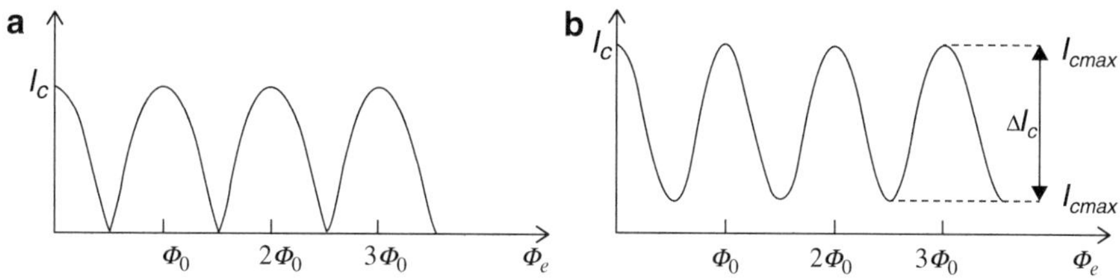

Fig. 17.2 Critical current of a DC-SQUID versus applied magnetic flux: (**a**) general case; (**b**) for the case when $LI_{crJ} \ll \Phi_0$

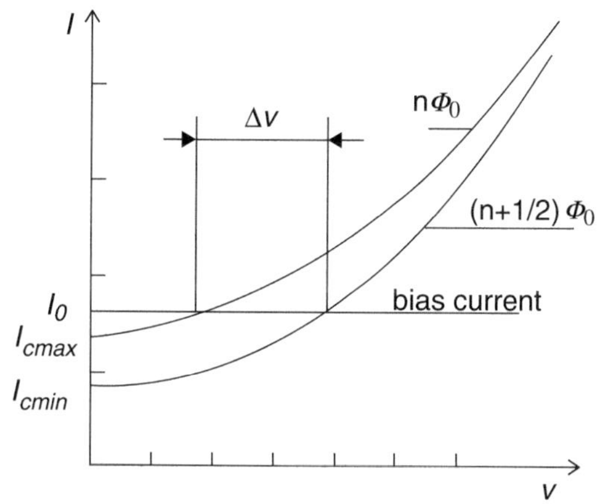

Fig. 17.3 Current-voltage characteristic of a DC-SQUID

The bias current I_0 of a DC-SQUID is a direct current exceeding the maximum critical current $I_{c\max}$. When the applied magnetic flux changes from $n\Phi_0$ to $(n + 1/2)\Phi_0$, i.e. by $0.5\Phi_0$, the voltage at the detector decreases by ΔV. In the currently manufactured DC-SQUIDs the value of ΔV ranges from 20 to 100 μV, which corresponds to a flux to voltage conversion coefficient

$$k_\Phi = \Delta V/0.5\Phi_0 = (20 \div 100)\,\mu\text{V}/(0.5 \times 2.07 \times 10^{-15}\,\text{Vs}) \cong (2 \div 10) \times 10^{10}\,\text{Hz}.$$

Figure 17.4 presents a block diagram of a measurement system with a DC-SQUID. Since the DC-SQUID is a detector of magnetic flux a measurement setup with a plain DC-SQUID (without input coil) will be a quantum magnetometer. The system depicted in Fig. 17.4 allows for high-sensitivity measurements of any physical quantity that can be converted to electric current in the input coil.

In real measurement setups SQUIDs are very carefully screened from the external magnetic field, since the Earth's magnetic field, as well as the field generated by electric devices such as electric motors, transformers or power lines, will interfere strongly with an unshielded SQUID (see Fig. 17.5) to make the measurement practically impossible. Interference is prevented by a superconducting shielding, typically of lead or niobium. Due to the perfect diamagnetism of superconductors

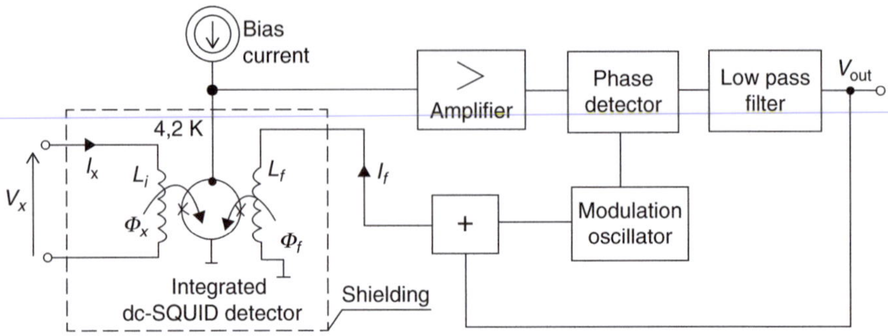

Fig. 17.4 DC-SQUID measurement system: DC-SQUID, bias current source, input coil L_i, 4 – modulation coil L_f, preamplifier, phase detector, low-pass filter, modulation signal oscillator

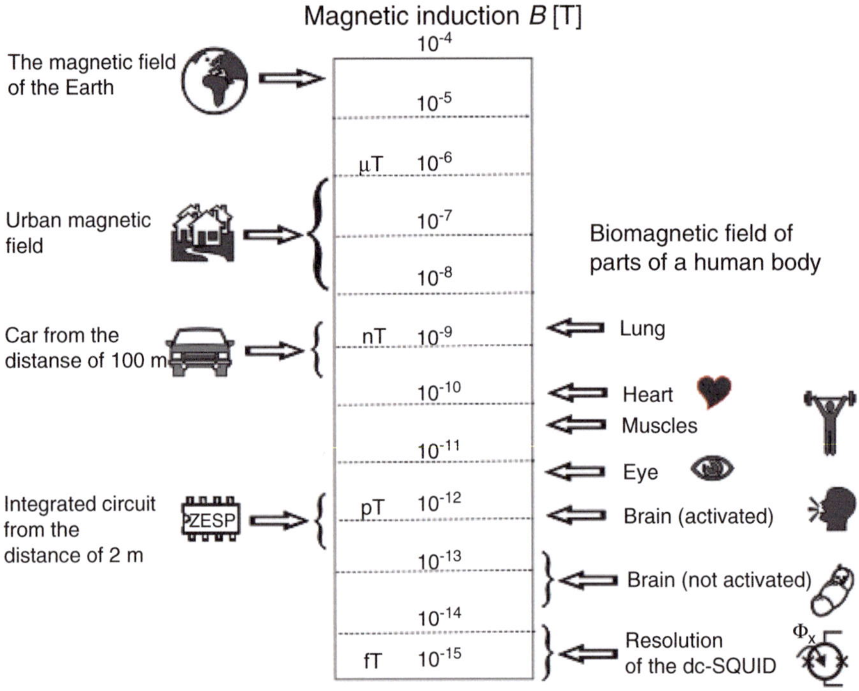

Fig. 17.5 Magnetic induction generated by the Earth, technical devices and parts of the human body

the external field is repulsed from the superconducting shielding. As a result the shielded detector is only exposed to the magnetic flux generated by the current I_x in the input coil.

Measuring the signal I_x at the terminals of the input coil, the system operates in a magnetic feedback loop with magnetic flux modulation [1]. The direct current source provides the bias current to the DC-SQUID. The total magnetic flux acting on the magnetic field sensor includes:

- the magnetic flux Φ_x from the input coil L_i to which the signal is brought,
- the negative feedback flux Φ_f induced by the modulation coil L_f and proportional to the modulated signal generated by the modulation signal generator.

The alternating voltage generated at the terminals of the detector is amplified by the preamplifier. After the phase detection in the phase detector the signal is filtered by the low-pass filter. The current in the modulation coil is proportional to the voltage V_{out} at the output of the low-pass filter. The measured magnetic flux Φ_x is automatically compensated by the feedback flux Φ_f in the measurement system. The greater the signal I_x in the input circuit, the greater the current I_f in the modulation coil generating the feedback flux. The voltage V_{out} at the output of the low-pass filter is the output voltage of the whole measurement system.

The system allows to measure the current I_x or any physical quantity convertible to current in the input coil. The most important parameters of a SQUID measurement setup include the noise level, the energy resolution, the frequency bandwidth of the transmitted signal and the slew rate. The total noise of a SQUID system is usually the noise level of the DC-SQUID, while the dynamic properties are determined by the electronic units of the feedback system.

17.3 Magnetic Interference and Its Reduction

Measurements of weak magnetic fields require the reduction of the interference caused by the magnetic field of the Earth or that generated by technical devices. Figure 17.5 shows the level of the magnetic induction generated by different sources against the biomagnetic field of different parts of the human body.

The magnetic field measured by a SQUID magnetometer is sensed by the detector indirectly via a superconducting magnetic flux transformer (see Fig. 17.6). In the measurement system shown in Fig. 17.6b the transformer is a first-order gradiometer. The measured field (magnetic flux Φ_{sig}) only acts on the first coil of the

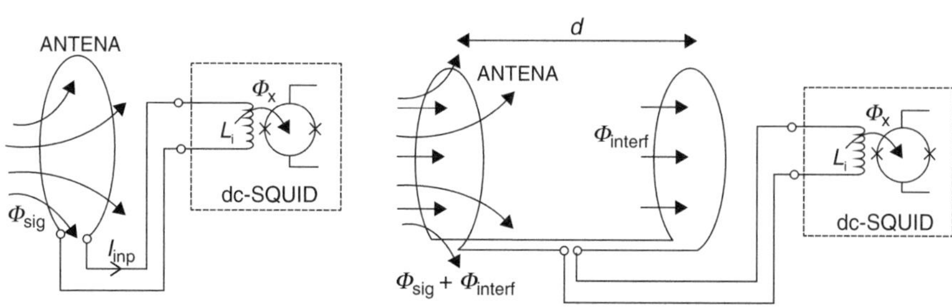

Fig. 17.6 An input part of a DC-SQUID measurement system: (**a**) a magnetometer; (**b**) a first-order gradiometer

gradiometer, while the interference (magnetic flux Φ_{interf}) acts on both coils. These are wound in opposite directions so that the currents induced in the coils of the gradiometer by magnetic interference have opposite directions and thus compensate each other.

Measurements of weak magnetic signals, e.g. biomagnetic measurements, with a SQUID setup are carried out in shielded rooms, usually with multilayer shields. Excellent shielding is provided by the shielded room in Physikalisch-Technische Bundesanstalt Berlin [2]. This 2.9 × 2.9 × 2.9 m room has an 8-layer shield that reduces the external electromagnetic interference with a coefficient of up to 170 dB (the reduction coefficient is a function of the frequency of the electromagnetic field).

17.4 Noise Level and Energy Resolution of DC-SQUIDs

The power spectral density S_v of the voltage white noise at the terminals of a DC-SQUID is described by the analytical formula found by Likharev [3]:

$$S_v = 8k_B T R_b \frac{1 + 0.5(I_C/I_0)^2}{1 + (I_C/I_0)^2}. \tag{17.2}$$

The power spectral density S_v of the voltage signal can be converted to the magnetic flux power spectral density Φ_r^2 in the equation (17.4) below. For low noise levels the optimal relation between the voltage change $\mathrm{d}V$ at the detector and the change $\mathrm{d}\Phi$ of the flux acting on the SQUID is [3]:

$$\frac{\mathrm{d}V}{\mathrm{d}\Phi} = \frac{R_b}{L}. \tag{17.3}$$

The magnetic flux power spectral density, or the noise level of the DC-SQUID, is expressed by the following formula obtained from (17.2) and (17.3):

$$\Phi_r^2 = \frac{S_v}{\left(\frac{\mathrm{d}V}{\mathrm{d}\Phi}\right)^2} = \frac{S_v}{k_\Phi^2} = \frac{8k_B T L^2}{R_b} \frac{1 + 0.5(I_C/I_0)^2}{1 + (I_C/I_0)^2}, \tag{17.4}$$

where Φ_r is the noise level of the DC-SQUID, ε is its energy resolution, L its self-inductance, I_C its critical current, I_0 its bias current, R_b its shunt resistance, and T is the temperature.

The magnetic noise Φ_r of a DC-SQUID is the thermal noise generated by shunt resistors R_b. A shunt resistor R_b is connected to each Josephson junction to prevent the hysteresis of its I–V characteristic. By the formula (17.2) obtained by Likharev [3] the noise level depends on the temperature, the parameters (inductance L and shunt resistance R_b) of the detector, and the bias current I_0. The noise level of the best DC-SQUIDs (LTc SQUIDs cooled by liquid helium) is of the order of $10^{-7}\Phi_0\,\mathrm{Hz}^{-1/2}$; commercial detectors have a noise level of the order of $10^{-6}\Phi_0\,\mathrm{Hz}^{-1/2}$. The noise level of HTc DC-SQUIDs is nearly 100 times higher.

The energy resolution of a SQUID detector is the lowest amount of energy it can detect. This parameter allows to compare the sensitivity of detectors of different physical quantities, e.g. the sensitivity of a DC-SQUID with that of a photodetector. The energy resolution ε of a DC-SQUID detector is described by the formula:

$$\varepsilon = \frac{\Phi_r^2}{2k_i^2} = \frac{4k_B T L^2}{k_i^2 R_b} \frac{1 + 0.5\frac{I_c^2}{I_0^2}}{1 + \frac{I_c^2}{I_0^2}} \tag{17.5}$$

The formula (17.5) implies conditions to be fulfilled for the DC-SQUID to have a good resolution. The operating temperature of the SQUID is often fixed. Typically, the operating temperature is 4.2 K (the temperature of liquid helium) for LTc SQUIDs and 77 K (the temperature of liquid nitrogen) for HTc SQUIDs. The parameters required for a good energy resolution (a low value of ε) of the detector are: a low self-inductance L, a high coefficient k_i of the feedback between the inductance L of the SQUID 'ring' and the inductance L_i of the input coil, and a high shunt resistance R_b of the Josephson junction. The best energy resolution achieved so far is $\varepsilon = 0.5h$ (where h is Planck's constant) in the DC-SQUID without the input coil [4]. The energy resolution is determined indirectly by measuring the noise level Φ_r and the current sensitivity $\Delta I_i/\Phi_0$ of the DC-SQUID:

$$\varepsilon = \frac{1}{2} L_i \Phi_r^2 \left(\frac{\Delta I_i}{\Phi_0}\right)^2. \tag{17.6}$$

17.5 Applications of SQUIDs for Non-destructive Evaluation of Materials

A non-destructive evaluation (NDE) means looking for flaws (defects) in structural parts of a mechanical construction. The second task of NDE are measurements of a stress in beams to be tested. Sensitive SQUID systems are used for non-destructive testing and inspection of materials if good performance and very high reliability are required. The NDE methods are used in particular for testing and inspection of sheets (and other construction parts) of airplanes and spacecrafts, and for testing of construction parts of nuclear power stations and oil platforms [5].

Other methods of NDE are: using a human sight when a man is looking for a visible defect, using an ultrasonic or X-ray inspection. We can note that X-rays do not penetrate hard materials such as steel. Ultrasonic waves reflect not only on the flaw but also on the separating layer of the multi-layer sheet. Therefore such methods are not right for testing most types of materials for mechanical constructions mentioned above. In contrast to the NDE – a destructive evaluation means applying to the structural part such a great force that this part would be destroyed.

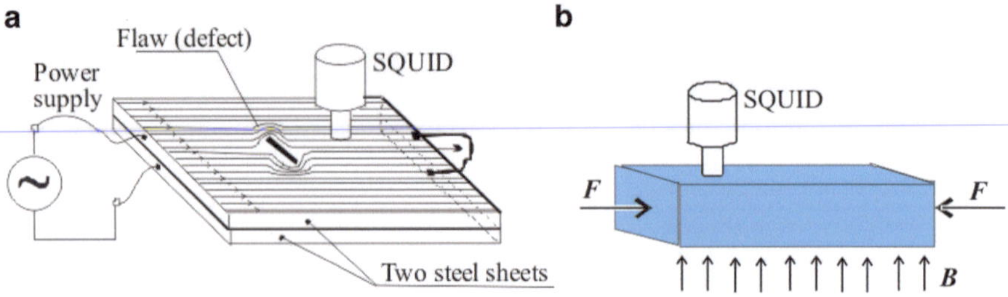

Fig. 17.7 Application of SQUID systems for NDE: (**a**) looking for inhomogeneous magnetic field in the multi-layer sheets; (**b**) measurements of mechanical stress in the construction beam

A setup for testing two-layer metallic sheets is shown in Fig. 17.7a (in general – for testing of multi-layer sheets). Current flow in a sheet without defects induces a homogeneous magnetic field over the entire surface of the plate. Scanning the magnetic field over the surface of the plate is carried out in order to detect places where the magnetic field is inhomogeneous. In such places, the material may be defective. This method can be applied for testing of sheets made of conducting materials, e.g. steel, titanium or copper plates.

Figure 17.7b illustrates a method for measuring strain in a tested beam. The magnetic field strength above-mentioned beam depends on the mechanical stress. In particular, using this method it is possible to detect the barrier between the stress causing elastic deformation and stress causing plastic deformation.

Advantages of non-destructive evaluation with SQUID are:

1. High sensitivity (B in order $10\div100\,\mathrm{fT/Hz^{1/2}}$),
2. Wide frequency band: from DC to about 10 kHz,
3. Quantitative evaluation.

High-temperature SQUID, operated at 77 K and cooled with liquid nitrogen, are used for NDE applications in industry. For measurements with high resolution (in x-y axis) of nanometers – nano-SQUID detectors can applied [6].

17.6 Other Applications of SQUIDs

Other fields of application of SQUIDs include [7, 8]:

- Medicine: magnetocardiography (MCG), magnetoencephalography (MEG) (see Fig. 17.8), magnetoneurography [9], magnetic markers, nuclear magnetic resonance (NMR) [10];
- Sensitive instruments for physics: NMR, noise thermometer [11];
- Tests of general relativity: the geodetic effect (gravity waves, curved spacetime due to the presence of the Earth); gyroscopes (a spinning niobium-coated sapphire sphere develops a magnetic dipole moment that can be sensed by a SQUID);

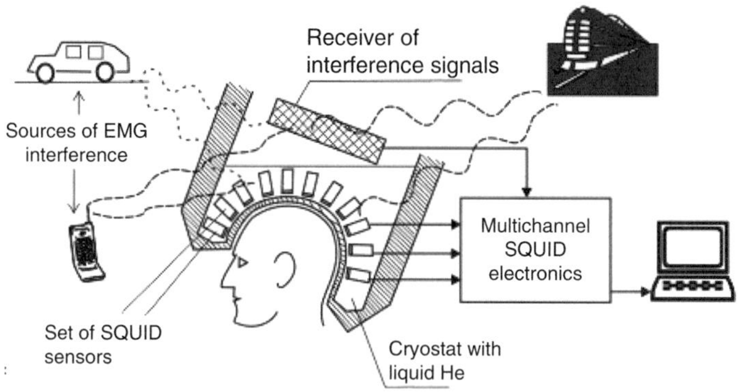

Fig. 17.8 Application of DC-SQUID measurement systems: magnetoencephalography

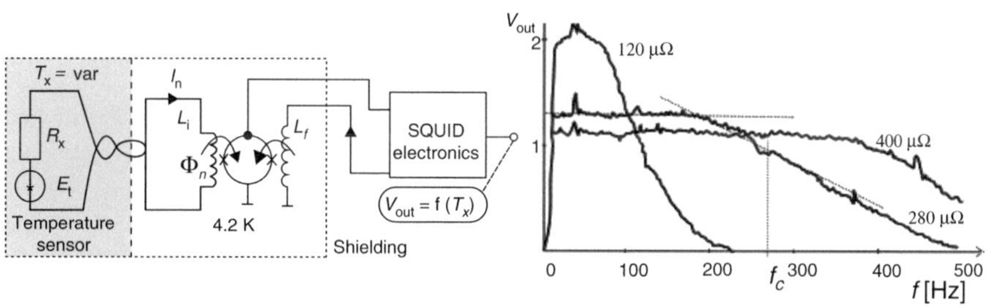

Fig. 17.9 Measurements of a thermal noise with a DC-SQUID: (**a**) the measurement system; (**b**) the thermal noise current measured by the system

- Geology: searching for mineral deposits [12];
- Military: monitoring of the coast in search of underwater bombs and submarines [13].

DC-SQUID detectors and measurement systems are the most widely used in biomedical studies. It is estimated that more than a half of the fabricated DC-SQUIDs are installed in magnetocardiographs and magnetoencephalographs (Fig. 17.7). DC-SQUIDs are also used for increasing the sensitivity of nuclear magnetic resonance (NMR) measurement systems used in medicine [6] and other fields. Another field of application of SQUID systems is non-destructive evaluation (NDE) of materials, mainly for testing sheets for aircraft and spacecraft siding. Complementary to ultrasonic testing, NDE tests performed with a SQUID system consist in measuring the distribution of the magnetic induction over the tested sheet (Fig. 17.8). If a defect occurs in the material the magnetic field is non-homogeneous, which allows the detection of the defect by a SQUID magnetometer.

Dc-SQUID measurement systems are also used for measuring weak electric signals. Figure 17.9a shows a DC-SQUID measurement setup used for measuring the resistor thermal noise current; the measurement data are presented in [14].

The mean square of the noise current is given by the Nyquist formula:

$$\hat{I}_x^2 = \frac{4k_B \Delta f T}{R}. \tag{17.7}$$

Figure 17.9b presents the thermal noise of 120 μΩ to 1 mΩ resistors over a bandwidth of $\Delta f = 100$ Hz measured at a temperature of 4.2 K. The noise level of the DC-SQUID system is 100 pA. The depicted measurement setup was subsequently used as a noise thermometer.

As a part of the cryogenic current comparator (CCC), the DC-SQUID is used for comparative measurements between the quantum Hall effect resistance and the resistance of the resistor under study [15].

17.7 Conclusions

SQUID detectors allow the measurement of weak electric and magnetic signals that cannot be measured by other methods. The high sensitivity of SQUIDs, however, implies the requirement of very effective shielding of the measurement system. SQUIDs are used in many fields, mainly in biomedical measurements. Still, the use of SQUID detectors and measurement systems is limited by their high price.

References

1. Clarke J, Goubau WM, Ketchen MB (1976) Tunnel junction dc-SQUID: fabrication, operation and performance. J Low Temp Phys 25:99
2. Burghoff M (2002) Proceedings of the 13th international conference on biomagnetism, Jena
3. Likharev K, Semenov V (1972) Fluctuation spectrum in superconducting point junctions. Pisma Zhurnal Eksperim i Teoret Fizyki 15:625
4. Ketchen MB (1992) Design and fabrication considerations for extending integrated dc-SQUIDs to the deep-submicron regime. In: Koch H, Lübbig H (eds) Superconducting devices and their applications. Springer, Berlin, p 256
5. Donaldson GD, Cochran A, McKirdy AD (1996) The use of SQUIDs for nondestructive evaluation. In: Weinstock H (ed) SQUID sensors. NATO ASI series. Kluwer, Dordrecht/Boston, p 599
6. Granata C, Vettoliere A (2015) Nano superconducting interference devices: a powerful tool for nanoscale investigations. ArXiv: http://arxiv.org/pdf/1505.06887
7. Nawrocki W (2016) Introduction to quantum metrology: quantum standards and instrumentation. Springer, Heidelberg, p 273
8. Kruchinin S, Nagao H (2012) Nanoscale superconductivity. Int J Mod Phys B 26:1230013
9. Andrä W, Nowak H (2006) Biomagnetic instrumentation. In: Magnetism in medicine, 2nd edn. Wiley-VCH, London
10. Chen H-H et al (2011) DC-SQUID based NMR detection from room temperature samples. J Appl Phys 110:093903
11. White DR et al (1996) The status of Johnson noise thermometry. Metrologia 33:325

12. Paik HJ (1996) Superconducting accelerometers, gravitational-wave transducers, and gravity gradiometers. In: Weinstock H (ed) SQUID sensors. NATO ASI series. Kluwer, Dordrecht, The Netherlands, p 569
13. Clen TR et al (1996) Superconducting magnetic gradiometers for mobile applications with an emphasis on ordnance detection. In: Weinstock H (ed) SQUID sensors. NATO ASI series. Kluwer, Dordrecht, The Netherlands, p 517
14. Nawrocki W, Berthel K-H, Döhler T, Koch H (1988) Measurements of thermal noise by dc-SQUID. Cryogenics 28:394
15. Rietveld G et al (2000) Accurate measurement of small currents using a CCC with DC-SQUIS. Sens Actuators A Phys 85:54

Chapter 18
Morphological Features of Nanostructured Sensor for X-Ray and Optical Imaging, Based on Nonideal Heterojunction

Ie. Brytavskyi, V. Smyntyna, and V. Borschak

Abstract A novel nanostructured sensor for X-ray and optical images obtaining is developed on the basis of nonideal heterojunction CdS-Cu_2S. Microscopy studies have been conducted to determine the optimum method of sensor manufacturing. The analysis of the original data set with microscopic image comparison of polycrystalline films on different technological parameters showed that the most homogeneous surface structure presents on samples, which CdS layer is obtained by vacuum thermal evaporation of CdS layer. This finding is consistent with results of current-voltage characteristics analysis, which showed the presence of high quality CdS-Cu_2S heterostructure, obtained by mentioned base layer forming methodics.

18.1 Introduction

The polycrystalline films containing Cu_2S-CdS heterostructure formed by dry deposition method [1] have been investigated in this work. The heterostructures were formed by employing a substitution technique where the nanolayer of copper sulfide is formed directly on the substrate layer of CdS during heat treatment using a pre-printed cuprous chloride film, formed by vacuum evaporation. This technology is promising due to low cost, easy processibility, and the possibility to fabricate large area films with satisfactory structural quality [2].

However, such a heterojunction consisting of p-type Cu_2S and n-type CdS is a rather complicated structure because of the interface between two materials with different electron affinities, band gaps, and polycrystalline structures. The lattice mismatch and interdiffusion of components might cause defects at or near the interface that strongly affect the junction properties [3].

Sensor based on nonideal heterojunction CdS-Cu_2S demonstrates the properties of internal amplification and signal accumulation effect [3–5], and substantially

Ie. Brytavskyi (✉) • V. Smyntyna • V. Borschak
Odessa I.I. Mechnikov National University, Dvorianska str. 2, 65028 Odessa, Ukraine
e-mail: brytav@ukr.net

J. Bonča, S. Kruchinin (eds.), *Nanomaterials for Security*, NATO Science for Peace and Security Series A: Chemistry and Biology, DOI 10.1007/978-94-017-7593-9_18

increased sensitivity compared to existing analogues. Practically valuable results towards the measuring devices and systems creation for using on new telecommunication and testing methods were obtained.

Actuality of designed sensor implementing is defined by opportunity of creation on its basis a large sensing area devices that can be used in various security systems. Developed sensor has outstanding practical importance because it can be used for creation new classes of nanostructured sensors based on nonideal heterojunction which are the part of functional nanoelectronics devices.

18.1.1 Sensor Samples Obtaining

A transparent SnO_2 layer with 200 nm thickness was deposited on a glass substrate in vacuum to serve as the back electrode. The CdS layer was then deposited by one of two used methodics: vacuum thermal evaporation [6] (VTE) of pure CdS powder at the rate of ~0.3 μm/min at 200 °C or electrohydrodynamical spraying [7] (EHDS) at room temperature under applied voltage about 35 kV. The evaporation time was varied from 60 to 80 min to get a CdS layer thick enough (15–26 μm) and to prevent shunting due to diffusion of copper ions, as a Cu_2S layer is subsequently formed on the top of the CdS layer. Thickness increase of the CdS layer results in an unacceptable increase of the base region resistivity. Thickness of 20–24 μm was found to be optimal for the CdS base layer. Spraying time varied in range 3–10 min to get a continuous layer of CdS with thickness up to 5 μm.

The top layer and heterostructure were formed by a substitution reaction technique [1] where a layer of copper sulfide is formed by reacting the base CdS layer [3] with a subsequently deposited layer of CuCl. The samples with CuCl film on the surface were heated at 200 °C for 4 min in a vacuum chamber to initiate and maintain a solid-phase substitution reaction (18.1).

$$CdS + 2CuCl \rightarrow Cu_2S + CdCl_2. \tag{18.1}$$

Thickness of the Cu_2S layer was controlled by varying the CuCl deposition time. The metallic top electrode was made by evaporation technique or using pressed needle-type contact to the surface of Cu_2S layer.

18.2 Microscopic Study of CdS-Cu_2S Heterostructure Film Layers

Investigation of surface morphology of photosensitive thin film samples were carried out using two instruments: atomic force microscope (AFM) NTEGRA Prima NS-150 using a silicon cantilever with a 10 nm radius tip of the probe and a scanning electron microscope (SEM) Inspect F50 with a maximum resolution of 1.0 nm at 30 kV operating potential on the cathode.

18.2.1 Atomic Force Microscopy

Scanning of sensor samples surface occurred in static contact mode, image processing and statistical analysis were performed using Nova 1138 and Gwyddion software. The main goal here was to compare changes in the characteristics of the surface morphology of investigated CdS-Cu_2S thin film structures in the manufacturing process, including observation and analysis of changes in the structural characteristics of the sample surface during the formation of the layer Cu_2S on CdS film.

Comparing the morphology of samples prepared by different technologies, using methods of atomic force microscopy, was impossible due to the peculiarities of surface structure of CdS base layer films, manufactured by electrohydrodynamic spraying. When trying to scan these samples, the probe signal almost immediately underwent to irreversible distortions. Further use of the probe turned out to be impossible. In the study by electron microscope it was established the presence of CdS microparticles, that were adhering to the probe tip and made impossible its further operation.

For samples obtained by vacuum thermal evaporation, scanning was performed successfully and typical distributions of surface morphology presented on Fig. 18.1.

Used AFM technique allowed the quantitative analysis of morphology on a number of parameters such as roughness R_a (a set of surface irregularities with small steps at the base length), waviness W_a (a set of repetitive irregularities in which the distance between adjacent elevations or valleys exceeds the basic length), the relative length of the profile L_r (the ratio of the length of the profile to the basic length) and average grain size.

This enabled the comparison of thin film sample surface microgeometry before and immediately after the formation of Cu_2S layer on CdS. Thus, for the same

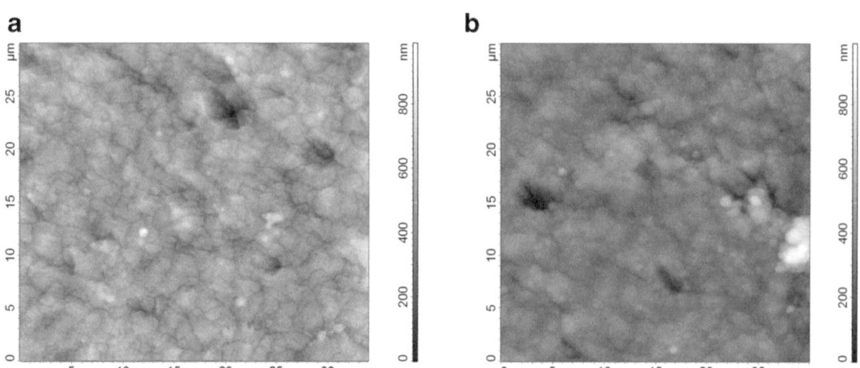

Fig. 18.1 AFM image of the sample surface before and after the heterojunction formation. (**a**) – the surface of CdS film, obtained by VTE, (**b**) – the surface Cu_2S layer, formed on the CdS. Scans are in the same scale

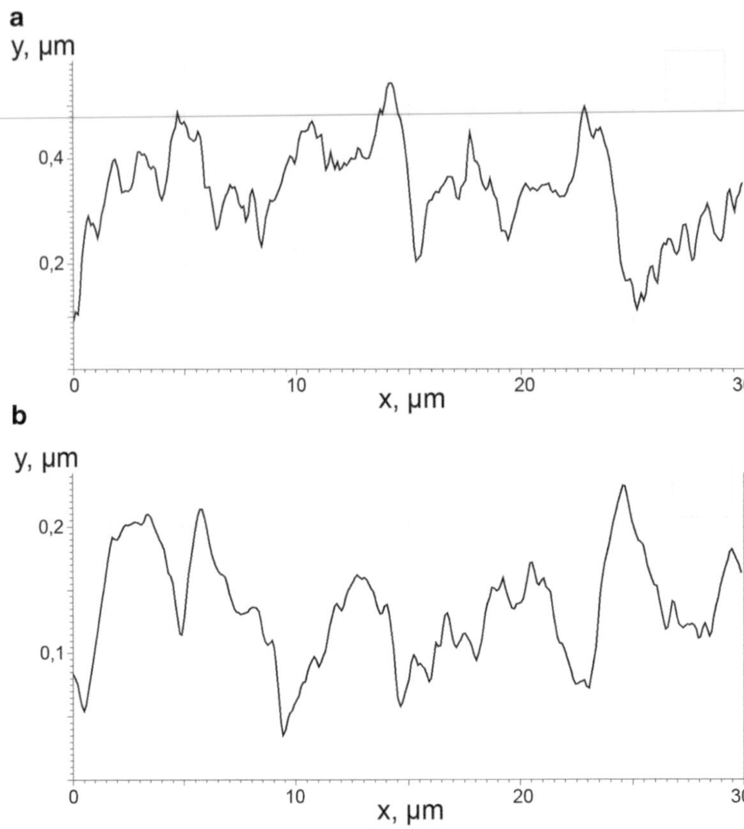

Fig. 18.2 Linear profiles of the surface for the base CdS layer, obtained by VTE (**a**) and for Cu_2S layer, formed on the surface of the same sample (**b**)

Table 18.1 The morphological parameters of the surface of CdS and Cu_2S layers

Parameter	CdS	Cu_2S
Average roughness R_a, μm	0.050	0.027
Average waviness W_a, μm	0.364	0.193
Relative length of profile L_r, a.u.	1.018	1.003

sample the sets of linear surface profiles for CdS layer and then for later deposited Cu_2S layer were obtained. Typical morphological profiles of the samples are shown on Fig. 18.2 (a – CdS, b – Cu_2S).

The values of morphological parameters numerically obtained from the profiles are presented in Table 18.1. It is seen that formation layer reduces the expression of Cu_2S surface microrelief, thereby the values of roughness and waviness decrease. The relative length L_r of the profile also reduced. Considering that the actual surface area is L_r^2 times more, then geometric one, so the deposition of Cu_2S layer reduces the effective area of the sample in the L_r^2 (CdS) – L_r^2 (Cu_2S) = 0.003, i.e. 3 %.

18.2.2 Scanning Electron Microscopy

The using of scanning electron microscopy allowed making the qualitative and quantitative comparative analysis of the surface microstructure for both types of samples, obtained by different base layer forming technologies.

The images of surface of sensor elements obtained using methods EHDS and VTE are shown in the same scale on the Fig. 18.3. The comparing of the surface images for CdS (Figs. 18.3a, c) clearly shows that there is more textured surface for EHDS samples, the granular structure of the base material is clearly seen. It was found that the CdS film, obtained by EHDS have a polycrystalline structure, orientation of polycrystallines is chaotic, average size of crystallites in the

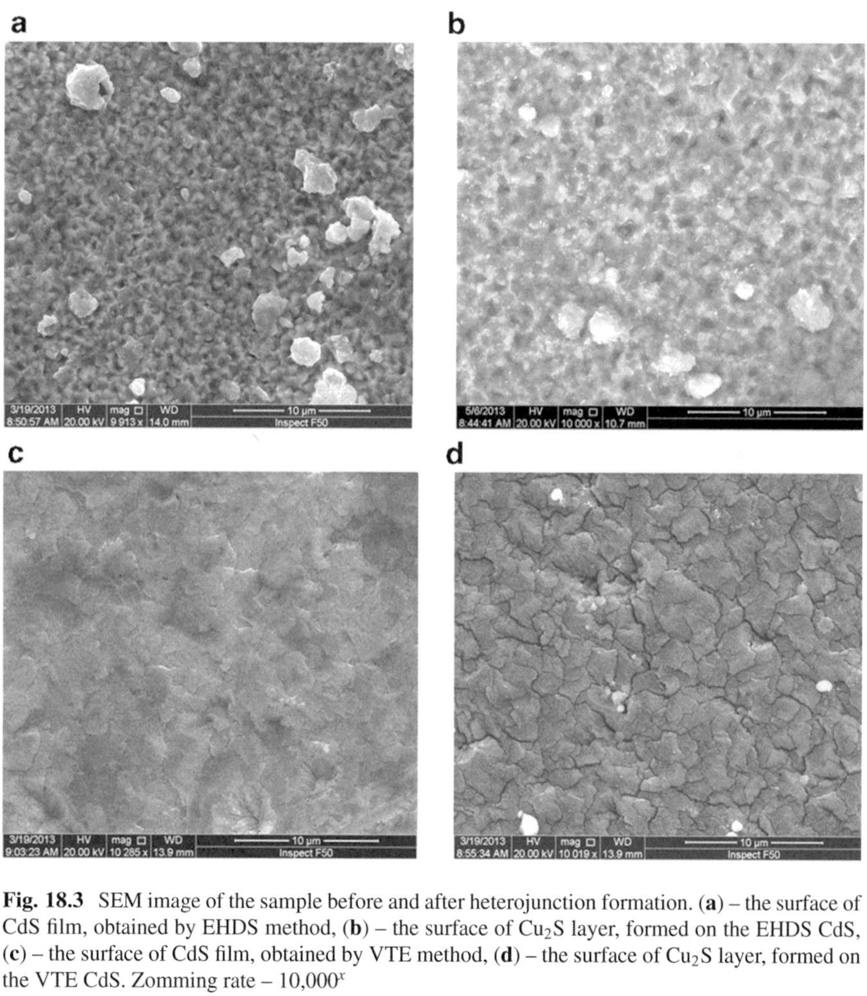

Fig. 18.3 SEM image of the sample before and after heterojunction formation. (**a**) – the surface of CdS film, obtained by EHDS method, (**b**) – the surface of Cu_2S layer, formed on the EHDS CdS, (**c**) – the surface of CdS film, obtained by VTE method, (**d**) – the surface of Cu_2S layer, formed on the VTE CdS. Zomming rate – $10{,}000^x$

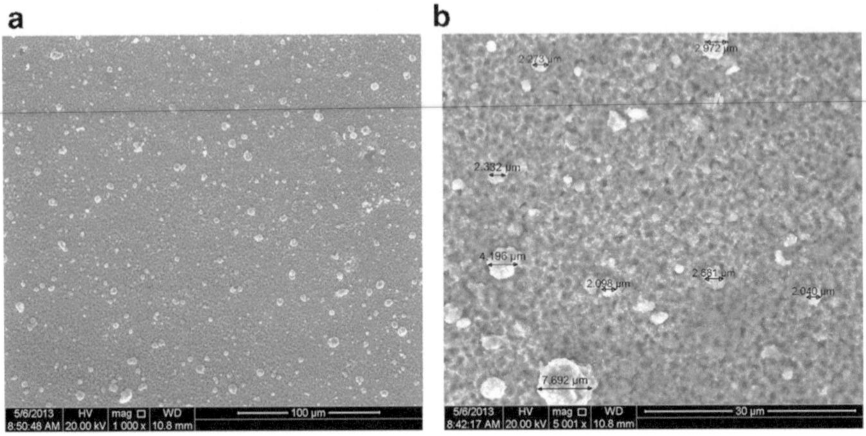

Fig. 18.4 SEM image of CdS film surface, obtained by EHDS method: (**a**) – 1000^x zoom (**b**) – 5000^x zoom. Measured linear dimensions of observed microparticles are marked

measurements on the SEM scan was 0.8 μm. The surface of VTE samples shows opposite – the relative homogeneity and less expressed relief.

Repeated scan series of sensor samples were made after forming the copper sulfide film on CdS layer by a substitution reaction in the solid phase. Both types of heterojunction samples, EHDS (Fig. 18.3b) and VTE (Fig. 18.3d) were investigated. This made it possible to compare the morphology of the surface as before and after the formation of the upper layer Cu_2S, and as between technological types of samples.

The splitting of Cu_2S layer surface to separate flakes about 5 μm^2 was observed for VTE samples (Fig. 18.3d) such features were not observed on the surface of EHDS samples (Fig. 18.3b).

SEM studies revealed the presence of particles on the surface that prevent the use of AFM techniques for scanning EHDS samples (Fig. 18.4a). The size of these particles ranged from 1 to 8 μm (Fig. 18.4b). They can easily separate from the surface of CdS by passing of AFM probe tip and stick on it, and that process make it impossible to use atomic force microscopy techniques for this type of samples. The formation of these particles is likely due to the presence of very large drops in the spray flare. When passing the reaction (18.1) the crystalline conglomerates on the substrate are formed of these droplets, and they are observed in SEM images as close to spherical shape particles on the surface of the sample.

A separate important task using microscopic methods was to study not only the surface but also the bulk characteristics of CdS-Cu_2S sensor heterostructures. For this part of the samples was taken away from the set to the cleavage procedure: cutting of the glass substrate and subsequent breakage by the incision line. In the majority of cases obtained cut was strictly perpendicular to the plane of the substrate glass plate on the side, where the heterojunction layers were deposited.

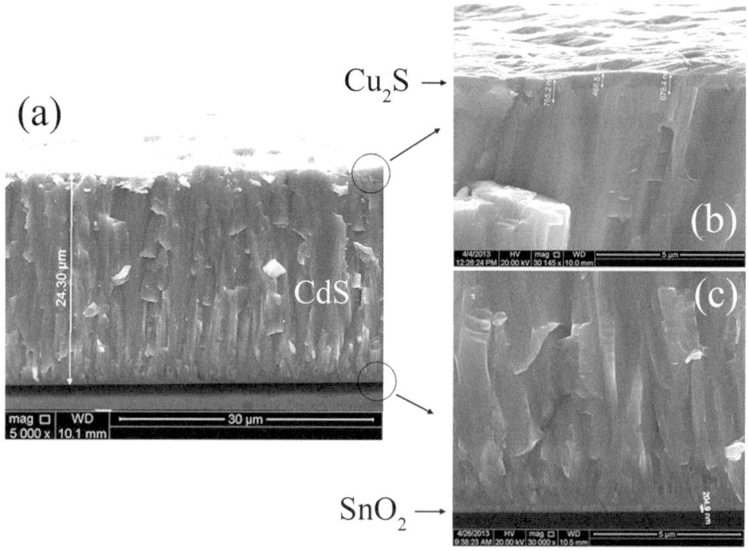

Fig. 18.5 SEM image of thin film heterojunction cross-section: general view at 5000^x zoom (**a**), the surface region with Cu_2S layer (**b**) and contact area with SnO_2 layer on a glass substrate (**c**) at 30,000^x zoom

The control of perpendicular matching of the cut line of the surface was held in electron microscope camera using a mobile holder, which allowed to make sample scanning from any side.

General view of CdS-Cu_2S nanoheterostructure cross-section with the base layer obtained by VTE is shown on Fig. 18.5. For this type of samples the columnar structure of crystals location in cadmium sulfide was revealed, which is consistent with the model of diffusion of copper to base layer by deep intercrystalline layers [1].

The precise measurements of geometric parameters of sensor samples, such as the thickness of the layers of the heterostructure, were obtained. The equipment allowed to carry out measurements directly on the sample in the scan mode. Zooming up to 30,000 times made it possible to distinguish tin oxide contact layer (Fig 18.5c). The thickness of SnO_2 for all samples was identical and uniform within 200 ± 20 nm (Fig. 18.6). Thickness of base layers CdS obtained by VTE method was in range 15–25 µm (Fig. 18.5a) for EHDS method – 1–4 µm (Fig. 18.4b). These data correspond well to the results of previous indirect measurements [8] of the thickness of the base layer (interferometry, chemical etching, film transparency measuring), but unlike them, method used in this work is a source of direct and most accurate values of the thickness of the layers CdS and Cu_2S (actual accuracy to 3 nm is limited by resolution of the device).

SEM cross-section image of EHDS base layer is presented on Fig. 18.7. CdS layer thickness is about 3 µm. However, in this case, the columnar structure of

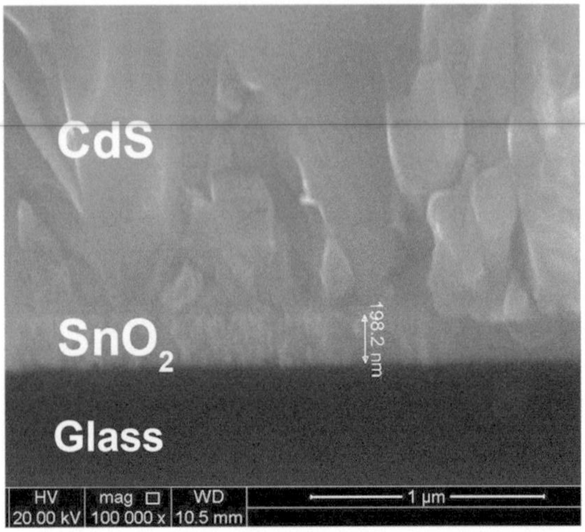

Fig. 18.6 SEM image of contact area cut with CdS base layer on glass substrate at 100,000^x zoom. The *arrow* indicates the measured thickness of SnO_2 contact layer

Fig. 18.7 SEM image of cross-section of sample formed by EHDS technology at 5000^x zoom

crystallites was not observed. Instead, visible grains location disorder, significant surface roughness and thickness variation comparable to the value of layer thickness were evident.

According to obtained linear measurements the characteristic curve of layer thickness dependence on film deposition time for VTE technology was built (Fig. 18.8). The difference between previously obtained and a new curve is due to insufficient precision of previous indirect measurement techniques [8] (electrochemical etching and CdS film optical transparency). A new curve shows also previously unmarked shift from linear dependence h (t_{dep}) to sublinear for larger time values. Thus, a decrease in growth rate of the film over time in the later stages of the CdS layer forming is observed, which can be explained by a decrease in the intensity of source material evaporation in a vacuum chamber.

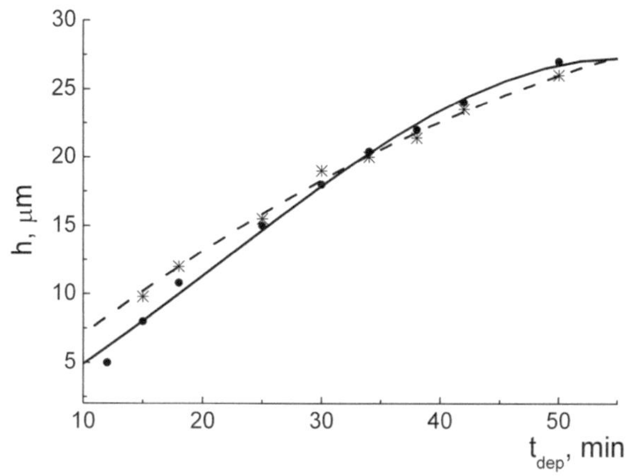

Fig. 18.8 The dependence of thickness (h) of CdS film on time (t_{dep}) of its deposition by VTE method. Thickness values calculated from experimental measurements of transparency coefficient (*stars*, *dotted line*) and measured directly on the SEM image of sectioned sample (*circles*, *solid line*)

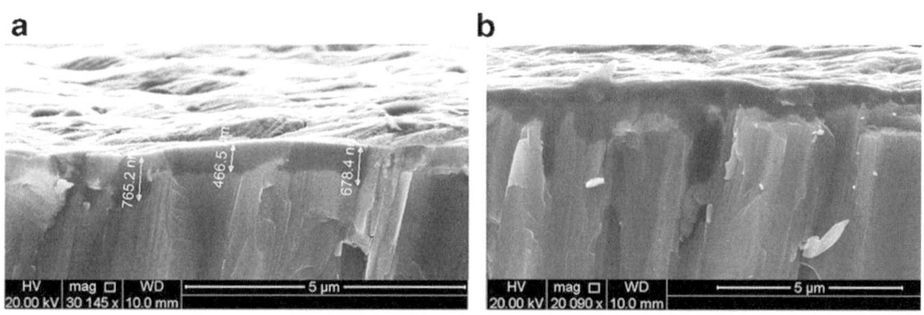

Fig. 18.9 SEM image of heteojunction surface area cross-section with 30,000^x zooming (**a**) and with 20,000^x zooming (**b**). The *arrows* indicate measured thickness of Cu_2S layer

The big difference between the values of conductivity of layers CdS and Cu_2S, and therefore between the intensity of electron scattering of scanning beam, allowed to observe copper sulfide layer directly formed by substitution reaction on the surface of CdS (Fig. 18.5b). Thus, the opportunity was used to receive the original results for investigated structures concerning the distribution of thickness h_1 of the light-absorbing upper layer Cu_2S. Thus, a detailed scan revealed a significant heterogeneity in thickness of copper sulfide film, as shown in Fig. 18.9. Measured range in thickness was 0.3–2 μm with an average 0.75 μm. Considering that most of the long-wavelength light is absorbed by a copper sulfide film, we can speak about significant local inhomogeneities in the intensity of light absorption in Cu_2S caused by varying of layer thickness.

Additional scanning of CdS layer cross-section revealed the presence of surface cracks with depth h_2 up to 1.5 μm (Fig. 18.10). These cracks can play the role of effective channels for Cu_2S layer thickening with subsequent diffusion of copper atoms deep into CdS film. Filling the cracks by copper sulfide, that is formed during the substitution reaction, changes the sensor surface roughness that was observed during the study of samples microrelief by AFM. Typical local increase in Cu_2S

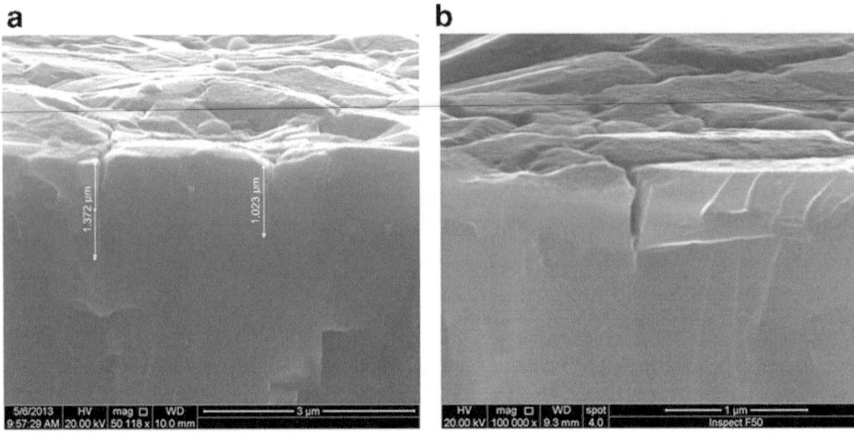

Fig. 18.10 SEM image the surface area cross-section of the base layer CdS, obtained by VTE, with 50,000^x (**a**) and 100,000^x (**b**) zooming. The *arrows* indicate measured depth of detected intercrystalline cracks

thickness was in rage of 1.5–2 μm (Fig. 18.10b), which is completely consistent with the possibility of its formation due to surface cracks in CdS.

Thus, in the use of VTE method of CdS base layer obtaining with subsequent substitution reaction to create a surface Cu_2S layer a "vertical" heterojunction is formed, when copper sulfide layer gets deeper in the base region due to the presence of surface cracks. This increases the heterojunction effective area and therefore leads to increasing of short circuit current and efficiency of photoconversion sensor.

According to obtained SEM microscopy results the model of surface microstructure of studied heterostructure was developed ant its visualization is given on Fig. 18.11, which shows an expected technological thickness $h_1$of Cu_2S layer and thickness irregularities caused by cracks in the surface layer of CdS with depth h_2.

The presence of cracks with 2 μm depth in the surface layer of cadmium sulfide formed by vacuum thermal evaporation was revealed. It was proved that these intercrystalline cracks promote the penetration of Cu_2S compound into base layer depth when forming heterojunction, as it was evidenced by local heterogeneity of measured Cu_2S thickness. This fact can explain observed variations in sensor surface photosensitivity [9].

18.3 Conclusion

Novel results concerning CdS-Cu_2S heterojunction surface morphology and layers geometrical distribution were obtained. In particular, the question of observed variation of surface photosensitivity caused by irregularities in thickness of light-absorbing Cu_2S layer were clarified. Also the comparison of samples formed

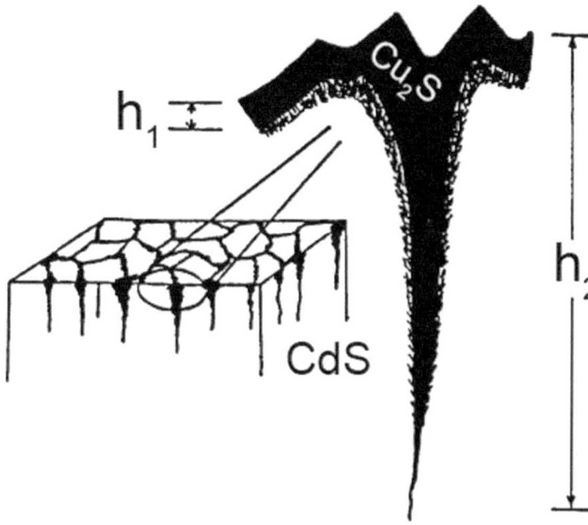

Fig. 18.11 Illustration of CdS-Cu_2S sensor heterostructure surface morphology according to performed SEM research, h_1 – thickness of Cu_2S; h_2 – depth of cracks in CdS surface layer

by two different methodics (electrodynamical spraying and vacuum evaporation techniques) was made. Obtained results enable to understand better the features of layers forming procedure, the junction components surface interaction and influence of external environment on sensor samples surface characteristics.

Reliable characteristic dependence of CdS thickness on evaporation time was also obtained, which enables more accurate process of films geometric parameters control during the formation of heterojunction layers and enables the doping of individual layers of the base layer by introducing an impurity substance at a certain time of CdS film depositing.

Developed sensor prototype demonstrates high stability parameters and shows prospectivity for comprehensive studies of inorganic, organic and biological objects. The originality of designed nanostructured sensors consists in the fact that they are effective at low doses of X-ray radiation, which makes it possible to use this material not only in controlling production processes, but in places of public use, such as security systems and customs control.

References

1. Goldenblum A, Oprea A (1994) Photocapacitance effects in dry processed Cu_2S-CdS heterojunctions. J Phys D 27:582–586
2. Vanhoecke E, Burgelman M, Anaf L (1986) Reactive sputtering of large-area Cu_2S/CdS solar cells. Thin Solid Films 144:223–228
3. Borschak VA, Brytavskyi Ie V, Smyntyna VA, Lepikh Ya I, Balaban AP, Zatovskaya NP (2012) Influence of internal parameters on the signal value in optical sensor based on the non-ideal heterostructure CdS-Cu_2S. Semicond Phys Quantum Electron Optoelectron 15(1):41–43

4. Borschak VA, Smyntyna VA, Brytavskyi Ie V, Karpenko AA, Zatovskaya NP (2013) Open-circuit voltage of an illuminated nonideal heterojunction. Semiconductors 47(6):838–843
5. Gaubas E, Borschak V, Brytavskyi I, Ceponis T, Dobrovolskas D, Jursenas S, Kusakovskij J, Smyntyna V, Tamulaitis G, Tekorius A (2013) Non-radiative and radiative recombination in CdS polycrystalline structures. Adv Condens Matter Phys 2013:1–15
6. Goldenblum A, Popovici G, Elena E, Oprea A, Nae C (1986) All-evaporation-processed Cu_2S/CdS solar cells with improved characteristics. Thin Solid Films 141:215–221
7. Nascu C, Pop I, Ionescu V, Indrea E, Bratu I (1997) Spray pyrolysis deposition of CuS thin films. Mater Lett 32(2):73–77
8. Vassilevski DL (1999) Physical principles of graphic registration by non-ideal heterojunction. Sens Actuators A 55:167–172
9. Borschak VA, Smyntyna VA, Brytavskyi Ie V, Balaban AP, Zatovskaya NP (2011) Dependence of conductivity of an illuminated nonideal heterojunction on external bias. Semiconductors 45(7):894–899

Chapter 19
Hetero-Carbon Nanostructures as the Effective Sensors in Security Systems

G. Kharlamova, O. Kharlamov, M. Bondarenko, and O. Khyzhun

Abstract Hetero-carbon (heteroatomic derivatives of carbon, in which one or several atoms of carbon are replaced on atoms of another non-metals) will a basic component a new generation of nanomaterials. Carbon nitride g-C_3N_4 it is possible to consider as the limiting saturated by nitrogen a hetero-carbon, that the monolayer $((C_6N_7)\text{-}N)_n$ should be named (on an analogy with azafullerene) azagraphene. Carbon nitride as well as azafullerenes, azananotubes (N-dope nanotubes) and N-doped graphene are used as very sensitive nanosensors. Modifying, in particular, oxidized derivatives of carbon nitride are not studied practically. Here the products of a new route of a pyrolysis of melamine which as against known methods is carried out at the presence of oxygen are described. New compound as carbon nitride oxide (g-C_3N_4)O was obtained. Its structure is analogue of graphite oxide. Nanosized powder of (g-C_3N_4)O is easily exfoliated and is dissolved in water with formation of a flake-like suspension. This suspension can contain nanosheets from several heptazine $((C_6N_7)\text{-}N)_n$ monolayers or azagraphenes. Alongside with carbon nitride oxide at pyrolysis of melamine at the presence of oxygen in a one step the O-doped ($\sim$8.1 %) carbon nitride (O-g-C_3N_4) is also formed.

19.1 Introduction

Up to date, after the brilliant opening of the new state of the matter, in particular, in the form of monatomic monolayer closed molecules (fullerenes and single-walled nanotubes (SCNT)) and planer structures (graphene) as well as the different morphologies of multi-walled nanostructures (multilayer carbon nanotubes (MCNT), onions, filaments and fibers) the most problem is the creation of the preparation the methods of hetero-carbon or heteroatomic carbon derivatives. Are most perspective, but are least investigated heteroatomic derivatives of carbon, in which one or several

G. Kharlamova (✉)
Taras Shevchenko National University of Kyiv, 64 Volodymyrsska str., 03001 Kyiv, Ukraine
e-mail: akharlamova@ukr.net

O. Kharlamov • M. Bondarenko • O. Khyzhun
Frantsevich Institute for Problems of Materials Science of NASU, Krzhyzhanovsky St. 3, 03142 Kyiv, Ukraine

J. Bonča, S. Kruchinin (eds.), *Nanomaterials for Security*, NATO Science for Peace and Security Series A: Chemistry and Biology, DOI 10.1007/978-94-017-7593-9_19

atoms of carbon are replaced on atoms, for example, nitrogen (aza-derivatives (azafullerenes)), boron (boron-derivatives (borafullerene, (azabora[60]fullerene)) or silicon. Endohedral fullerenes as special type of fullerene derivatives that have atoms of metals ($Na@C_{60}$) and non-metals, ($N@C_{60}$) or molecules ($H_2O@C_{60}$), ($(NH_3)_2@C_{60}$)) incapsulated into fullerene cage are studied enough detailed.

There is, naturally, a question: why so are interesting heteroatomic derivatives of molecules and nanostructures of carbon? It is supposed, that aza-, azabora[60]-fullerenes and N-doped nanotubes and onions because of brightly expressed asymmetry of the distribution of electronic density in a graphene layer will have the unusual electronic characteristics, that will allow to create the new effective carriers and catalysts, is functional active adsorbents and perspective nanomaterials for application them as a new generation of nanosensors.

Hetero-carbon can be a main component of a new generation of nanomaterials useful in information technology, where rapid communication processes are based on optical methods rather than electronics. Studies of B-, N-, and BN-substitutional fullerenes shown that the doping enlarges the second hyperpolarizability by several orders of magnitude compared to the pure fullerenes and, consequently, may suppose that hetero-fullerenes will be unique candidates for photonic devices. Indeed, BN-doped fullerene (azabora[60]fullerene) serve as a nanoscale sensor to differentiate an adenine and its rare isomer Cu(II) complexes which their chemical identification practically impossible [1]. (Adenine ($C_5H_5N_5$) is a part of many vital compounds in living organisms, such as adenosine, adenozinfosfotazy, adenozinfosfornye acids, nucleic acids, adenine nucleotides, etc.)

Nitrogen-doped graphene is sensitive for probing Rhodamine B (RhB) molecules [2]. (Rhodamine B ($C_{28}H_{31}ClN_2O_3$) often is used as a tracer, biomarker in oral rabies vaccines.)

Sensors based on N-doped CNTs (azananotubes) for the detection of various gases (NH_3, NO_2, O_2, N_2, CO_2, CH_4, H_2O, H_2, Ar) and the vapours of the solvents (ethanol, acetone, chloroform, gasoline, pyridine, benzene) show the advantages over undoped nanotubes because they have higher sensitivity and response times (0.1–1 s).

Graphite-like carbon nitride g-C_3N_4, as a matter of fact, it is possible to consider as the limiting saturated by nitrogen hetero-carbon, in which the degree of a replacement (C/N) of atoms of carbon by atoms of nitrogen is maximal and corresponds to 0.75. Carbon nitride has a wide area of the homogenity concerning a nitrogen [3–5], within the limits of which a graphite-like structure is kept, but the ratio C/N can change from 0.75 (C_3N_4) up to 3.0 (C_3N). Here it is important to note, that carbon nitride can be, from the chemical point of view, only graphite-like one with sp^2 atoms of carbon. The assumptions based only on geometrical parameters of atoms of carbon and nitrogen about an opportunity of formation, in particular, "superhard" hexagonal phase of carbon nitride β-C_3N_4 similar to silicon nitride (β-Si_3N_4) are in the chemical relation absolute unreasonable [4]. Unfortunately, subsequent numerous barothermal influences on transformation g-C_3N_4 in imaginary β-C_3N_4 with sp^3 hybridized atoms of carbon have appeared unsuccessful, but quite natural.

The atom of silicon as against atom of carbon is capable to locate (and locates, for example, in a flat configuration of trisilylamine $N(SiH_3)_3$) on empty low energetic 3d-orbitals not binding electrons of atom of nitrogen with formation the additional π-bond N–Si. This additional interaction how the unique multiple bond between atoms of nitrogen and silicon promotes the greater strengthening σ-bond N–Si. Therefore theoretically pyramidal skeleton NSi_3 in a molecule $N(SiH_3)_3$ gets a flat configuration. A case this is not unique: the interaction of atom of silicon with oxygen ($O(SiH_3)_2$) and phosphorus ($P(SiH_3)_3$) is characterized also additional double pπ-dπ-bond. In structure β-Si_3N_4, probably, multiple interaction N–Si also is realized, where not binding electrons of atom of nitrogen are located on low energetic 3d-orbitals of atom of silicon and in addition this bond are strengthened. The atom of carbon has not at all d-orbital, therefore at formation σ-bond C–N not binding electrons of atom of nitrogen will delocalizated and can only weaken this bond. Therefore molecule of trimethylamine $N(CH_3)_3$ as analogue of a flat molecule $N(SiH_3)_3$ exists only in a pyramidal state. Hence, β-C_3N_4 as against β-Si_3N_4 cannot in general exist because of absence of an opportunity by atom of carbon to locate not binding electrons of atom of nitrogen.

On logic of a choice superhard β-C_3N_4 to have only to be surprised, why up to date the researches on obtaining of cubic aluminium nitride (cAlN) as an analogue of existed cubic boron nitride (cBN) or graphite-like silicon nitride (g-Si_3N_4) as an analogue of g-C_3N_4 are not carried out. The atoms of boron and silicon in sp^3 – hybridizated state form the bonds with nitrogen accordingly in cBN and β-Si_3N_4 for the same reason: both atoms as against atoms of carbon and aluminium have the vacant energetically low orbitals. The atom of boron (as against an atom of aluminium) has energetically low vacant pπ-orbital, therefore can locate and locates in cBN not binding electrons of an atom of nitrogen with the formation of an additional pπ-pπ-interaction.

Dispergation of g-C_3N_4 up to a several (6–12) monolayer and its modifying by various elements, in particular, oxygen, allows essentially to increase a sensitivity of nanosensors on the basis of a nitrogen-carbon material. So, on a basis graphitic carbon nitride (g-C_3N_4) fluorescent nanosensor for high selective and sensitive detection of mercury ions and L-cysteine [6] was created. g-C_3N_4 is used as a nonmetal biosensor for a selective detecting of heparin with a limit of the detection 18 ng/ml [7]. Nanosensor on a basis of nanostructurized carbon nitride has a high sensitivity and allows to detect up to 2.8 μA/% of relative humidity [8]. Moreover, the fluorescent nanosensor based on g-C_3N_4 nanosheets was applied to detect tap water and the well of water. G-C_3N_4 is an attractive material as the candidate for integrated humidity sensors because of its high melting point and thermal stability [9–12]. Nanostructured carbon nitride is used as an alternative (traditional on the basis of diamond doped by boron (BDD)) nanosensor for simultaneous definition of dopamine (DA) and an ascorbic acid (AA) in synthetic biological samples [13]. The limits of the detection established for simultaneous definition of these analytes with use of a nitride-carbon electrode were 0.0656 $\mu mol{\cdot}L^{-1}$ for DA and 1.05 $\mu mol{\cdot}L^{-1}$ for AA, while with an electrode BDD these meanings were 0.283 $\mu mol{\cdot}L^{-1}$ and 0.968 $\mu mol{\cdot}L^{-1}$ accordingly. Nanosensor on a basis of nanosheets of g-C_3N_4 intercalated LiCl has superhigh time of the response (~0.9 s) and recovery (~1.4 s)[9] at the detecting of a humidity.

Consequently, the increased interest in the development of novel effective methods of dispergation and modifying of carbon nitride, in particular, by oxygen is wellfounded. Therefore, we can assume that the existing route of the preparation of graphene from graphite

$$\text{graphite} \rightarrow \text{graphite oxide} \rightarrow \text{graphene oxide} \rightarrow \text{graphene} \tag{19.1}$$

may be also most suitable for the synthesis from carbon nitride of its nanodimensional monolayer consisting of tri-*s*-triazine or heptazine (C_6N_7) fragments linked nitrogen bridges (C_6N_7)–N–(C_6N_7). By analogy with azafullerene ($C_{59}N$), wherein in a fullerene structure only one carbon atom replaced by a nitrogen atom, hypothetical heteroatomic monolayer $((C_6N_7)\text{–}N)_n$ from the structure of carbon nitride can be called an azagraphene (or (heptazinic graphene). Hence, the synthesis of the hypothetical azagraphene from graphitic carbon nitride can be represented by a scheme (19.2) similar to the scheme (19.1):

$$\begin{aligned} &\text{carbon nitride } (g - C_3N_4) \rightarrow \text{carbon nitride oxide } ((g - C_3N_4)O) \\ &\rightarrow \text{azagraphene oxide } (((C_6N_7) - N_n)O) \rightarrow \text{azagraphene } ((C_6N_7) - N_n). \end{aligned} \tag{19.2}$$

However, in present time is reliably synthesized only a graphite-like carbon nitride, which in the majority of the known solvents (behind exception DMSO) is practically insoluble [14]. Therefore the heterophase method of obtaining of graphite oxide from graphite in water solutions of strong oxidizers is represented also quite logical and for the synthesis of oxide of carbon nitride from carbon nitride. It is known [15] that carbon nitride unlike graphite is partially oxidized in the aqueous solution of hydrogen peroxide (H_2O_2). So, a carbon nitride with the contents of nitrogen close to the formulaic composition (C/N = 0.751), which was obtained at 550 °C (in during 5 h) from dicyandiamide, subjected to processing by a hydrothermal method: a powder g-C_3N_4 dispergated in 30 % a water solution of H_2O_2 in teflon germetic autoclave at 140 °C in during 10 h. The resulted product after five cycles of centrifugation and washing by water [15] contained according to the data XPS till 7.98 % oxygen. Therefore currently doped oxygen carbon nitride prepared by only two stage method at an oxidation by H_2O_2 of previously synthesized g-C_3N_4. However, obtained by such method of doped by oxygen of carbon nitride, as well as g-C_3N_4, in water solutions of strong oxidizers in addition is not oxidized with the formation of oxide of carbon nitride as analogue of oxide of graphite. The ability and an ease (in comparison with carbon nitride) graphite to intercalate water and oxidize up to graphite oxide, probably, is caused a somewhat larger (on 0.016 nm) in it an interplanar distance 002.

Noticeable dissolution and stratification of g-C_3N_4 nanosized particles (to form ultrathin nanosheets) have been performed only in concentrated alkaline solutions at 50–90 °C and in the absence of reducing agents by "green" peeling liquid [16]. (Note that this method is used to obtain stable suspensions of graphene from exfoliated graphite oxide.) Compared with the insoluble carbon nitride its nanosheets have increased photoabsorption and photoresponse, are used as non-metal biosensor. The information on obtaining carbon nitride oxide to date in the literature absents.

We believe that the carbon nitride oxide may be formed in parallel with the synthesis of carbon nitride by thermal transformation of melamine (cyanamide or dicyandiamide) molecules (or fragments of their degradation) into heptazine layers $((C_6N_7)–N)_n$.

In this paper results of a research of melamine pyrolysis at 500–580 °C and at the presence of the small fixed quantity of oxygen are submitted. The aim of this study to obtain partially oxidized derivatives of nitride carbon and, in particular, oxide of carbon nitride and by oxygen doped carbon nitride.

19.2 Experimental

First of all, it is important here to remind, that partially (11.1 mass % O) oxidized carbon nitride with the small contents of nitrogen (O-C_3N) was obtained by us earlier [4, 5] at pyridine pyrolysis [17–19]. Onion nanostructures of turbostratic carbon nitride have been formed in the reactionary vapour-gas stream of the products of pyridine molecules destruction and have been deposited far from the most high-temperature zone of reactor. We believe that in a various degree of the oxidized forms of carbon nitride can also be synthesized by means of gas-transport reactions at pyrolysis of molecules, for example, melamine (cyanamide or dicyandiamide) at the presence of oxygen. Graphite-like structure of carbon nitride from heptazine fragments (C_6N_7)–N can be formed at the expense of consecutive escalating of twosized heptazine monolayers from molecules (and its fragments C_3N_6, CN, C_2N) evaporating melamine. Formed an electrondonor (connected from π-bonds) heteroatomic structure (C_6N_7)–N is extremely inclined to oxidation and at the presence of oxygen it can easy oxidize with the formation of partially oxidized carbon nitride. However it is usual the pyrolysis of the data precursors is carried out, as consider in [3, 14], in conditions of the solid-phase reactions excluding the presence of oxygen and precursor sublimation [20, 21].

So, the testes on the preparation from melamine carbon nitride of a formulaic composition (g-C_3N_4) have been carried out in conditions practically excluding of the presence of oxygen: are used a closed vacuum ampoules and superpure inert gas. In [22] melamine pyrolysis was carried out by a solvothermal method ($P = 2.5$ GP, $T = 800$ °C) in a solution of hydrazine. The authors [20, 21] for the prevention of a sublimation and oxidation of melamine at temperatures 520–550 °C used a special sheeting of a crucible from aluminium oxide. The products obtained in similar reactionary conditions contain nitrogen, carbon and hydrogen as well as very small (2–3 %) amount of oxygen from adsorbed water.

By us the pyrolysis of melamine to produce partially an oxidized product is carried out at the presence of a fixed volume of an air. Melamine in open ceramic crucible is located in a tubular quartz reactor with a fixed air volume and, consequently, with a fixed quality of oxygen. The testes are carried out at heating of ceramic crucible up to 500–580 °C and at the variation of temperature, melamine amount and the duration of its thermal treatment. The products of reaction are condensed and are deposited outside of a zone of precursor localization and mainly in low temperature zones of quarts reactor. The powdery products obtained at fixed reactionary regimes previously estimated on their chemical and XRD analysis. For the further investigation the samples were selected which according to their diffractogram contained g-C_3N_4 and did not contain an initial reagent (melamine).

Synthesized substances have been investigated by methods chemical and X-ray phase (DRON UM-1 with Cu K_α radiation and nickel filter) analyses, XPS and IR spectroscopy, as well as temperature-programmed desorption mass spectrometry (TPDMS) formed products. X-ray photoelectron spectra (XPS) of the samples were measured with photoelectron spectrometer manufactured by SPECS Surface Nano Analysis Company (Germany), with a PHOIBOS semi-spherical analyzer. IR spectra were measured in the reflection mode in the spectral range 4000–400 cm^{-1} with a resolution of 8 cm^{-1} with a Nicolet Nexus FTIR spectrometer (Thermo Scientific). The samples under study were compacted in pellets with potassium bromide in a ratio of 1:10.

The mass-spectrometric analysis at m/z of up to 200 was made on an MX-7304A monopole mass spectrometer (Sumy, Ukraine) with electron ionization (EI) [23]. The samples (3 mg) were placed on the bottom of a quartz-molybdenum ampule and the ampule was evacuated at room temperature to approximately 5 × 10–5 Pa. The programmed linear heating of the sample to 750 °C was performed at a rate of 0.15°/s [23]. Volatile thermolysis products were delivered through a vacuum valve with a diameter of 5.4 mm into the ionization chamber of the mass spectrometer and ionized by electrons.

19.3 Results and Discussion

19.3.1 X-Ray Diffraction and Chemical Analysis of Products of Melamine Pyrolysis

In particular amount of melamine by varying of the temperature and duration of pyrolysis was determined the temperature interval in which the resultant product of the reaction is substantially (according to X-ray diffraction) contained no a phase of precursor. In our experimental conditions the conversion of melamine in the carbon nitride is begun at 520 °C: in the X-ray spectrum (Fig. 19.1, curve 4) of a resultant product N0 of pyrolysis on a background of intense lines belonging melamine are presented noticeable enough reflexes at $2\theta = 12.40$ and 27.49 °C,

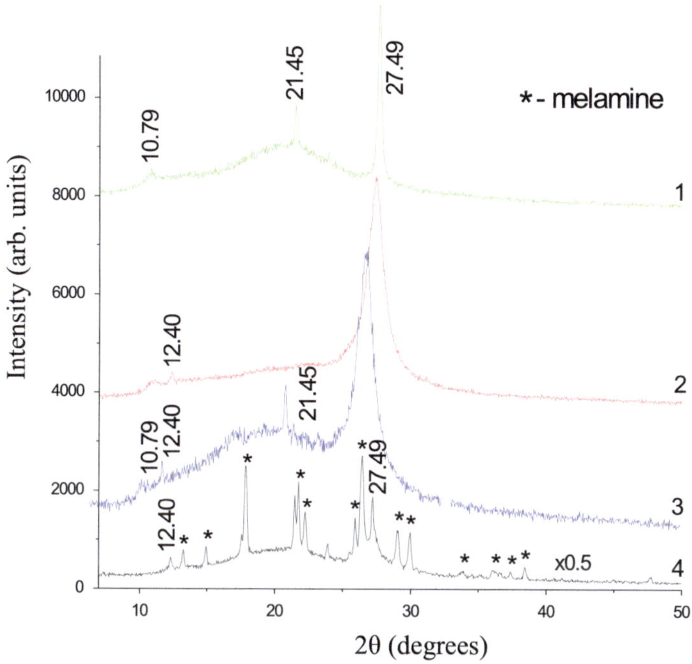

Fig. 19.1 XRD patterns of samples O2, N2, N1 and N0 (curves 1, 2, 3 and 4 respectively)

those are characteristic for graphitic carbon nitride. At temperatures above 550 °C phase of melamine into pyrolysis products N1 (Fig. 19.1, curve 3) and O2 is not detected at all. However, from the X-ray diffraction patterns may see that the phase composition of obtained at different temperatures resultant products are markedly different and, hence, the route of melamine conversion depends on its pyrolysis the temperature.

A brown powdery product N1 (Fig. 19.1, curve 3) is formed at melamine pyrolysis at 550 °C, in which X-ray spectrum clearly are detected four reflex, two of which ($2\theta = 12.40°$ and $2\theta = 27.49°$) completely are characteristic for g-C_3N_4, respectively for the planes 100 ($d = 0.714$ nm) and 002 ($d = 0.324$ nm). The presence in the X-ray spectrum of the sample N1 of additional low-intensity peaks at $2\theta = 21.45°$ and 10.79° may indicate on the presence in the product of the pyrolysis of other O, C, N- containing phases. Indeed, the sample N1, unlike carbon nitride of a formulaic composition partially dissolved in hot water to form an intensely colored yellow solution O2. Diffractogram of treated by water insoluble brown powder N1 (Fig. 19.1, curve 3, Fig. 19.2a), on which there are two characteristic for carbon nitride reflex, fully consistent with both the calculated and experimental [3, 14] X-ray spectra of the g-C_3N_4. Indeed, the sample N1 according to an elemental chemical analysis contains 35.9 mass % carbon, 54.7 mass % nitrogen, 1.3 mass % hydrogen and 8.1 mass % oxygen, which may correspond to carbon nitride ($C_3N_{3.92}H_{1.30}O_{0.51}$) with a ratio C/N (0.765) close formulaic (0.75). Consequently, during the pyrolysis of melamine doped by oxygen carbon nitride is formed.

Sample O2, which is formed after evaporation of water, has a yellow color. Its diffraction pattern there are three reflex at $2\theta = 10.79$, 21.45 and 27.49°, which are probably inherited from the original sample N2. We can assume that in this sample is realized unidirectional shift both interplanar spaces in the direction of their increase in comparison with the carbon nitride sample N2. Consequently, during the pyrolysis of melamine, together with oxygen-doped carbon nitride is formed a water-soluble substance O2, which may correspond to carbon nitride oxide, the X-ray spectrum (Fig. 19.1, curve 1) which is presented here for the first time.

At higher temperatures (particularly at 580°C) pyrolysis of melamine is formed yellow product O1 (Fig. 19.2b), which has XRD patterns identical to the diffraction pattern of the sample O2: three reflexes at $2\theta = 10.79$, 21.45 and 27.49° are

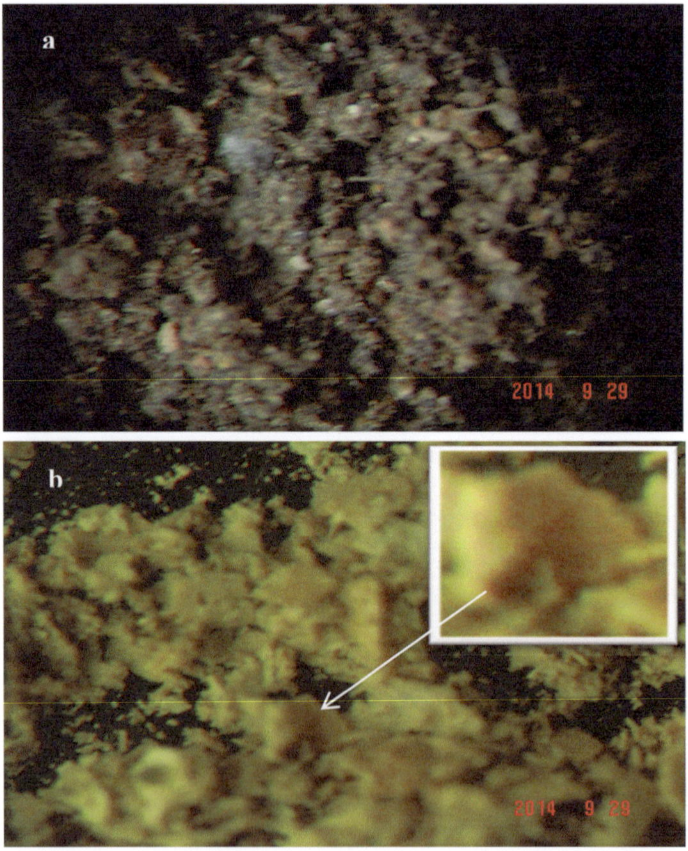

Fig. 19.2 Optical microscopy images of the sample N1 (**a**) and the flake-like sample O1 (**b**)

characteristic for sample O2. In accordance with the chemical composition (49.6 % nitrogen, 1.5 % hydrogen, 16.1 % oxygen and 32.8 % carbon) sample O1 can be represented as formula $C_3N_{3.89}H_{1.65}O_{1.09}$. It can be seen that the product O1 compared to sample N2 contains twice more oxygen, while the ratio C/N in which approximately comparable, 0.771 and 0.765 respectively. Sample O1 unlike N1 completely dissolves in hot water to form a flocculent yellow solution from which after an evaporation of water yellow powder O3 is precipitated. (The flocculent solution containing nanosheets (2–4 nm thick) of carbon nitride is formed by ultrasonic peeling in the acid powdery g-C_3N_4 [16]). Thus, three reflex at 2θ = 10.79, 21.45 and 27.49° in the diffraction patterns of recrystallized yellow powder O3 coincide with the reflections in the diffraction patterns of samples O1 and O2. Sample O2, which is formed after evaporation of water, has a yellow color.

When comparing the diffraction patterns of water-insoluble oxygen-doped carbon nitride $C_3N_{3.92}H_{1.30}O_{0.51}$ (sample N2, 8.1 mass % O) and water-soluble substantially more oxidized carbon nitride (sample O1 (16.1 mass % O) as well as O2) clearly may see that the structure of their X-ray spectra differ markedly. The spectrum of sample O1 there is a significant shift of the signal corresponding to the interaction between heptazine fragments in a monolayer, with 12.40°–10.79°. Thus, the most intense signal at $2\theta = 27.49°$, characterizing interlayer interaction g-C_3N_4, is maintained. Furthermore, there is a new reflex at $2\theta = 21.45°$ in XRD patterns O1, not present in the spectrum of the sample N2 (and g-C_3N_4). It can be assumed that the obtained water-soluble product O1, containing 16.1 % oxygen and having an X-ray spectrum differs from the spectrum of g-C_3N_4, corresponds to a new compound, carbon nitride oxide ((g-C_3N_4)O). We believe that the observed changes (compared to carbon nitride) in structure of carbon nitride oxide associated with the features (compared to the structure of graphite oxide) of location of heteroatoms (N and O) in the plane of heptazine layer.

Each plane of graphite oxide is a continuous graphene network and any oxygen-containing groups (hydroxyl or epoxy) can only be located between the graphene layers, which lead to such a large increase in the interlayer spacing from 0.34 nm in the graphite to about 0.7 nm in the graphite oxide. Therefore in the X-ray spectrum of graphite oxide the signal at $2\theta = 26°$ characteristic for graphite virtually disappears, and the halo appears at the $2\theta = 10$–15°. However heteroatomic (nitrogen-carbon) plane of g-C_3N_4 has not continuous but “openwork network” because it consists of heptazine C_6N_7 fragments connected nitrogen bridges $((C_6N_7)–N)_n$. Thus, in a plane between three neighbouring heptazine fragments exists the hollow with diameter 0.717 nm (Fig. 19.3a). In the space of this hollow can be freely placed two oxygen containing groups (Fig. 19.3d), in which the bond lengths are ~0.143 nm (C–O) and ~0.091 nm (O–H). (Note that in the carbon nitride oxide (g-C_3N_4)O to six carbon atoms heptazine fragments have two oxygen atoms). Moreover, the oxidation of carbon nitride may also accompanied by some gap between heptazine bonds and nitrogen-bridged fragments (Fig. 19.3b), which leads to an increase of the distance between the heptazine blocks with 0.717 ($2\theta = 12.40°$) to 0.818 ($2\theta = 10.79°$) nm.

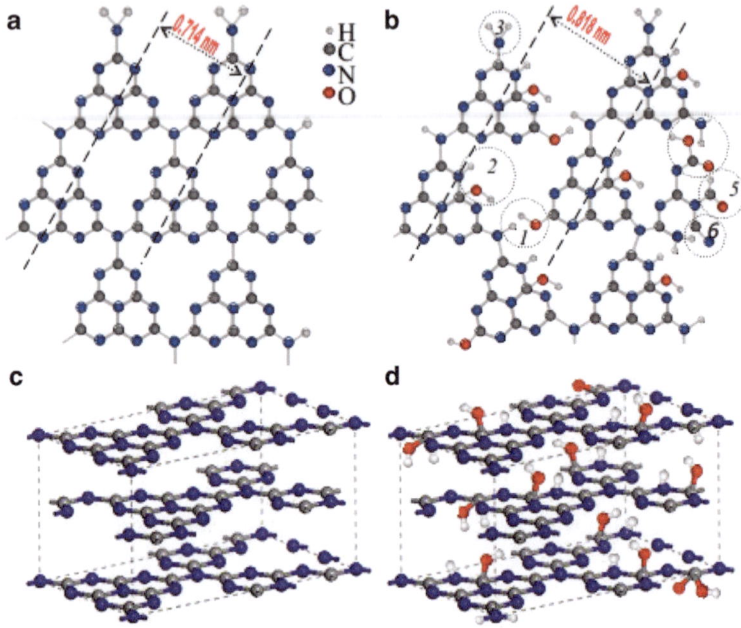

Fig. 19.3 Schematic on-plane atomic model of g-C_3N_4 (**a**), (g-C_3N_4)O (**b**); schematic interplanar atomic model of g-C_3N_4 (**c**), (g-C_3N_4)O (**d**)

However, in the carbon nitride oxide interlayer spacing close to the characteristic (0.324 nm ($2\theta = 27.49°$)) for carbon nitride can be partially preserved in some parts of heptazine planes due to the distortion of the planarity in the opening of the double bonds (–C=N– → –C(OH)–NH–) in heterocycles. This "fluting" heptazine cell in carbon nitride oxide may be the cause of additional peak at $2\theta = 21.45$, corresponding substantially greater (compared with carbon nitride) interplanar spacing, 0.414 nm.

There is a natural question that if an oxide of carbon nitride is the structural analogue of graphite oxide than why in the diffractogram of graphite oxide in comparison with the diffractogram of graphite of an occurrence of similar additional peak is not observed?

Indeed, the structures of X-ray spectra of carbon nitride and carbon nitride oxide differ more essentially than known structures of X-ray of spectra of graphite and it oxide. In X-ray spectrum of graphite oxide in comparison with graphite the shift of a characteristic signal with 26° up to 10–15° is fixed only, as corresponds to an increase of an interplanar distance 002 with 0.340 up to ~0.7 nm. In a spectrum of oxide of carbon nitride (samples O1, O2) the change (an increase) both 002 and 100 the interplanar distances is fixed according to 0.324 ($2\theta = 27.49°$) up to 0.414 ($2\theta = 21.45°$) and with 0.714 ($2\theta = 12.40°$) up to 0.818 nm ($2\theta = 10.79°$). Besides in oxide of carbon nitride the interplanar distance 0.324 nm, characteristic for g-C_3N_4, also partially is kept (Fig. 19.3c, d).

Main difference between the structures oxide of graphite and oxide of carbon nitride is caused by a principal distinction in an arrangement of monoatomic (carbon) and heteroatomic (nitrogen-carbon) hexagons. Each plane of oxide of graphite represents a continuous graphene network and anyone formed at oxidation of graphite of oxygen containing groups (hydroxyl or epoxy) can be located exclusively between the graphene layers. Therefore an interplanar distance 002 in oxide of graphite is so considerably increased.

Heteroatomic heptazine plane $(C_6N_8)_n$ in g-C_3N_4 represents not continuous but "openwork" network (Fig. 19.3a, c) as everyone 3 heterocycles (C_6N_7) of the monolayer connected one with another by means of tertiary nitrogen atoms (C_6N_7)–N–(C_6N_7)–N–(C_6N_7)–N form a large hollow (diameter 0.714 nm) in this monolayer. Thus, carbon nitride oxide due to the hollows in heptazine planes and localization features therein oxygen (or –OH groups) interplanar distance in (g-C_3N_4)O slightly (by 0.09 nm) (Fig. 19.3a, b) increases only in certain areas of these planes.

By the method X-ray photoelectron spectroscopy can estimate the chemical composition of the sample O2, to determine the status of the elements and a character of bonds between the atoms of these elements. We also hope, that IR spectrum of $(C_3N_{3.89})O_{1.09}H_{1.65}$ will contain both signals characteristic for vibrations of the bonds in a heptazine layer of carbon nitride and signals of the oxygen containing functional groups that usually present in the IR spectrum of graphite oxide, but absent in a spectrum of g-C_3N_4.

19.3.2 The Study of the Oxide of Carbon Nitride $(C_3N_{3.89})O_{1.09}H_{1.65}$ by X-Ray Photoelectron Spectroscopy

First of all, it is necessary to note, that earlier by method XPS were investigated the various samples of carbon nitride obtained from various precursors [3, 14, 24–26] as well as the samples of O-doped carbon nitride [15]. General and characteristic for the investigated samples both g-C_3N_4 and O-g-C_3N_4 is the indispensable presence at a spectrum of carbon of an intensive line at ~288.0 eV. It is accepted to consider that this line corresponds to sp^2-bonded carbon to three of neighboring nitrogen atoms $(N{=}C\text{–}(N)_2)$ inside the aromatic heptazine structures. In low-nitrogen O-doped carbon nitride (O-g-C_3N) [4, 5], consisting of mainly not from heptazine (C_6N_7) but pyrazine and pyrimidine (C_4N_2) of structures and, hence, containing $(N{=}C\text{–}(C)_2)$, instead of $(N{=}C\text{–}(N)_2)$ fragments the line C1s at of ~288.0 eV is low intensive (Fig. 19.4, inset). At the same time, in a spectrum C1s of O-g-C_3N distinctly it is possible to see a line at 284.5 eV characteristic for atom of carbon in hexagon of graphite $(C{=}C\text{–}(C)_2)$, which also distinctly is visible in C1s spectra of some samples g-C_3N_4 [3, 14, 24–26] and O-doped g-C_3N_4 [15].

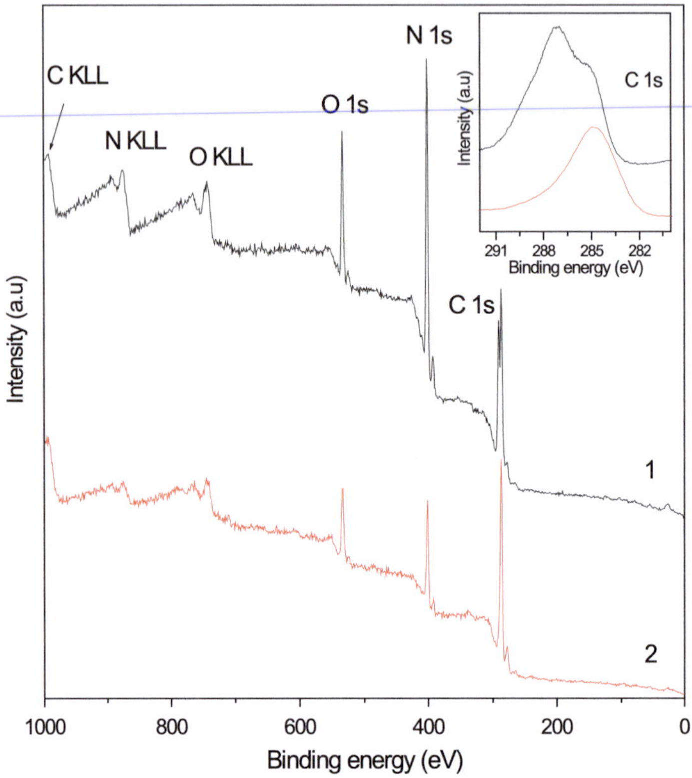

Fig. 19.4 XPS survey spectra of the samples (g-C_3N_4)O (curve 1) and O-g-C_3N (curve 2) with C1s spectrum of the samples (g-C_3N_4)O (curve 1) and O-g-C_3N (curve 2) in the *inset*

Distinctive feature of O1s spectra g-C_3N_4 from partially oxidized samples of carbon nitride (O-g-C_3N_4 and O-g-C_3N) is the absence in the first intensive line O1s at 531.5 eV. This intensive line in spectra both oxidized samples of carbon nitride testifies to presence at these samples bond C–O.

It is possible to see (Fig. 19.4) the survey scan XPS spectra of oxide of carbon nitride (sample O1) and O-g-C_3N are similar: both spectra contain (as well as a sample O-g-C_3N_4 [15]) three intensive peaks C1s, N1s and O1s and, hence, samples contain carbon, nitrogen and oxygen.

Wide and assimetric peaks C1s and N1s can specify the existence in openwork networks (g-C_3N_4)O and O-g-C_3N of atoms of nitrogen and carbon with the various valence status: their individual lines, undoubtedly, distinctly will be visible at detailed deconvolution of these peaks. At the same time, in a spectrum C1s of $(C_3N_{3.89})O_{1.09}H_{1.65}$ (Fig. 19.4, inset) it is possible see the two intensive peaks at ~284.5–285.3 and ~288–290 eV. These peaks are characteristic both g-C_3N_4 (284.7–285.7 and 287.8–288.2 eV [14, 24–26], and O-g-C_3N_4 (284.6 and 288.0 eV [15]). Asymmetry of peak C1s at 288–290 eV can be caused by a superposition of

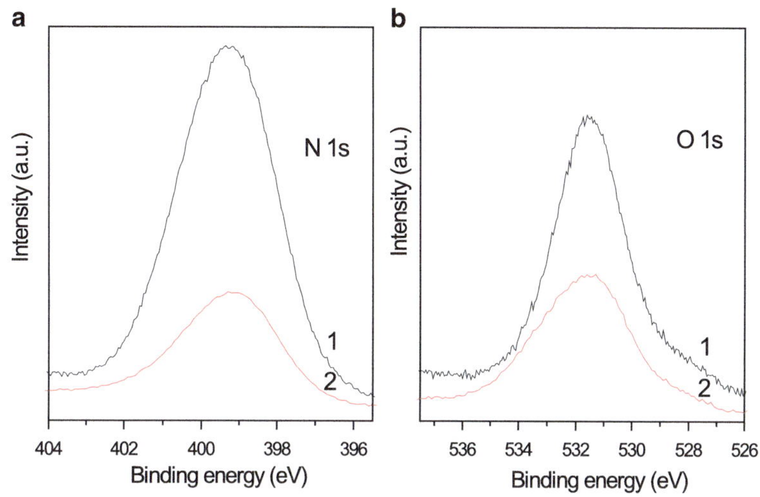

Fig. 19.5 XPS N1s (**a**) and O1s (**b**) spectra of the samples (g-C_3N_4)O (curves 1) and O-g-C_3N (curves 2)

two also of intensive peaks: carbon (at ~288 eV) connected with three atoms by nitrogen in a heptazine fragment (N = C–$(N)_2$), and carbon (at ~290 eV) connected with oxygen of hydroxyl group. Indeed, under the deconvolution of an asymmetric spectrum C1s of a sample O-g-C_3N_4 [15] the lines distinctly were visible both at 288.0 eV, and at 289.0 eV. Perhaps, that the asymmetric peak C1s at ~289–291 eV displays the contribution of weaker signals of carbonyl (C = O) and carboxyl (O–C = O) terminal functional groups, which, as will be shown later, distinctly are detected in the IR spectrum of a sample $(C_3N_{3.89})O_{1.09}H_{1.65}$.

Asymmetric N1s peaks of the XPS spectra (Fig. 19.5a) of the samples (g-C_3N_4)O and O-g-C_3N at ~398.5–400 eV are similar and, probably, are the superposition of two basic peaks with the values of binding energies ~398.5–398.7 and ~399.8–400.1 eV. These values it is accepted to correspond[21–23], to two types of nitrogen, bonded to two (C–N = C) and three (C_6N_7)–N–$(C_6N_7)_2$ neighboring carbon atoms, respectively. Appreciable asymmetry of N1s peaks at a few large value of binding energies (400–401 eV), perhaps, testify to the presence at structure of the researched samples also of nitrogen of amino group (C–NH_x), which at 401.0 eV is detected both in g-C_3N_4 [14, 24–26], and in O-g-C_3N_4 [15].

Figure 19.5b shows that O1s peaks for the samples (g-C_3N_4)O and O-g-C_3N are rather symmetric with the values of binding energy ~531.5 eV. In XPS the spectrum of O-g-C_3N_4 (~7.98 mass %) the peak O1s with the binding energy 531.6 eV corresponds, in opinion [15], to the bond N–C–O. Is remarkable, that in XPS a spectrum of oxygen-free g-C_3N_4 [3, 14, 26] there is only of weak intensity a line O1s with binding energy of 532.9 eV, characteristic for adsorbed H_2O.

So, the XPS spectrum of the water-soluble sample $(C_3N_{3.89})O_{1.09}H_{1.65}$ as well as the known spectra of O-doped carbon nitride (O-g-C_3N_4 [15] and O-g-C_3N [4, 5])

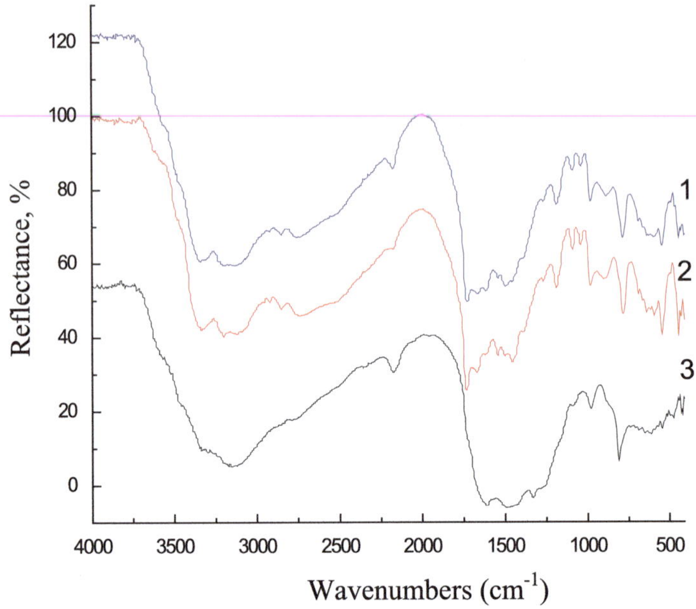

Fig. 19.6 FTIR spectra of samples O3, O1 and N2 (curves 1, 2 and 3 respectively)

contains intensive lines of carbon, nitrogen and oxygen. Valence status of elements in $(C_3N_{3.89})O_{1.09}H_{1.65}$ practically coincides with the binding energies of C, N and O in the samples of oxygen-free and O-doped carbon nitride.

Consequently, under thermal transformation of melamine molecules at the presence of small quantity of oxygen heptazine monolayers of carbon nitride partially are oxidized with formation of two new earlier not detected at melamine pyrolysis of products: a carbon nitride oxide $(C_3N_{3.89})O_{1.09}H_{1.65}$ and doped by oxygen (~8.1 %) carbon nitride $(C_3N_{3.92}O_{0.51}H_{1.30})$.

19.3.3 IR Spectra of the Samples of O-Doped Carbon Nitride (N1) and Water-Soluble Oxide of Carbon Nitride (O1 and O3)

IR spectra of the samples N1, O1 and O3 (Fig. 19.6, curves 3, 2 and 1, respectively) contain a number of distinct intense absorption bands in the region 1200–1650 cm^{-1}, which correspond to characteristic for g-C_3N_4 stretching vibrations of aromatic CN-bonds in the heptazine (C_6N_7) fragments of carbon nitride [14]. The spectra of all three samples contain the characteristic peak at 810 cm^{-1}, corresponding to the "breathing" vibration of triazine ring (C_3N_3). In contrast the IR spectra of samples g-C_3N_4, obtained by known methods, in the IR spectra of N1, O1 and O3 are present the absorption bands of hydroxyl group. The absorption band at 1135 cm^{-1} corresponds to C–O bond of hydroxyl group, but broad strong absorption bands at about 3200 and 3300 cm^{-1} correspond to the stretching vibrations of –OH and –NH groups respectively.

It is important to note that in almost identical IR spectra of water-soluble samples (O1 and O3) of carbon nitride oxide presents at 1728 cm^{-1} intense absorption band carbonyl (carboxyl) terminal groups, whereas in the IR spectrum of O-g-C_3N_4 (N1), which contains in 2 times (8.1 %) less amount of oxygen, this absorption band completely absents. In the presence of carboxyl groups in IR spectra of samples O1, O3 indicates also broadened band at ~2700 cm^{-1}, characteristic for –O–H bonds of the carboxyl group. Furthermore, in the IR spectra of samples O1, O3 weaker signals present at ~1270, 970 and 780 cm^{-1}, which may correspondent to the characteristic stretching vibrations of the epoxy groups. It is noteworthy that the absorption bands characteristic for given oxygen-containing groups (hydroxyl, epoxy, carbonyl and carboxyl) always present in the IR spectra of graphite oxide. Consequently, the IR spectrum of the sample synthesized carbon nitride oxide is original since it contains additional bands which are absented both in the known samples of g-C_3N_4 and oxygen doped carbon nitride.

19.3.4 Mass Spectrometric Study of Thermal Stability of the Powdery Samples Oxide of Carbon Nitride and O-g-C_3N_4

First of all, it is necessary to note, that the researches of the thermal stability of partially oxydized, in particular the doped by oxygen, samples of carbon nitride were not yet realized at all. The study of the thermostability of g-C_3N_4 (as extremely important characteristic for many areas of its application) with detailed mass spectrometric detecting of the gas-like products of the thermolysis also are quite unique. In [27] is shown, that the thermolysis products of g-C_3N_4 contain mainly of (CN)-fragments, such as $C_3N_2^+$ (m/z 64), $C_2N_3^+$ (m/z 66), CN_4^+ (m/z 68), $C_3N_3^+$ (m/z 78), $C_3N_4^+$ (m/z 92). The thermal decomposition of g-C_3N_4 is started at 500 °C, and the complete destruction of structure of g-C_3N_4 is finished at 850°C [27]. According to [28], carbon nitride obtained from dicyandiamide or melamine, is completely decomposed in an atmosphere of nitrogen in an interval of temperatures 600–800 °C by an evolution of smaller (CN) – fragments: C_2N_2, CNH_2, NH_3.

The synthesized sample of the water-soluble carbon nitride oxide compared to carbon nitride of formulaic composition naturally is less thermally stable. The main feature of the thermal decomposition of carbon nitride oxide (Fig. 19.7), as well as carbon nitride doped with oxygen is the presence in the composition of the gaseous products of their thermolysis both nitrogen-carbon (NH_3, HCN, C_2N_2), and oxygen containing (NCO, NCOH, CO, CO_2) fragments.

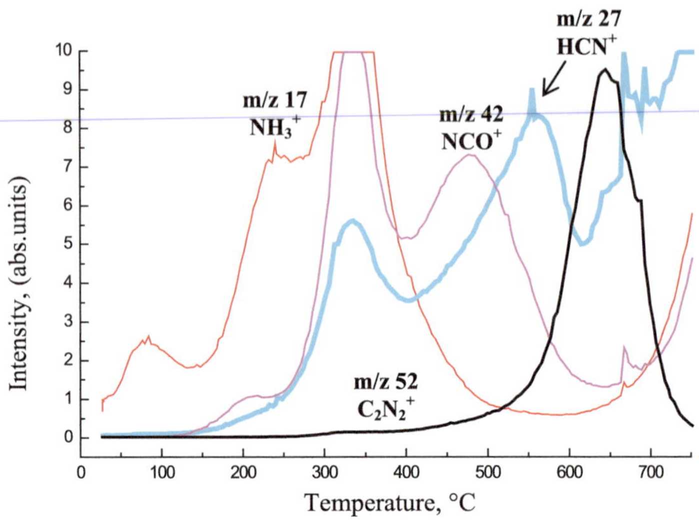

Fig. 19.7 Selected thermodesorption curves of sample O1

The data fragments that are detected at a temperature of thermolysis ~300 °C are formed, probably, due to the destruction of functional groups arranged in layers defects. Evacuation of cyanogen at thermolysis of (g-C_3N_4)O and O-g-C_3N_4 indicates on complete destruction of heptazine nitride-carbon layer, which is begun (according to mass spectrometric data) for both samples at ~600 °C. It should to note that known oxygen-free examples of carbon nitride (obtained from precursors such as dicyandiamide or melamine [28]) is completely decomposed in a nitrogen atmosphere in the temperature range 600–800 °C with evacuation of C_2N_2, CNH_2 and NH_3. However ions of CNH_2^+ and NH_3^+ are detected in the products of thermolysis of undoped carbon nitride already at about 450 °C.

Distinctive (from g-C_3N_4) feature of the thermal destruction of carbon nitride oxide and doped by oxygen of carbon nitride is that in products of them thermolysis alongside with nitrogen-carbon (NH_3, HCN, C_2N_2) present and nitrogenoxygen fragments (NCO, NCOH). The decrease of the thermal stability of compounds in a range g-C_3N_4 – O-g-C_3N_4 – (g-C_3N_4)O of an increase of the contents in the substances of oxygen is completely logical.

So, the study of melamine pyrolysis at the presence of oxygen has shown that the products of its transformation are essentially new substances. First of all, earlier the unknown substance as carbon nitride oxide $(C_3N_{3.89})H_{1.65}O_{1.09}$ as well as partially oxidized or doped by oxygen carbon nitride($C_3N_{3.92}O_{0.51}H_{1.30}$) are formed from melamine molecules in presence of oxygen. Let's note that is traditional at melamine pyrolysis in oxygen-free atmosphere were obtained only and only carbon nitride practically formulaic composition, g-C_3N_4. Notably, the ratio N/C in the synthesized samples $(C_3N_{3.89})H_{1.65}O_{1.09}$ and ($C_3N_{3.92}O_{0.51}H_{1.30}$) is also rather high, close to a ratio N/C in g-C_3N_4. It is possible to assume, that at the pyrolysis of melamine at presence a small quantity of oxygen the construction of heptazine monolayers $((C_6N_7)–N)_n$ of g-C_3N_4 from melamine molecules ($C_3N_6H_6$) or fragments ((C_3N_6), C_2N and CN) of them destruction is accompanied by parallel partial oxidation of a resultant "openwork" network.

The researched in this paper products of reaction, in particular, O-doped carbon nitride and it oxide are formed only in more low temperature (<400 °C) zone of reactor, far from a place of the localization of precursor. Therefore the reactions of the formation of $(C_3N_{3.89})H_{1.65}O_{1.09}$ and $(C_3N_{3.92}O_{0.51}H_{1.30})$ from melamine at the presence of oxygen are gas-transport. Melamine in most high-temperature (~550 °C) zone of reactor not only evaporates from a surface of a sample in the crucible, is occurs also the destruction of its molecules. Molecules of melamine from a vapour-gas phase are condensed in more low-temperature (<400 °C) reactor zone. Certainly, some part of melamine molecules is exposed the destruction and in a vapour-gas phase. Moreover, in a vapour-gas phase the formation not only heptazine fragments $((C_6N_7)–N)$ can be performed, but also larger fragments $((C_6N_7)–N)_n$ of "openwork" network, which are deposited, naturally, in a low-temperature zone of a reactor. However, as is established, the most part of a product as "openwork" network partially oxidized carbon nitride is formed in a low-temperature zone of a reactor. The formation of a heptazine fragment (C_6N_7) as basic element of construction of a nitrogen-carbon monolayers $((C_6N_7)–N)_n$ can be fulfilled at the expense of an interaction of melamine molecule with carbon-nitrogen radicals C_2N, CN, C_2N_3 and CN_3 as products of disintegration of a molecule $C_3N_6H_6$. The reaction of creation, for example, one fragment (C_6N_7) assumes the formation of four additional bonds C–N. Such reaction is high-exothermal. The escalating of a number of fragments $((C_6N_7)–N)$ in a "openwork" network reminds the growth on more high-temperature (in comparison with general temperature in the reactor) of thread-like crystals, for example, carbide of silicon from the vapours of silicon and carbon [29]. The high-temperature centre of growth of nanofilaments and nanotubes of carbon is created also at the expense of huge exoeffect of the escalating of the bonds C–C [30].

An addition of oxygen atoms to resultant fragments $((C_6N_7)–N)_n$ is fulfilled in a place of the destruction of the bonds C–N mainly between a tertiary amine nitrogen and a carbon of a heterocycle $((C_5N_7)C)–N$. The oxidation of heptazine fragments is extremely exothermal reaction. Therefore only in lack of oxygen their oxidation is not complete and is realized by the formation of partially oxidized forms of carbon nitride, in particular, doped by oxygen of carbon nitride $C_3N_{3.92}O_{0.51}H_{1.30}$ and oxide of carbon nitride $(C_3N_{3.89})O_{1.09}H_{1.65}$.

In an oxide of carbon nitride the oxygen-containing groups are placed mainly in the hollows of a plane of a heptazine monolayer limited by three of the next heptazine fragments (C_6N_7), whereas in graphite oxide as its structural analogue they are placed only between monolayers

Therefore an interplanar distance 0.324 nm ($2\theta = 27.49°$), characteristic for g-C_3N_4, in oxide of carbon nitride (g-C_3N_4)O also is kept. An occurrence in (g-C_3N_4)O of additional a little bit greater interplanar distance 0.414 nm ($2\theta = 21.45°$) is caused by a distortion of the planarity, "corrugating" of partially dearomatizated at oxidation of some heterocycles (–C = N– → –C(OH)–NH–) of "openwork" network. Besides, in oxide of carbon nitride is observed the increase (in comparison with g-C_3N_4) a distance between heptazine cycles in the "openwork" network with 0.714 ($2\theta = 12.4°$) up to 0.818 nm ($2\theta = 10.79°$), whereas in graphite oxide the structure of graphene layer is kept similar to graphite. This increase can be caused by a destruction of the bonds C–N between some heptazine fragments and tertiary amine nitrogen.

Consequently, at the conversation of carbon nitride to it oxide the more appreciable change is observed not in an interplanar space 002, as it is characteristic for graphite oxide, but in a plane 100 of a heteroatomic monolayer. First, the distance between the nearest heptazine fragments (C_6N_7)–(C_6N_7) in a monolayer is increased a little in the greater degree (on 0.104 nm), than between the heptazine monolayers (on 0.090 nm). Secondly, the oxygen containing groups are located mainly not in an interplanar spacing 002, as it is observed in graphite oxide, but in the increased hollows of a plane 100. Therefore in comparison with graphite oxide the ability for an exfoliation oxide of carbon nitride can be shown to a lesser degree. According to the described structure of oxide of carbon nitride there appeared quite logical results [16], in which the exfoliation of carbon nitride in rather rigid acid conditions is finished by the formation only nanosheets from several heteroatomic monolayers. It is possible only to hope, that the heptazine monolayer $((C_6N7)\text{–}N)_n$ (or azagraphene) will be successfully reduced and stabilized from a water-soluble oxide of carbon nitride as perspective material for nanosensors.

19.4 Conclusions

1. For obtaining of oxidized derivatives of carbon nitride as the perspective nanosensors is developed the special method of pyrolysis of melamine: first the pyrolysis of melamine was performed at the presence of oxygen.
2. New substance as carbon nitride oxide $(C_3N_{3.89})H_{1.65}O_{1.09}$ as well as doped by oxygen carbon nitride $(C_3N_{3.92}O_{0.51}H_{1.30})$ are formed at melamine pyrolysis at the presence of small quantities of oxygen. Earlier at melamine pyrolysis in oxygen-free atmosphere were obtained only the samples of practically oxygen-free carbon nitride.
3. Oxide of carbon nitride is the structure analogue of both carbon nitride, and carbon oxide. Perhaps, a water-soluble oxide of carbon nitride will be the precursor of hetero-atomic heptazine graphene or azagraphene.
4. Of yellow colour of the powder $(C_3N_{3.89})H_{1.65}O_{1.09}$ as against $(C_3N_{3.92}O_{0.51}H_{1.30})$ is easily dissolved in water with the formation of an yellow flake-like solution. It is possible, that these flakes represent ultra-thin nanosheets from the azagraphene layers.

References

1. Krainara N, Illas F, Limtraku J (2012) Interaction of adenine Cu(II) complexes with BN-doped fullerene differentiates electronically equivalent tautomers. Chem Phys Lett 537:88–931
2. Lv R, Li Q, Botello-Méndez AR et al (2012) Nitrogen-doped graphene: beyond single substitution and enhanced molecular sensing. Sci Rep 2(586):1–8
3. Thomas A, Fischer A, Goettmann F et al (2008) Graphitic carbon nitride materials: variation of structure and morphology and their use as metal-free catalysts. J Mater Chem 18:4893–4908
4. Kharlamov AI, Bondarenko ME, Kharlamova GA (2014) New method for synthesis of oxygen-doped graphite-like carbon nitride from pyridine. Russ J Appl Chem 87(9):1284–1293
5. Kharlamova G, Kharlamov O, Bondarenko M (2015) O-Doped Carbon Nitride (O-g-C_3N) with high Oxygen content (11.1 mass %) synthesized by pyrolysis of pyridine. In: Camesano TA (ed) Nanotechnology to aid chemical and biological defense. NATO science for peace and security series A: chemistry and biology, chapter 9. Springer, Dordrecht, pp 129–145
6. Zhang H, Huang Y, Hu S et al (2015) Fluorescent probes for "off–on" sensitive and selective detection of mercury ions and L-cysteine based on graphitic carbon nitride nanosheets. J Mater Chem C 3:2093–2100
7. Ma TY, Tang Y, Dai S, Qiao SZ (2014) Proton-functionalized two-dimensional graphitic carbon nitride nanosheet: an excellent metal-label-free biosensing platform. Small 10: 2382–2389
8. Lee SP, Lee JG, Chowdhury S (2008) CMOS humidity sensor system using Carbon Nitride film as sensing materials. Sensors 8(4):2662–2672
9. Zhang Z, Huang J, Yuan Q, Dong B (2014) Intercalated graphitic carbon nitride: a fascinating two-dimensional nanomaterial four an ultra-sensitive humidity nanosensor. Nanoscale 6: 9250–9256
10. Kruchinin S, Pruschke T (2014) Thermopower for a molecule with vibrational degrees of freedom. Phys Lett A 378:157–161
11. Ermakov V, Kruchinin S, Pruschke T, Freericks J (2015) Thermoelectricity in tunneling nanostructures. Phys Rev B 92:115531
12. Ermakov V, Kruchinin S, Fujiwara A (2008) Electronic nanosensors based on nanotransistor with bistability behaviour. In: Bonca J, Kruchinin S (eds) Proceedings NATO ARW "Electron transport in nanosystems". Springer, Berlin, pp 341–349
13. Medeiros RA, Matos R, Benchikh A et al (2013) Amorphous carbon nitride as an alternative electrode material in electroanalysis: simultaneous determination of dopamine and ascorbic acid. Anal Chim Acta 797:30–39
14. Dai H, Gao X, Liu E et al (2013) Synthesis and characterization of graphitic carbon nitride sub-microspheres using microwave method under mild condition. Diam Relat Mater 38: 109–117
15. Li J, Shen B, Hong Z, Lin B et al (2012) A facile approach to synthesize novel oxygen-doped g-C_3N_4 with superior visible-light photoreactivity. Chem Commun 48:12017–12019
16. Zhang X, Xie X, Wang H et al (2013) Enhanced photoresponsive ultrathin graphitic-phase C_3N_4 nanosheets for bioimaging. J Am Chem Soc 135:18–21
17. Kharlamov AI, Kharlamova GA, Bondarenko ME (2013) New products of a new method for pyrolysis of pyridine. Russ J Appl Chem 86(2):167–175
18. Kharlamov AI, Kharlamov GA, Bondarenko ME (2013) Preparation of onion-like carbon with high nitrogen content (~15 %) from pyridine. Russ J Appl Chem 86(10):1493–1503
19. Kharlamova G, Kharlamov O, Bondarenko M et al (2013) Hetero-carbon: heteroatomic molecules and nano-structures of carbon. In: Vaseashta A, Khudaverdyan S (eds) Advanced sensors for safety and security, NATO science for peace and security series B: physics and biophysics, Part VII. Springer, Dordrecht, pp 339–357
20. Xin G, Meng Y (2013) Pyrolysis synthesized g-C_3N_4 for photocatalytic degradation of methylene blue. J Chem 2013:1–5

21. Kentaro K, Naoya M, Chen Y et al (2013) Development of highly efficient sulfur-doped TiO_2 photocatalysts hybridized with graphitic carbon nitride. Appl Environ Microbiol 142–143: 362–367
22. Montigaud H, Tanguy B, Demazeau G et al (2000) C_3N_4: dream or reality? Solvothermal synthesis as macroscopic samples of the C_3N_4 graphitic form. J Mater Sci 35(10):2547–2552
23. Kulik TV, Barvinchenko VN, Palyanitsa BB et al (2007) A desorption mass spectrometry study of the interaction of cinnamic acid with a silica surface. Russ J Phys Chem A 81:83–90
24. Kim M, Hwang S, Yu JS (2007) Novel ordered nanoporous graphitic carbon nitride with C_3N_4 stoichiometry as a support for Pt–Ru anode catalyst in DMFC. J Mater Chem 17:1656–1659
25. Zhang Y, Liu J, Wu G, Chen W (2012) Porous graphitic carbon nitride synthesized via directly polymerization of urea for efficient sunlight-driven photocatalytic hydrogen production. Nanoscale 4:5300–5303
26. Dong F, Wu L, Sun Y et al (2011) Efficient synthesis of polymeric g-C_3N_4 layered materials as novel efficient visible light driven photocatalyst. J Mater Chem 21:15171–15174
27. Pozdnyakov OF, Pancakes LN, Arif M et al (2005) Mass spectrometry of carbon nitride C_3N_4. Tech Phys Lett 31:17–23
28. Fischer A (2008) "Reactive Hard Templating" from carbon nitrides to metal nitrides. Ph.D. thesis, Potsdam .
29. Kharlamov AI, Kirillova NV, Kaverina SN (2002) Hollow silicon carbide nanostructures. Theor Exp Chem 38(4):237–241
30. Kharlamov AI, Kirillova NV (2009) Fullerenes and fullerenes hydrides as products of transformation (polycondensation) of aromatic hydrocarbons. Rep Acad Sci Ukr 5:112–120 (Russian)

Chapter 20
Characterization of SnO_2 Sensors Nanomaterials by Polarization Modulation Method

V.S. Grinevych, L.M. Filevska, V.A. Smyntyna, M.O. Stetsenko, S.P. Rudenko, L.S. Maksimenko, and B.K. Serdega

Abstract The polarization characteristics for tin dioxide cluster films are studied by the method of modulation polarization spectroscopy. The galvanic conductivity presence in the films is the basis for registration in it the surface plasmon resonances. The spectral characteristics analysis by means of expansion in terms of Gauss components for the Stocks vector Q component of the probe radiation gave the parameters of the revealed resonances. The resonance excitation of polaritons and localized surface plasmons is established. The dispersion characteristics of non-radiative modes of surface plasmons are obtained which matches the cluster film structure. The numerical values comparison for resonances relaxation constants leads to the conclusion of application of one of them in sensors.

20.1 Introduction

The unique chemical resistance in an aggressive medium, specific adsorptive-catalytic, electrophysical and optical properties, cheapness in production and possibility to create Tin dioxide nano size forms secures its leading position among materials used for ecological monitoring [1, 2]. Nevertheless it is not the only sphere for application of this material. The resistance sharp decrease property for sensitive elements based on SnO_2 films at pressure decrease is successfully used in pressure transducers [2]. The conductive SnO_2 films are actively used as transparent electrodes in sun cells [3].

V.S. Grinevych (✉) • L.M. Filevska • V.A. Smyntyna
Department of Experimental Physics I.I. Mechnikov, Odessa National University (ONU), Dvoryanskaya str., Odessa 65082, Ukraine
e-mail: grinevich@onu.edu.ua

M.O. Stetsenko • S.P. Rudenko • L.S. Maksimenko • B.K. Serdega
V.E. Lashkarev Institute of Semiconductors Physics, National Academy of Sciences of Ukraine (NASU) Nauky Prospect, 45, Kyiv 03680, Ukraine

J. Bonča, S. Kruchinin (eds.), *Nanomaterials for Security*, NATO Science for Peace and Security Series A: Chemistry and Biology, DOI 10.1007/978-94-017-7593-9_20

The nonstoichiometric oxides, SnO_x is among them, may become the basic material in the nearest future for new elements of computer energy independent memory-memristors, which modify their electric resistance under the current influence [4]. The application of nanoparticles of the dioxide composite material built in graphene nanoribbons, as anodes in Lithium Ions batteries in mobile electron technique were recently reported [5]. This may seriously enhance the productivity and effectiveness of such batteries and to achieve high initial charge and discharge potentials [6].

The use of optical methods of gas and biological objects detection in liquid and gaseous mediums became the new stage in nanosized tin dioxide application [7].

The tin-dioxide physical properties are principally dependent on the methods of its production. Among them are expensive ion-plasma, electron – beam deposition, magnetron spray and more cheap chemical methods – electric spray pyrolysis, chemical deposition from vapour phase, sol-gel method [8–10]. These methods are used for obtaining thin SnO_2 films of high defectness and as a result of high conductivity.

Free electrons presence in them secures the surface plasmon resonance (SPR) effects developing. The surface plasmons excitation by electromagnetic radiation (EM) is specific for nanosized structures, containing nanoparticles (NPs) of noble metals and may be registered by the plasmon adsorption in visible, nearest UV and IR wave length range [11]. The spectral and angular SPR characteristics of nanostructured metal-dielectric films are very sensitive to the dielectric permittivity variation of the contacting with them outer medium [12]. This is widely used for monitoring. Addition of gold (Au) on the SnO_2 surface in [13] enhances a sensor best performance due to response (s) and time of response (t).

Previously in our [14] the SPR existence in SnO_2 films without Au NPs was shown. At the same work the effectiveness of modulation-polarisation spectroscopy method (MPS) was shown, as an alternative method for diagnostics and characterization of SnO_2 films. The present work is a continuation of SnO_2 films characterization by MPS method aiming the determination of optimal resonance-optical SPR parameters in dependence on the resonance type (localized or polariton one) and also the structure morphology specificity aiming future application in sensors.

20.2 Samples and Experimental Details

Samples for the investigation were prepared by the technique described in detail in [15]. Bis(acetylacetonato)dichlorotin (BADCT) was used as a tin dioxide precursor [16]. Freshly prepared BADCT was dissolved in acetone at different concentrations, then equal volumes of each solution were mixed with the same volumes of polyvinylacetate (PVAC) solutions in acetone prepared at different concentrations. The mixtures were then sprayed onto the microscope cover glass of $22 \times 22\,mm^2$ size. Samples were kept at room temperature for about 15 min to allow the acetone

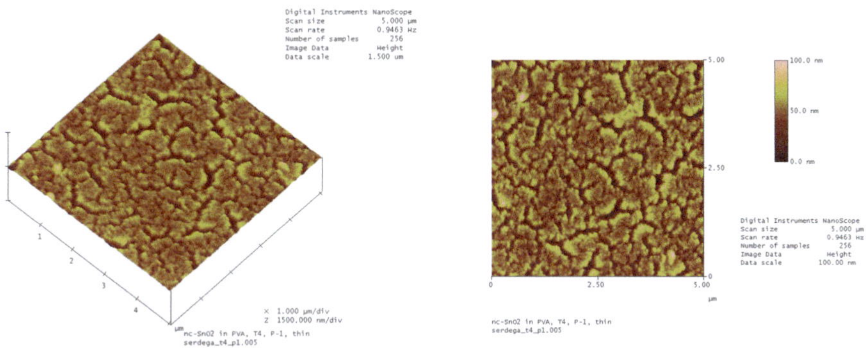

Fig. 20.1 AFM images of SnO_2 films

removing prior to annealing them at 600 °C for 6 h in air to achieve the thermal decomposition of the film organic components (BADCT and PVAC) and subsequent removing decay products. The organic components removing were confirmed by our thermogravimetric studies of the precursor [16] and by the data on the PVAC decomposition at temperatures above 200 °C, particularly in the presence of catalytic oxides (tin dioxide in our case). After annealing, the tin dioxide film was left on the substrate. PVAC was employed to structurize the film during removing its decay products. Force Microscopy (AFM) method (NanoScope IIIa, Digital Instruments). The tapping mode measurements were carried out with use of a silicon probe with nominal radius of about 10 nm. The AFM phase topology image of the sample T4.P1 (4 wt. % BADCT and 1 wt. % PVAC in the initial solution) is shown in Fig. 20.1a.

The optical scheme of experimental technique for spectral and angular SPR characteristics measurement based on the Kretchman [17] geometry and improved by the MPS method is described in details in [18]. The method is based on modulation of the EM radiation polarization. The idea of modulation is the periodical variation of probing radiation polarized state with constant intensity, frequency, phase and wave vector. The magnitude of the registered signal at the modulation frequency ($f = 60\,\text{kHz}$) is equal to the intensities difference of the reflection coefficients s and p polarized radiations $\rho(\lambda, \theta) = R_s^2 - R_p^2$ – the polarization difference. According to the conventional terminology in polarimetry [19] this parameter is a Q-component of Stocks vector within the EM radiation which is reflected in this case by SnO_2 film fixed on a semicylinder surface and sensible to resonance. The phase locking detection of angular and spectral polarization characteristics increases the information ability of $\rho(\theta, \lambda)$ parameter, both due to the dynamic range extension of the measured quantity, and due to different signs of amplitude signals. The parameter ρ is more convenient in detection with high sensitivity to the peculiarities morphology of structure due to simultaneous measuring of R_s^2 and R_p^2 components under interaction between radiation and sample.

20.3 Results and Discussions

The angular characteristics of the internal reflection coefficients $R_s^2(\theta)$ and $R_p^2(\theta)$ and their polarization difference $\rho(\theta)$ at $\lambda = 500$ nm for SnO_2 films are shown at Fig. 20.2. For the incidence angles θ, more than the critical angle of total internal reflection $\theta_{cr} = 43°$, the presented characteristics minima, are caused by the surface plasmons excitation for both *s*- and *p* polarized radiation.

The effectiveness of the resonance interaction of *s*-polarization in the present case exceeds over *p*-polarization (R_{sp}^p and R_{sp}^s are oscillators' forces, excited by corresponding polarized radiation). The negative values of $\rho(\theta)$ characterize the SnO_2 films structure as cluster one and nonuniform, which corresponds to the results of investigation of ultra thin and cluster gold films in [20].

In dependence on the structure peculiarities of the films studied, the SPR excitation is subdivided to several types: the excitation of the surface plasmon-polaritons on the infinite plane boundary of the metal-dielectric surface and the excitation of surface plasmons, localized on separate not interacting NPs or between NPs of metal due to dipole-dipole interaction. The spectral characteristics of the polarization difference $\rho(\lambda)$ at different incidence angles $\theta = 45°, 55°, 65°$ are shown at the Fig. 20.3a for characterization of SnO_2 films, and SPR types definition. The complicated character of the parameter $\rho(\lambda)$ spectral contour is connected with different types of SPR in the films. The variable in signs form of curves $\rho(\lambda)$ caused by the resonance response from both polarizations at the phase-locked detection allows to interpret each extremum as a result of the resonant excitation of the surface plasmons of different nature.

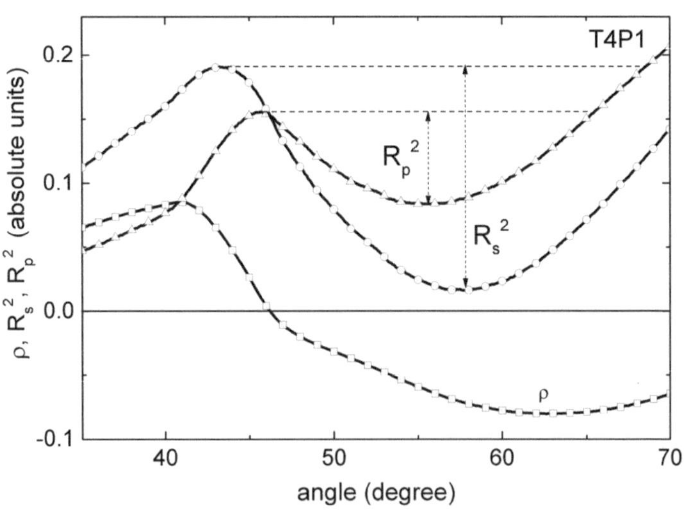

Fig. 20.2 Angular dependences of the internal reflection coefficients $R_s^2(\theta)$, $R_p^2(\theta)$ and polarization difference $\rho(\theta)$ at $\lambda = 500$ nm

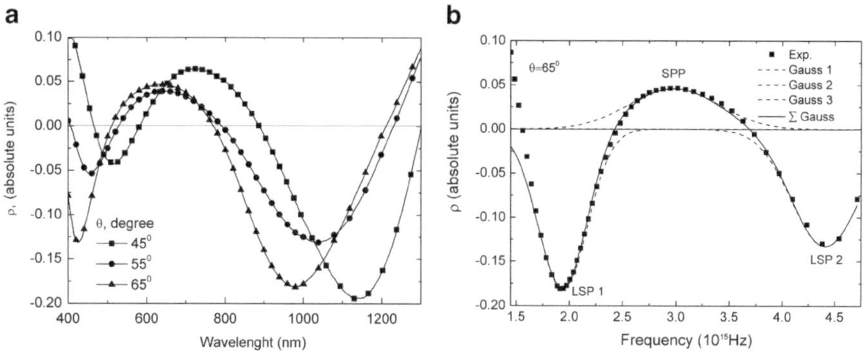

Fig. 20.3 (**a**) – spectral dependences of polarization difference $\rho(\lambda)$ for SnO_2 films at incident angles of $\theta = 45$ (*square*), 55 (*circle*), 65 (*triangle*), degrees. (**b**) – frequency dependence of polarization difference $\rho(\omega)$ at $\theta = 65°$ (*marks*) recalculated from the corresponding dependence of $\rho(\lambda)$ in comparison with the Gaussian distribution function (*solid line*) expanded into components (*dashed line*)

The negative sign extrema in $\rho(\lambda)$ spectra are conditioned by the resonant excitation of the localized surface plasmons in a short wavelength region ($\lambda_{LSP1} \sim 500$ nm) on clusters or surface structure inhomogeneity, but in a long wavelength region ($\lambda_{LSP2} \sim 1000$ nm) between clusters due to electrodynamic/ dipole-dipole interaction. In both cases, the simultaneous resonant interaction of both *s*-, and *p*-polarized radiation with SnO_2 film occurs. Extrema with positive $\rho(\lambda)$ values at $\lambda_{SPP} \sim 650$ nm are conditioned by the surface plasmon-polariton excitation in the interface film-air, consequently, as a result of resonant interaction only for p-polarized radiation. It is to be mentioned, that the "blue shift" occurs for corresponding extrema, which is conditioned by the considerable thickness $d = 230$–400 nm of the SnO_2 films. Such peculiarities of $\rho(\lambda)$ curves are specific for cluster type and inhomogenous structure films [21].

Aiming the definition of every SPR parameters, the $\rho(\lambda)$ characteristics were studied by means of Gaussian functions approximation. As a result, the amplitude and relaxation resonance parameters were obtained. The frequency dependence of $\rho(\omega)$ of one of $\rho(\lambda)$ curves at $\theta = 65°$ was shown at Fig. 20.3b. The following principal frequencies and FWHM parameters (full width at a half maximum) are obtained: $\omega_{LSP1} = 1.9 \times 10^{15}$ Hz, $\gamma_{LSP1} = 0.73 \times 10^{15}\,s^{-1}$; $\omega_{SPP} = 2.86 \times 10^{15}$ Hz, $\gamma_{SPP} = 0.994 \times 10^{15}\,s^{-1}$; $\omega_{LSP2} = 4.43 \times 10^{15}$ Hz, $\gamma_{LSP2} = 0.544 \times 10^{15}\,s^{-1}$. In all cases considered the surface plasmons excitation has non radiative nature and needs to secure phase-synchronous conditions, when both frequency (ω) and wave vector (k) of light excitation matches to those of the surface plasmon's frequency and wave vector.

The resonances frequencies values, obtained from the components' extrema (dash lines), both with the wave vectors magnitudes, considering the incidence angle (θ), permit to built the dispersion characteristics $\omega(k)$ of the surface plasmons (Fig. 20.4). The dispersion curves shape, their position in $\omega(k)$ coordinates, and

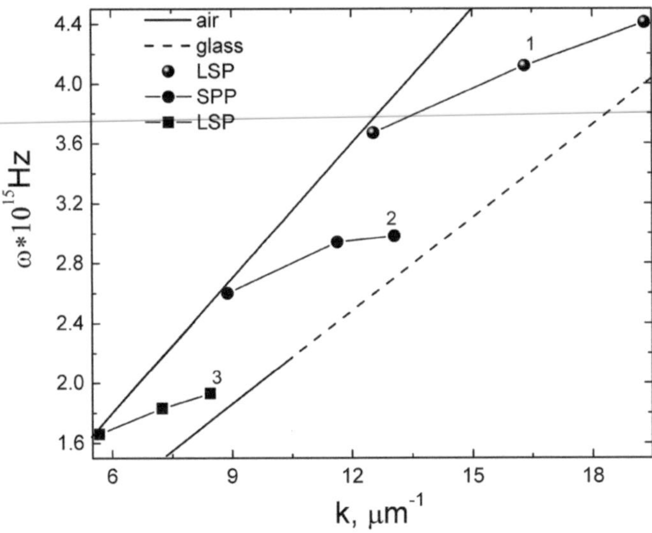

Fig. 20.4 Dispersion characteristics $\omega(k)$ of surface plasmons localized on isolated clusters (1), between clusters caused by dipole fields' interactions (3), and surface plasmon-polaritons (2) for SnO_2 films

also AFM images, makes it possible to identify two of them, correspondingly, as excitation of the surface plasmons, localized on separate, not interacting clusters, and interclusters interaction and the third – the excitation of surface plasmon-polaritons.

The angle dependencies of the relaxation parameters, γ, connected with SPR type is presented at Fig. 20.5. It is evident, that the biggest γ values corresponds to the surface plasmon-polariton excitation. According to the least γ parameter value rule, the SPR intercluster interaction dependence is the most appropriate for sensors applications. It is evident, the parameter value $\gamma_{LSP2} = 0.472 \times 10^{15}\,s^{-1}$ at the angel $\theta = 45°$, provides the maximum $\rho(\lambda)$ characteristic slope. It's shift as a result of the film interaction with an external detected medium, secures the maximum effectiveness in the sensors applications.

20.4 Conclusions

The characterization of resonant-optical parameters and morphology features of sensor – SnO_2 films is fulfilled by modulation – polarization spectroscopy method. Two SPR types are registered, which are conditioned both by the surface plasmon-polaritons excitation and localized surface plasmons on separate not interacting clusters and between clusters due to dipole interaction. The polarization analysis of SnO_2 research showed the inhomogeneity and cluster type of the structure

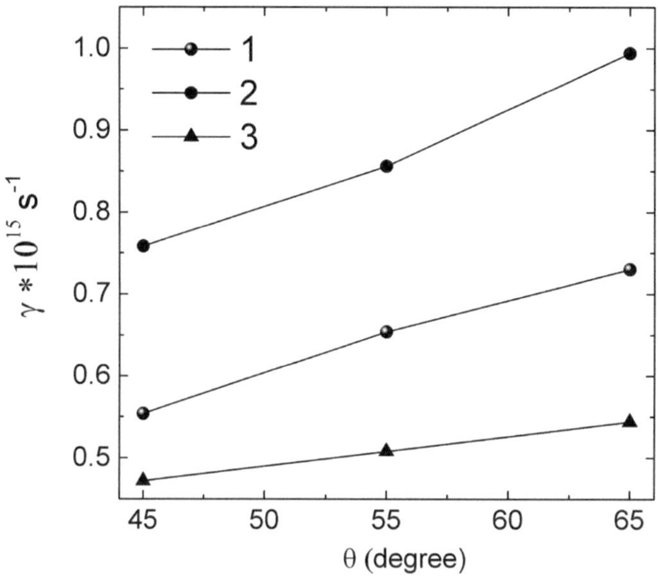

Fig. 20.5 Angular dependences of the relaxation parameters of SPR characteristics for SnO_2 films

morphology. The optimal resonance-optical parameters were obtained for SPR intercluster inter action, which supposes and secures their angular and spectral characteristics maximum effectiveness in the sensors applications.

References

1. Batzill M, Diebold U (2005) The surface and materials science of tin oxide. Prog Surface Sci 79:47–154
2. Ganesh EP, Kajale DD, Chavan DN, Pawar NK, Ahire PT, Shinde SD, Gaikwad VB, Jain GH (2011) Synthesis, characterization and gas sensing performance of SnO_2 thin films prepared by spray pyrolysis. Bull Mater Sci 34(1):1–9
3. Ginley DS, Hosono H, Paine DC (2010) Handbook of transparent conductors. Springer, New York
4. Ryabtsev SV (2011) Electrical and optical properties of various tin oxide nanoforms. Doctor's dissertation Thesises, Voronezh
5. Lin J, Peng Zh, Xiang Ch, Ruan G, Yan Zh, Natelson D, Tour JM (2013) Graphene nanoribbon and nanostructured SnO_2 composite anodes for lithium ion batteries. ACS Nano 7(7):6001–6006
6. Xia G, Li N, Li D, Liu R, Wang Ch, Li Q, Lü X, Spendelow JS, Zhang J, Wu G (2013) Graphene/Fe_2O_3/SnO_2 ternary nanocomposites as a high-performance anode for lithium ion batteries. ACS Appl Mater Interfaces 5(17):8607–8614
7. Wu R, Chen X, Hu J (2012) Synthesis, characterization, and biosensing application of ZnO/SnO_2 heterostructured nanomaterials. J Solid State Electrochem 16(5):1975–1982. doi:10.1007/s10008-011-1590-6

8. Joshi BN, Yoon H, Yoon SS (2013) Structural, optical and electrical properties of tin oxide thin films by electrostatic spray deposition. J Electrost 71(1):48–52
9. Inchidjuy P, An KS, Pukird S (2013) Growth and characterization of SnO_2 nanostructures by vapor transport technique. Adv Mater Res 677(3):94–97
10. Gnanam C, Rajendran V (2010) Synthesis of tin oxide nanoparticles by sol-gel process: effect of solvents on the optical properties. J Sol-Gel Sci Technol 53(3):555–559
11. Maier SA (2007) Plasmonics: fundamentals and applications. Springer, New York, p 221
12. Orfanides P, Buckner TF, Buncick VC, Meriaudeau F, Ferrell TL (2000) Demonstration of surface plasmons in metal island films and the effect of the surrounding medium – An undergraduate experiment. Am J Phys 68(10):936–942
13. Bakrania SD, Wooldridge MS (2010) The effects of the location of Au additives on combustion-generated SnO_2 nanopowders for CO gas sensing. Sensors 10(7):7002–7017
14. Grinevich VS, Filevska LM, Matyash IE, Maximenko LS, Mischuk ON, Rudenko SP, Serdega BK, Smyntyna VA, Ulug B (2012) Surface plasmon resonance investigation procedure as a structure sensitive method for SnO_2 nanofilms. Thin Solid Films 522:452–456
15. Filevskaya LN, Smyntyna VA, Grinevich VS (2006) Morphology of nanostructured SnO_2 films prepared with polymers employment. Photoelectronics 15:11–14
16. Ulug B, Türkdemir HM, Ulug A, Büyükgüngör O, Yücel MB, Grinevich VS, Filevskaya LN, Smyntyna VA (2010) Structure, spectroscopic and thermal characterization of bis(acetylacetonato)dichlorotin(IV) synthesized in aqueous solution. Ukr Chem J 7:12–17
17. Kretzschmann E, Raether H (1968) Radiative decay of non-radiative surface plasmons by light. Z Naturforsch A 23:2135
18. Berezhinsky LJ, Maksimenko LS, Matyash IE, Rudenko SP, Serdega BK (2008) Polarization modulation spectroscopy of surface plasmon resonance. Opt Spectrosc 105(2):257–264
19. Gerrard A, Burch JM (1975) Introduction to matrix methods in optics. Wiley, London
20. Serdega BK, Rudenko SP, Maksimenko LS, Matyash IE (2011) Plasmonic optical properties and the polarization modulation technique. In: Mishchenko MI, Yatskiv YS, Rosenbush VK, Videen G (eds) Polarimetric detection, characterization and remote sensing. Springer, Dordrecht, pp 473–500
21. Vozny AA, Stetsenko MO, Kosyak VV, Maksimenko LS, Opanasyuk AS, Serdega BK (2015) Detection of structural characteristics of nanosized SnxSy film by the modulation-polarization spectroscopy of plasmon resonance. Proc NAP 4, 02 NAESP06

Chapter 21
Diagnostic of Resonant Properties of Au-PTFE Nanostructures for Sensor Applications

L.S. Maksimenko, S.P. Rudenko, M.O. Stetsenko, I.E. Matyash, O.M. Mischuk, Yu.V. Kolomzarov, and B.K. Serdega

Abstract Diagnostic of nanocomposite films formed by gold nanoparticles embedded in a polytetrafluorethylene matrix has been performed utilizing modulation polarization spectroscopy technique. The dependencies of annealing influence on morphology structure and LSPR parameters of Au-PTFE films were studied and discussed. Effective optical parameters of Au-PTFE films were obtained. Two types of LSPR were detected: the first is on isolated non-interacting Au NPs and the second – between Au NPs caused by dipole fields' interactions. The experimental data were confirmed by theoretical calculations. Radiative and non-radiative modes of surface plasmons were studied. Plasma frequencies were obtained for Au-PTFE films with unheated structure and annealed at temperatures of 300 and 520 °C by measuring of spectral characteristics of the angle of isotropic reflection $\theta_{\rho=0}(\lambda)$.

21.1 Introduction

Properties and behavior of small gold nanoparticles (Au NPs), such as their melting point, density, lattice parameter, and optical properties, are significantly different from those of bulk materials. Gold is technologically important among of many elements and compounds from which NPs can be made, because of its low electrical resistivity, more biocompatible, and stability into environment [1]. Numerous practical applications of metal NPs require them to be well dispersed into or on surface of various substrates and matrices without formation of large aggregates at low chances to coagulating. Polymers are the best host materials, which commonly are used for improvement of mechanical stability of NPs. Polytetrafluorethylene (PTFE) films possess a high thermal stability and also have a high degree resistance to any chemical attack and excellent transparency. They are generally low-cost

L.S. Maksimenko • S.P. Rudenko (✉) • M.O. Stetsenko • I.E. Matyash • O.M. Mischuk •
Yu.V. Kolomzarov • B.K. Serdega (✉)
V.E. Lashkaryov Institute of Semiconductors Physics, National Academy of Sciences of Ukraine (NASU) Nauky Prospect, 45, Kyiv 03028, Ukraine
e-mail: bserdega@isp.kiev.ua; rudenko.svetlana@gmail.com

J. Bonča, S. Kruchinin (eds.), *Nanomaterials for Security*, NATO Science for Peace and Security Series A: Chemistry and Biology, DOI 10.1007/978-94-017-7593-9_21

materials, versatile, and can easily be processed into any shape including thin films. PTFE can be used in waveguide as materials with low and high refractive indices [2–4].

Nanocomposite Au-PTFE film is a combination of AuNPs embedded in PTFE matrix. In these systems small particle sizes enhance the optical properties while the polymer matrix materials will act to stabilize the particle size and growth besides providing the required transparent medium. Polymer-metal nanocomposites are promising candidates for many applications [5] as in chemical catalysis, biochips, surface-enhanced Raman spectroscopy, optical and photonic sensor devices for environment monitoring including hazardous processes in industry [6–8].

The sensing principle relies on the high sensitivity of the localized surface plasmon resonance (LSPR) spectral shifts caused by the surrounding dielectric environmental change in a binding event [9, 10]. LSPR is one of the signature optical properties of noble metal NPs, which arises when the incident photon frequency is resonant with the collective oscillation of the conduction electrons in the metal NPs. The bandwidth, intensity and position of the LSPR wavelength of nanocomposites depend on the amount, shape and size of the metal clusters in the film, as well as on the interparticle separation [11, 12]. The great significance of the polymer-metal nanocomposites is the ability to tune and extend the plasmon resonance wavelength, which is highly desirable in various applications of metallic NPs.

In the field of sensor security, much attention is paid to new approaches of control and diagnostics for the resonant parameters, morphology structures, the concentration and manipulation of light to improve the absorption and/or transmission for energy of sensor plasmonic films [13]. The aim of this study is to characterize and determine of optical and polarization parameters of LSPR, which are necessary for sensor applications of Au-PTFE nanocomposites. In this context, we investigated the influence of different morphology structure on resonant parameters of LSPR for Au-PTFE films utilizing the modulation polarization spectroscopy (MPS) technique. The effectiveness of MPS technique was demonstrated in our previous studies of both continuous and porous nanosized Au and Ag films [14, 15] as well as nanocomposite structures containing Au NPs [16, 17]. Two type of LSPR (surface plasmon's excitations on isolated non-interacting Au NPs and between adjacent Au NPs due to dipole-dipole interactions) were studied by means of analysis of angular and spectral characteristics of polarization difference $\rho(\lambda, \theta) = R_s^2 - R_p^2$ of the internal reflection coefficients of s- and p-polarized radiation (perpendicular R_s^2 and parallel R_p^2 to the incident plane, respectively). Radiative and non-radiative modes of surface plasmons were considered. The obtained experimental results contain the information about the morphological peculiarities of Au-PTFE films and parameters of LSPR, such as amplitude, frequency position, and FWHM (full width at half maximum). Plasma frequency oscillations were determined for each Au-PTFE films. Effective optical parameters of Au-PTFE films were obtained. These studies allowed performing rapid diagnosis of Au-PTFE nanocomposites with different morphology structure and receiving their resonant and optical parameters, which are necessary in further sensor applications.

21.2 Samples

Nanocomposite films formed by AuNPs embedded in PTFE matrix were made by thermal vacuum co-deposition (pressure of 4×10^{-2} Pa). One-millimeter-thick quartz plates were used as substrates. PTFE was evaporated with vapor activation by electron cloud. The heated tungsten boat was used for gold evaporation. Further details can be observed in previous works [18–22]. Deposition rate was 6 nm/min. Gold concentration in Au-PTFE films was 27.4 % of the mass or 4.12 % from the film volume. The amount of deposited Au and PTFE was controlled in situ with a quartz thickness gage (Sigma Instruments SQM-242), frequency shift by the quartz sensor readings with the gold was 747.6 Hz, and with the teflon – 1983.7 Hz. The film thickness was determined by means of the height of a step profile from an Atomic force microscopy (AFM) Nanoscope IIIa (Digital Instrument, USA). The morphology and thickness were studied using a periodic contacts regime with silicon AFM probes (a rated end point radius up to 10 nm). Annealing was carried out in a homemade oven in air for various time intervals at temperatures of 300 and 520 °C.

The AFM images for Au-PTFE films are demonstrated in Fig. 21.1. The unheated film (sample 1) shows that shape of Au NPs was somewhat spheroidal, but not

Fig. 21.1 AFM images of Au-PTFE films: sample 1 – unheated; sample 2 – annealed at 300 °C; sample 3 – annealed at 520 °C

completely spherical. Although, in order to achieve a high degree of accuracy, one should suppose that Au NPs has ellipsoidal shapes. A higher Au concentration is characterized by a bigger NP size. At the higher concentration NPs aggregation is presented on the image. PTFE matrix prevents formatting of smooth Au film. Also Au NPs keeps formation of PTFE phase in thermodynamically disequilibrium conditions. It is amorphous. The root-mean-square (RMS) roughness equals to ~4.65 nm. The mass thickness of sample 1 is ~24 nm with average particle size of ~59 nm at a height drop of ~13.5 nm. After annealing at 300 °C the film (sample 2) exhibits more uniform structure with smaller size of Au NPs at approximately the same mass thickness of gold. The RMS roughness and mass thickness are decreased to ~1.01 and ~22 nm, respectively, at NPs size of ~10 nm and a height drop of ~7.1 nm. Similar results of influence of annealing on the surface morphology of Au films at certain heat temperature were observed in the works of Romanyuk et al. [23] and Worsch et al. [24].

Annealing at higher temperature up to 520 °C lead to a drastic change in morphology structure (sample 3) at increasing of Au NPs-size of 223 nm with a slight difference in height drop of ~3.7 nm. In this case the RMS roughness and mass thickness are decreased to ~0.77 and ~13 nm, respectively. The clusters with Au NPs can be merged into bigger aggregates, which were made possible due to a pyrolysis of PTFE at $T > 500\,°C$. As shown in Madorsky et al. [25], the pyrolysis of PTFE occurs relatively slowly only up to 420 °C. Moreover the size and inter-particle distance of Au NPs depend on annealing time and temperature. High temperature annealing tended to increase the size and inter-islands distance of Au islands [26].

21.3 Experimental Details

Polarization characteristics of Au-PTFE films were measured in Kretschmann geometry using the MPS technique. Setup scheme was described in detail in the work of Berezhinsky et al. [27]. The probe radiation in case of modulation of polarization state had constant intensity, frequency, phase, and wave vector, with polarization state as the only variable. A diffraction monochromator MDR-4 (with a halogen tube KGM-150 at the input and Franck–Ritter polarizer (FRP) at the output) served as a source of spectral radiation in the wavelength range $\lambda = 0.4$–$1\,\mu m$. A photoelastic polarization modulator PEPM [28] acted as dynamic phase plate. Alternating phase incursion was caused by compression/expansion of the quartz plate. A quarter-wave mode was selected by a proper supplying voltage. As a result, linear polarization was transformed into alternating right-to-left circular polarization. A stationary quarter-wave phase plate (PP) was placed after the PEPM. By rotating PP around the optical axis a position of PP was selected at which polarization azimuths of radiation after PP alternated between parallel and perpendicular positions relative to the incident plane (p- and s-polarization, respectively). The radiation with modulated polarization state is directed onto the system of quartz half-cylinder with the Au-PTFE film. The output radiation was directed at a photodetector PD (silicon photodiode). Reflected light was a measure of the difference of orthogonally polarized intensities, which was transformed by a PD into alternating signal. This signal was registered by a selective amplifier equipped with a phase-lock detector (lock-in-voltmeter) tuned to the modulation frequency ($f = 60\,\text{kHz}$). The variable component value is a magnitude of the difference of internal reflection coefficients of s- and p-polarizations $\rho = (R_s^2 - R_p^2)\sin(\omega t)$. The resulting value is the polarization difference $\rho(\lambda,\theta) = R_s^2 - R_p^2$. If necessary, the angular and spectral dependences of $R_s^2(\lambda,\theta)$ and $R_p^2(\lambda,\theta)$ reflection coefficients separately were measured. The magnitude of polarization difference $\rho(\theta)$ equals zero at particular incident angles. In that case, the reflection coefficients of s- and p-polarized radiation have equal amplitude values, i.e. $R_s^2(\theta) = R_p^2(\theta)$. Hence, light reflection occurs regardless of polarization state at the angles of isotropic reflection $\theta_{\rho=0}$ as defined in the work of Serdega et al. [14]. The parameter ρ is the analog of Q-component of the Stokes parameters and also describes the amplitude characteristics of samples. The parameter ρ is more convenient in detection with high sensitivity to the peculiarities morphology of structure due to simultaneously measuring of R_s^2 and R_p^2 components under interaction between radiation and sample.

21.4 Experimental Results and Discussion

In Fig. 21.2 angular characteristics of polarization difference $\rho(\theta)$, internal reflection coefficients of p-polarized $R_p^2(\theta)$ and s-polarized $R_s^2(\theta)$ radiation at a wavelength $\lambda = 600\,\text{nm}$ for Au-PTFE films are shown. In the range of incident angles more than the critical angle of total internal reflection, $\theta > \theta_{\text{cr}} \sim 43°$, the resonant parts in the dependencies of $R_p^2(\theta)$ and $R_s^2(\theta)$ are very poorly visible and their detection can be difficult. In our previous works [29, 30] the negative sign of $\rho(\theta)$ in the range of $\theta > \theta_{\text{cr}} \sim 43°$ indicates to the presence of a medium with high optical density (e.g. Au-PTFE film) when anomalous reflection manifests itself $R_s^2(\theta) < R_p^2(\theta)$.

Negative values and positive curvature of $\rho(\theta)$ indicates in the resonant excitations of surface plasmons by both s- and p-polarized radiation, which correspond to the clustered structure. Increasing of amplitude value of $\rho(\theta) < 0$ in the direction from sample 1 toward more annealed sample 3 is due to changing of morphology structure up to clusters aggregation. These are correlated to decreasing of mass thickness of films from 24 nm (1), 22 nm (2) to 13 nm (3) and corresponded to changing of optical parameters. Note that all curves of $\rho(\theta)$ intersect x-axis (axis of angles) close to the θ_{cr}. A small shift takes place for sample 3 and indicates in significant changes in morphology structure due to annealing at a drastically high temperature. Agreement between theoretically calculated and experimentally measured data was obtained due to a fitting procedure in which the Fresnel formulas

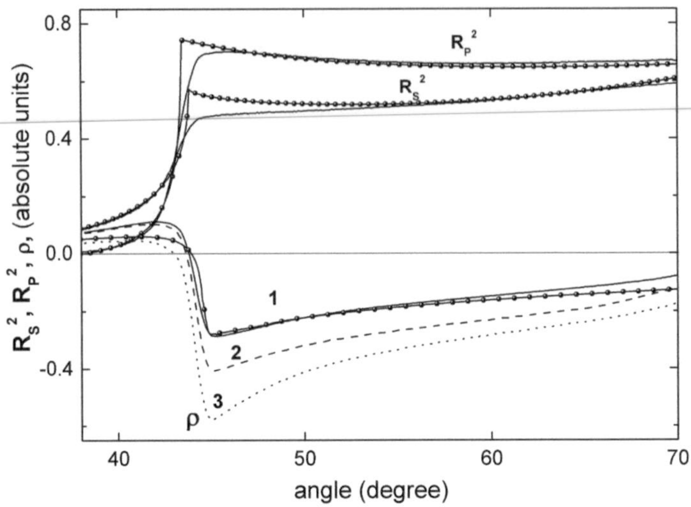

Fig. 21.2 Angular dependences of the internal reflection coefficients $R_s^2(\theta)$, $R_p^2(\theta)$ and polarization difference $\rho(\theta)$ at $\lambda = 600$ nm: experimental *curves* for samples: 1 – *solid line*, 2 – *dashed line*, and 3 – *dotted line* in comparison with calculated (marks) characteristics

were used. The modeling was performed by means of multiparameter fitting of the calculated curves of $\rho(\theta)$, $R_s^2(\theta)$, and $R_p^2(\theta)$ to the experimental ones. The following effective refractive n and absorption k indexes of Au-PTFE films were used for modeling: thickness $d = 24$ nm, $n = 1.5$, $k = 0.3$ (sample 1); $d = 22$ nm, $n = 1.65$, $k = 0.39$ (sample 2); $d = 13$ nm, $n = 1.8$, $k = 0.75$ (sample 3) at a wavelength of $\lambda = 600$ nm and refractive index of a half-cylinder $N_0 = 1.458$. The difference between obtained optical parameters is caused by differences in morphology structure of samples 3 in comparison with 1 and 2. In spite of the large amplitude of $\rho(\theta)$ for samples 3, the samples 1 and 2 are most suitable for sensor applications due to their stable structure.

The spectral characteristics of $\rho(\lambda)$ in the range of $\theta > \theta_{cr}$ for sample 1 are shown in Fig. 21.3a in order to investigate the LSPR in Au-PTFE films. Such behavior of curves is typical for nanosized films with clustered structure. Asymmetrical contour in characteristics of $\rho(\lambda)$ can be caused by presence of two partially overlapping resonances in the films. On the one hand extremum in a short-wavelength range is caused by resonant excitations of surface plasmons on isolated non-interacting Au NPs. On the other hand extremum in a long-wavelength range is caused by resonant excitations of surface plasmons between NPs due to electrodynamic/dipole-dipole interactions of adjacent NPs. As a result, the surface plasmon oscillations in a single AuNPs induces the surface plasmons oscillations in adjacent Au NPs when both frequency (ω) and wave vector (k) of light excitation matches to those of the surface plasmon's frequency and wave vector. In other words, the phase synchronism condition is satisfied, the surface plasmons resonance manifests itself. The average size of Au NPs is much less the radiation wavelength over the whole range of

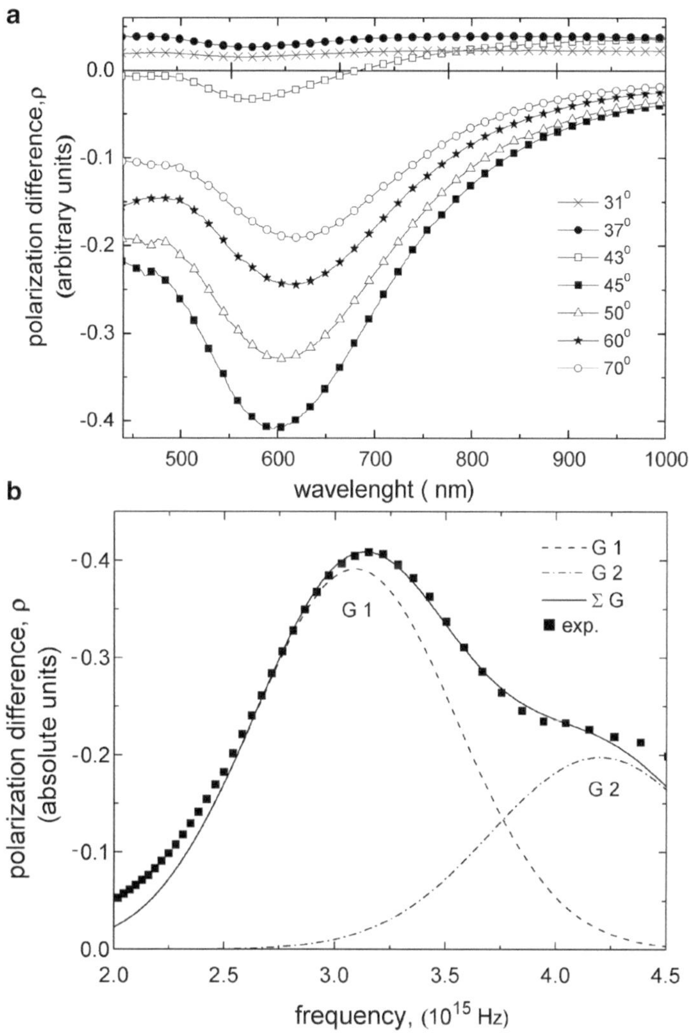

Fig. 21.3 (**a**) – spectral dependences of polarization difference $\rho(\lambda)$ for sample 1 at incident angles of $\theta = 31$ (×), 37 (•), 43 (□), 45 (■), 50 (△), 60 (⋆), and 70 (○) degrees. (**b**) – frequency dependence of polarization difference $\rho(\omega)$ at $\theta = 45°$ for sample 1 (marks) recalculated from the corresponding dependence of $\rho(\lambda)$ in comparison with the Gaussian distribution function (*solid line*) expanded in two components (*dashed line*)

measurement. The band of LSPR is broadened due to wide size distribution with increasing 'mean free path of an electron-to-size of Au NPs' ratio. Similar results were obtained in the absorption spectra of thin Au films in research laboratories by J. Siegel's group [31], where it was also noted that for Au NPs resonant absorption band shifts to long-wavelength range as cluster density increases. One can see that for $\theta > \theta_{cr}$, the curves of $\rho(\lambda)$ have their extrema around $\lambda \sim 600$ nm, which slightly red-shifts with increasing incident angle.

In Fig. 21.3b the experimental dependence of $\rho(\lambda)$ at $\theta = 45°$ was converted to the corresponding frequency dependency and decomposed into elementary components that are described by Gaussian function with fundamental frequencies ω_1, ω_2 and oscillation relaxation times γ_1, γ_2, respectively. Their sum determines a calculated characteristic, which has a quite satisfactory agreement with the experimental one. In this case two broad extrema of $\rho(\lambda)$ corresponds to two types of resonant mechanisms of surface plasmon's excitations for Au-PTFE films. first extremum in the angle range of $45° < \theta < 60°$ at a frequency of $\omega_1 = 3.09$–3.03×10^{15} Hz and $\gamma_1 = 1.08$–$1.12 \times 10^{15}\,s^{-1}$ is related to the LSPR due to dipole fields' interactions of adjacent Au NPs. The second extremum at a frequency of $\omega_2 = 4.19$–4.41×10^{15} Hz and $\gamma_2 = 1.16$–$0.8 \times 10^{15}\,s^{-1}$ is determined by surface plasmon's excitations on isolated non-interacting Au NPs.

In both cases the surface plasmons have non-radiative modes. The excitation of surface plasmons of two types of resonant mechanisms in the ultrathin metal films was confirmed by research results from D. Dalacu [32] and B.-h. Choi [33].

In Fig. 21.4a all curves of $\rho(\lambda)$ exhibit identical extrema at $\lambda = 568$ nm for the whole range of incident angles less than the critical one $\theta < \theta_{cr}$. The surface plasmon's oscillations of small metal NPs have radiative modes. In other words, the surface plasmons can be excited with light of appropriate wavelenhgt (and polarization) irrespective to the excitation light wave vector. For $\theta < \theta_{cr}$, the origin of extremum $\rho(\lambda)$ is determined by a number of metal centers that interact with electromagnetic wave that manifests itself in resonant excitations of surface plasmons on isolated non-interacting Au NPs at a wavelength close to the resonant wavelength of natural oscillations of the conduction electrons or plasma oscillations. The frequencies at which small Au NPs absorb light are near but not equal to those of gold material absorption [34]. In particular, from the similar measurements of $\rho(\lambda)$ for $\theta < \theta_{cr}$ the extrema at $\lambda = 568$, 560, and 585 nm are obtained for samples 1, 2, and 3, respectively.

The nanocomposite Au-PTFE films consist of AuNPs embedded in PTFE matrix. The matrix presence leads to shifting of the surface plasmons absorption band to long-wavelength range because of change of the Fröhlich conditions for surface plasmon's resonance in composite system. Moreover, if Au NPs are deposited on a glass substrate the shift-direction of resonant wavelength is determined by the relation between the refractive indices of the substrate and material of polymer substance (i.e., PTFE). These shift value depend on the permittivity of material of NPs as well as the dielectric properties and thickness of polymer substance [35]. Similar splitting of plasmon resonances, when new formed resonances are shifted in both red and blue spectrum directions, takes place as a result of the interaction of metal NPs with the substrate [36] and also is possible because of predominant orientation of Au NPs due to the structural particularities of thin films [37, 38].

Multifunctionality of the MPS technique enables one to perform diagnostic of Au-PTFE films by using a special parameter – angle of isotropic reflection for *s*- and

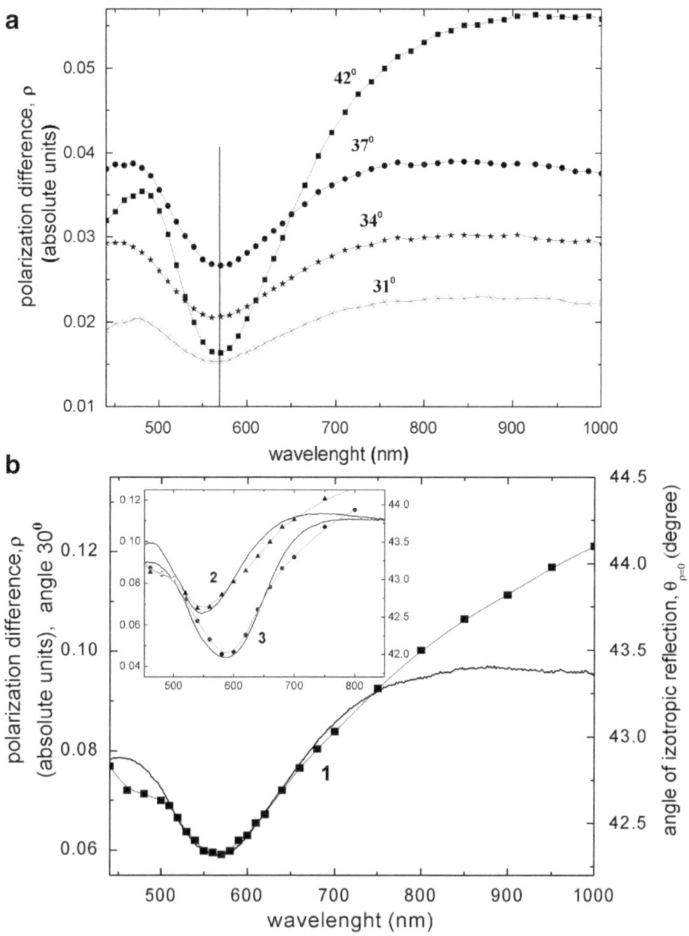

Fig. 21.4 (a) – spectral dependences of polarization difference $\rho(\lambda)$ for sample 1 at incident angles of $\theta = 31$ (×), 34 (⋆), 37 (•), and 42 (■) degrees. (b) – spectral dependences of the angle of isotropic reflection $\theta_{\rho=0}(\lambda)$ at $\rho(\theta) = 0$ (*filled marks*) in comparison with the dependence of $\rho(\lambda)$ at $\theta = 30°$ (*solid curve*) for sample 1. Inset: similar characteristics of $\theta_{\rho=0}(\lambda)$ and $\rho(\lambda)_{\theta=30°}$ for samples 2 and 3

p-polarized radiation, $\theta_{\rho=0}$. The above parameter $\theta_{\rho=0}$ is based on value equality of the reflection coefficients of s- and p-polarized radiation, i.e. $R_s^2(\theta) = R_p^2(\theta)$ and $\rho(\theta) = 0$ as defined in the work of Serdega et al. [14]. Here, light reflection is independent on the polarization state or, in other words, the polarization state of the electromagnetic wave is isotropic at angles of $\theta_{\rho=0}$. In Fig. 21.4b the spectral characteristics of the angles of isotropic reflection $\theta_{\rho=0}(\lambda)$ (dots) along with the spectral dependencies of $\rho(\lambda)$ (solid line) at $\theta = 30°$ for samples 1, 2 and 3 of Au-PTFE films are shown.

The presented curves are characterized by nonmonotonic behavior, with an extrema at $\lambda = 568, 560$, and $585\,\text{nm}$ for sample 1, 2 and 3, respectively, which are in complete agreement with extrema values of $\rho(\lambda)$ in the radiative region for $\theta < \theta_{\text{cr}}$. The light propagation in the direction of $\theta_{\rho=0}$ is defined by appropriate size of Au NPs. In this case surface plasmons are localized. These surface plasmons' excitations take place in small, non-spherical metal particles with irregular and uneven shape, when the surface charge distribution is essentially inhomogeneous. Such non-uniform distribution generates higher multipoles (quadrupole, etc.) even if the film thickness is much smaller than the wavelength. These higher multipoles are occurring when NPs size increases and interaction transition mode from quasi-static to radiation is observed [39]. Under such conditions, surface electromagnetic wave does not develop and its propagation is similar to a standing wave. Moreover, even small intensity of exciting electromagnetic wave can lead to strong oscillations, provided that the frequency of the incident radiation is in resonance with the frequency of collective oscillations of conduction electrons in Au films. The plasma frequency ωp depends on the electron concentration in each separated Au film which can deviate for small size of Au NPs [40, 41]. Parameter of $\theta_{\rho=0}(\lambda)$ is very sensitive to change of intensity towards one or another polarization state (s or p) of radiation during its interaction with Au NPs. Consequently, the angle of isotropic reflection $\theta_{\rho=0}(\lambda)$ is an additional registration method of plasma frequency for nanocomposited Au-PTFE films: $\omega_p = 3.32 \times 10^{15}$ Hz (1), 3.37×10^{15} Hz (2) and 3.22×10^{15} Hz (3). These values are increased according to increase of appropriate mass thickness d for Au-PTFE films.

In Fig. 21.5 spectral characteristics of $\rho(\lambda)$ at $\theta = 44°$ exhibit the LSPR for all samples under investigations 1, 2 and 3. The amplitude value of $\rho(\lambda)$ is gradually increased from sample 1 to sample 3 as in case of the angular characteristics of $\rho(\theta)$ in Fig. 21.2. The LSPR frequencies for the sample 1 and 2 are remained without changing $\omega_1 = 3.24 \times 10^{15}$ Hz ($\gamma_1 = 0.824 \times 10^{15}\,\text{s}^{-1}$) and slightly changed for the sample 3 $\omega_1 = 3.093 \times 10^{15}$ Hz ($\gamma_1 = 0.8 \times 10^{15}\,\text{s}^{-1}$). PTFE adding into nanosized Au films allows obtaining the nanocomposites with stable resonant parameters, which are demonstrated by similar frequency values of the LSPR. The annealing at delicate temperature of nanocomposite Au-PTFE films makes it possible to obtain the structures with increased amplitude properties of the LSPR.

In Fig. 21.6 the dispersion characteristics $\omega(k)$ of surface plasmons as a result of polarization analysis of spectral characteristics of $\rho(\lambda)$ in a wide incident angle range were obtained for samples 1, 2 and 3. Two dispersion branches are shown for non-radiative modes (to the right of the light straight line) where the phase synchronism condition is necessary for surface plasmon's excitations. The low-frequency branch corresponds to the surface plasmon's excitations between Au NPs due to dipole fields' interactions of adjacent NPs (the influence of neighbouring Au NPs is dominant). These LSPR frequencies are also increased according to increase of appropriate mass thickness d for Au-PTFE films. The high-frequency branch corresponds to the surface plasmon's excitations on isolated non-interacting Au NPs. Removal of branches from each other in frequency is associated with the dependence of the plasmon's frequency on Au NP size. In our case, such a size can

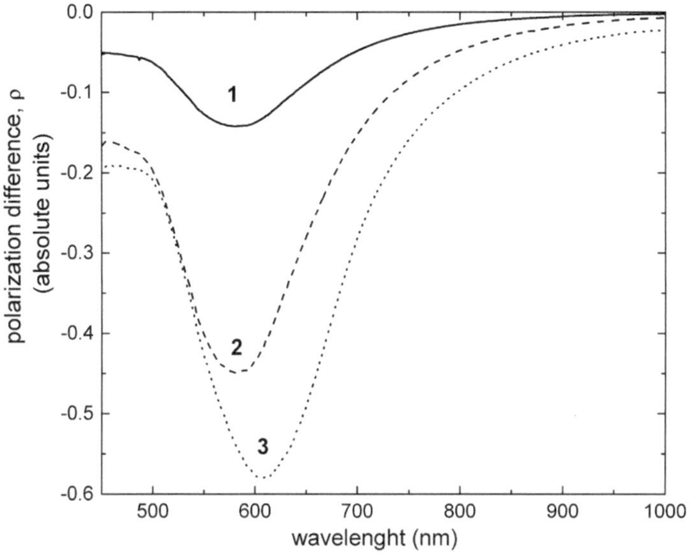

Fig. 21.5 Spectral dependences of polarization difference $\rho(\lambda)$ for samples 1 – *solid*, 2 – *dashed*, and 3 – *dotted* experimental curves at $\theta = 44°$

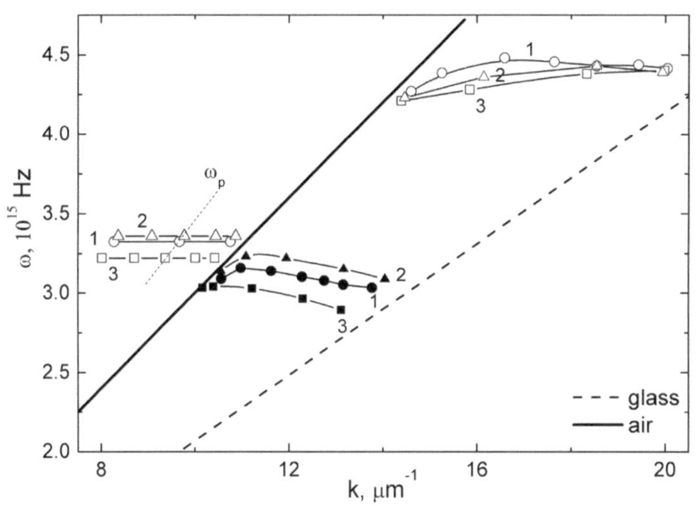

Fig. 21.6 Dispersion characteristics $\omega(k)$ of localized surface plasmons on isolated Au NPs (*opened*) and between Au NPs caused by dipole fields' interactions (*filled*) for Au-PTFE films

be particle cross-section (that depends on the incident angle) toward the electric field of electromagnetic wave and the distance between Au NPs. These characteristic features are associated with a spheroidal in form Au NPs, but not completely spherical for Au-PTFE film. The splitting of the dispersion characteristics into two branches is associated with the small thickness of the sample [42]. One more dispersion branch is shown for radiative modes (to the left of the light straight line). Constant frequency value of the LSPR at varying wave vector completely corresponds to the plasma frequency ω_p for Au-PTFE films. These values are also increased according to increase of appropriate mass thickness d of Au-PTFE films.

21.5 Conclusions

In summary, particularities of morphology structure due to annealing for Au-PTFE nanocomposites were studied by measuring angular and spectral dependencies of polarization characteristics utilizing developed modulation polarization spectroscopy technique. As a result of polarization analysis two types of the LSPR were detected: the first is on isolated non-interacting Au NPs and the second is between Au NPs caused by dipole fields' interactions. The obtained optical and resonant parameters of Au-PTFE nanocomposites are necessary characterization parameters for each individual film in case of their sensor applications. The paper discussed the following obtained results. Frequencies and damping parameters of the LSP resonances were determined. Dispersion characteristics $\omega(k)$ of LSP are shown for radiative and non-radiative modes. Plasma frequencies ω_p were obtained for Au-PTFE films with unheated structure and annealed at temperatures of 300 and 520 °C by measuring of spectral characteristics of the angle of isotropic reflection $\theta_{\rho=0}(\lambda)$. The correlation between plasma frequency and film thickness was discussed. The annealing at delicate temperature of nanocomposite Au-PTFE films makes it possible to obtain the structures with opportune morphology structure, optical and LSPR parameters. The annealing at high temperature leads to changing in morphology surface up to clusters aggregation and restructuring of films.

References

1. Kreibig U, Vollmer M (1995) Optical properties of metal clusters. Springer, Berlin
2. Takele H, Greve H, Pochstein C, Zaporojtchenko V, Faupe F (2006) Plasmonic properties of Ag nanoclusters in various polymer matrices. Nanotechnology 17:3499–3505
3. Siegel J, Kvítek O, Kolská Z, Slepicka P, Svorcík V (2012) Gold nanostructures prepared on solid surface. In: Pardhi Y (ed) Metallurgy – advances in materials and processes. InTech, pp 43–70. doi:10.5772/51617
4. Miyake S, Shindo T (2013) Deposition and tribological properties of multilayer and mixed films composed of gold and polytetrafluoroethylene. Thin Solid Films 527:210–221
5. Faupel F, Zaporojtchenko V, Strunskus T, Elbahri M (2010) Metal-polymer nanocomposites for functional applications. Adv Eng Mater 12(12):1177–1190
6. Zvatora P, Rezanka P, Procopec V, Siegel J, Svorcík V (2011) Polytetrafluorethylene-Au as a substrate for surface-enhanced Raman spectroscopy. Nanoscale Res Lett 6:366
7. Smausza T, Kecskemétia G, Csizmadiaa T, Benedekb F, Hoppa B (2013) Study on the applicability of polytetrafluoroethylene–silver composite thin films as sensor material. Appl Surf Sci 278:117–121

8. Mitsushio M, Nagaura A, Yoshidome T, Higo M (2015) Molecular selectivity development of Teflon AF1600-coatedgold-deposited surface plasmon resonance-based glass rod sensor. Prog Organ Coat 79:62–67
9. Zhao J, Zhang X, Yonzon Ch R, Haes AJ, Duyne RP (2006) Localized surface plasmon resonance biosensors. Nanomedicine 1(2):219–228
10. Onodera T, Toko K (2014) Towards an electronic dog nose: surface plasmon resonance immunosensor for security and safety. Sensors 14:16586–16616
11. Novotny L, Hecht B (2006) Principles of nano-optics. Cambridge University Press, New York
12. Maier SA (2007) Plasmonics: fundamental and application. Springer, New York
13. Biswas A, Aktas OC, Schurmann U, Saeed U, Zaporojtchenko V, Faupel F, Strunskus T (2004) Tunable multipole plasmon resonance wavelengths response from multicomponent polymer-metal nanocomposite systems. Appl Phys Lett 84(14):2655
14. Serdega BK, Rudenko SP, Maksimenko LS, Matyash IE (2011) Plasmonic optical properties and the polarization modulation technique. In: Mishchenko MI, Yatskiv Ya S, Rosenbush VK, Videen G (eds) Polarimetric detection, characterization and remote sensing (NATO Science for Peace and Security Series C: Environmental Security). Springer Science+Business Media B.V., pp 473–500
15. Kaganovich EB, Kravchenko SA, Maksimenko LS, Manoilov EG, Matyash IE, Mishchuk ON, Rudenko SP, Serdega BK (2011) Polarization properties of porous gold and silver films. Opt Spectrosc 110(4):513–521
16. Grynko DA, Barabash Yu M, Maksimenko LS, Matyash IE, Mishchuk ON, Rudenko SP, Serdega BK (2012) Modulation polarimetry of topological effects in films of gold-organic nanocomposites. Phys Solid State 54(11):146–153
17. Fedorenko L, Matyash I, Kazantseva Z, Rudenko S, Kolomiychenko Ya (2014) Laser-assisted implantation of gold nanoparticles, formed under surface plasmon-polariton resonant conditions in polymer layer. Appl Surf Sci 290:1–5
18. Gritsenko KP, Krasovsky AM (2003) Thin-film deposition of polymers by vacuum degradation. Chem Rev 103(9):3607–3650
19. Grytsenko KP (2008) Growth mechanism, properties and applications of vacuum-deposited PTFE films. Russ J Chem Soc 52(3):112–123
20. Grytsenko K, Schrader S (2005) Nanoclusters in polymer matrices prepared by co-deposition from a gas phase. Adv Colloid Interface Sci 116:263–276
21. Goncharenko AV, Grynko DO, Grytsenko KP, Lozovski VZ (2005) Preparation and optical properties of Au/teflon nanocomposites. J Nanosci Nanotechnol 5(11):1919–1924
22. Grytsenko K, Lozovski V, Strilchuk G, Schrader S (2012) Evaluation of the mechanism of the gold cluster growth during heating of the composite gold-polytetrafluoroethylene thin film. Nanomaterials 2(4):366–378
23. Romanyuk VR, Kondratenko OS, Fursenko OV et al (2008) Thermally induced changes in thin gold films detected by polaritonic ellipsometry. Mater Sci Eng B 149:285–291
24. Worsch Ch, Kracker M, Wisniewski W, Rüssel Ch (2012) Optical properties of self assembled oriented island evolution of ultra-thin gold layers. Thin Solid Films 520:4941–4946
25. Madorsky SL, Hart VE, Straus S, Sedlak VA (1953) Thermal degradation of tetrafluoroethylene and hydrofluoroethylene polymers in a vacuum. J Natl Bur Stand 51(6):327–333
26. Chen B, Mokume M, Liu C, Hayashi K (2014) Structure and localized surface plasmon tuning of sputtered Au nano-islands through thermal annealing. Vacuum 110:94–101
27. Berezhinsky LJ, Maksimenko LS, Matyash IE, Rudenko SP, Serdega BK (2008) Polarization modulation spectroscopy of surface plasmon resonance. Opt Spectrosc 105(2):257–264
28. Jasperson SN, Schnatterly SE (1969) An improved method for high reflectivity ellipsometry based on a new polarization modulation technique. Rev Sci Instrum 40(6):761
29. Berezhinski LI, Litvin OS, Maksimenko LS, Matyash IE, Rudenko SP, Serdega BK (2009) Size effects in the internal reflection in gold cluster films in polarization modulation experiments. Opt Spectrosc 107(2):264–269

30. Kaganovich EB, Kizyak SA, Manoilov EG, Maksimenko LS, Matyash IE, Rudenko SP, Serdega BK (2009) Polarization-modulation spectroscopy of the surface plasmon resonance in gold nanostructures obtained by the method of pulsed laser deposition. Ukr J Phys 54(6):621–626
31. Siegel J, Lyutakov O, Rybka V, Kolská Z, Švorčík V (2011) Properties of gold nanostructures sputtered on glass. Nanoscale Res Lett 6:96
32. Dalacu D, Martinu L (2001) Optical properties of discontinuous gold films: finite-size effects. J Opt Soc Am B 18(1):85–92
33. Choi B-h, Lee H-H, Jin S, Chun S, Kim S-H (2007) Characterization of the optical properties of silver nanoparticle films. Nanotechnology 18:075706
34. Dmitruk NL, Goncharenko AV, Venger EF (2009) Optics of small particles and composite media. Naukova Dumka, Kyiv
35. Zamkovets AD, Kachan SM, Ponyavina AN (2003) Optical properties of thin-film metal-dielectric nanocomposites. Phys Chem Solid State 4(4):627–631
36. Xu G, Tazawa M, Jin P, Nakao S (2005) Surface plasmon resonance of sputtered Ag films: substrate and mass thickness dependence. Appl Phys A 80:1535–1540
37. Hutter E, Fendler J (2004) Exploitation of localized surface plasmon resonance. Adv Mater 16(19):1685–1705
38. Gupta G, Tanaka D, Ito Shibata Y, Shimojo M, Furuya K, Mitsui K, Kajikawa K (2009) Absorption spectroscopy of gold nanoisland films: optical and structural characrerization. Nanotechnology 20:025703
39. Khlebtsov NG (2008) Optics and biophotonics of nanoparticles with a plasmon resonance. Quantum Electron 38:504–529
40. Maaroof AI, Gentle A, Smith GB, Cortie MB (2007) Bulk and surface plasmons in highly nanoporous gold films. J Phys D Appl Phys 40:5675–5682
41. Belahmar A, Chouiyakh A (2014) Influence of the fabrication conditions on the formation and properties of gold nanoparticles in alumina matrix produced by cosputtering. Int J Nano Mater Sci 3(1):16–29
42. Dmitruk NL, Litovchenko VG, Strizhevsky VL (1989) Surface polaritons in semiconductors and dielectrics. Naukova Dumka, Kiev

Chapter 22
Metal Oxide Based Biosensors for the Detection of Dangerous Biological Compounds

A.V. Tereshchenko, V.A. Smyntyna, I.P. Konup, S.A. Geveliuk, and M.F. Starodub

Abstract In this report an application of some metal oxide nanostructures as a biosensor platform for the detection of dangerous biological compounds (Bovine leucosis, Salmonella) have been discussed. The attention is paid to the TiO_2 nanoparticles and ZnO nanorods deposited on the flat surface. The changes in photoluminescense signal from nanostructured surface were applied as biosensor response to detect the analytes. The detection range of TiO_2 based biosensor for Bovine leucosis antibodies was in the range of 2–10 μg/ml. The detection range of ZnO based biosensor for Salmonella antigens was 10^2–10^6 cells/ml. The obtained results provide a good basis for the use of optical properties of metal oxide based semiconductor nanostructures in biosensor technology.

22.1 Introduction

During last decades metal oxide nanostructures based on TiO_2 and ZnO are the materials that attract a lot of attention due to their optical, catalytic and sensing applications. Physico-chemical properties of these nanomaterials that can be controlled and changed by growth methods or by modification of nanostructures can play crucial role in sensing application. For these metal oxides, quantum confinement effect caused by nanoscale size, have resulted not only in band gap increase and improved photocatalytic activity but also in photoluminescence peaks appearance at room temperature. Being wide band gap semiconductors that have a good affinity to biological compounds, TiO_2 and ZnO nanostructures are promising materials to be used as optical biosensor transducers [1–5].

A.V. Tereshchenko (✉) • I.P. Konup • S.A. Geveliuk
Odessa National I.I. Mechnikov University, Odessa, Ukraine
e-mail: alla_teresc@onu.edu.ua

V.A. Smyntyna
Department of Experimental Physics I.I. Mechnikov, Odessa National University (ONU), Dvoryanskaya str., Odessa 65082, Ukraine

M.F. Starodub
National University of Life and Environmental Sciences, Kyiv, Ukraine

J. Bonča, S. Kruchinin (eds.), *Nanomaterials for Security*, NATO Science for Peace and Security Series A: Chemistry and Biology, DOI 10.1007/978-94-017-7593-9_22

Among various types of biosensors, an immune biosensor is a type, based on specific interaction between antibody (Ab) – antigen (Ag) couple [6]. The reaction between this couples has high specificity and sensitivity to detecting analytes what makes immune biosensors suitable for accurate and precise tests with electrochemical, optical, magnetic and piezoelectric transducers [6–8]. As it is known from the range of works, optical methods of detection based on absorbance, reflectance and photoluminescence demonstrate simple, fast and accurate detection of target analyte [9–14]. In particular, photoluminescence from nanostructured metal oxides is a promising property that can be used for the detection of chemical and biologic compounds [4, 9, 10, 13, 14]. Therefore, the photoluminescence from TiO_2 nanoparticles was used for the detection of Bovine leucosis antibodies.

22.1.1 Application of Photoluminescence TiO_2 Nanoparticles for the Detection of Bovine Leucosis

Bovine leucosis virus (BLV) – is the highly foetal neoplasia of the cattle characterized by the abnormality maturation process of the blood cells [15]. Diagnosis of the BLV infection based on the clinical signs alone is difficult because of the wide range of symptoms. The traditional immune methods have high specificity and sensitivity, but they take a lot of time and require additional parameters such as labelled molecules [16]. To overcome such drawbacks we need to use the modern instrumental analytical devices based on the biosensor technology.

TiO_2 is a material, which is widely applied for different applications [17–22] including sensors and biosensors [20, 23–25]. TiO_2 shows good stability in aggressive environment what makes it attractive for chemical sensors applications [18, 20]. TiO_2 is a wide band gap semiconductor with indirect optical transitions [19]. TiO_2 has low isoelectric point pH $= 5.5$ what makes some advantages to protein immobilization on its surface [25].

22.2 Experimental

Anatase nanoparticles with mean size 32 nm were used as biosensor template. TiO_2 nanoparticles were solved in water to prepare sols (with concentration 0.05 mg/ml). TiO_2 layers were formed on glass substrates by dropping TiO_2 sols and drying at room temperature with post annealing treatment at 300 C for 1 h was provided to remove water from the samples. SEM measurements showed that nanoparticles formed high surface area porous structure (Fig. 22.1a) that is suitable platform for immobilization of biological species.

Raman spectrometer with Ar/Kr laser (Jobin Yvon-Labram 1B, $\lambda = 647.1$ nm) and spectral resolution 1 cm^{-1} were used for the study of Raman spectra. Raman spectrum of TiO_2 nanostructures, deposited on glass substrates is shown in Fig. 22.1b. The peaks were found at 392, 512 and 634 cm^{-1}, which correspond to B_{1g}, A_{1g} and E_g modes of anatase phase of TiO_2. Surface morphology of

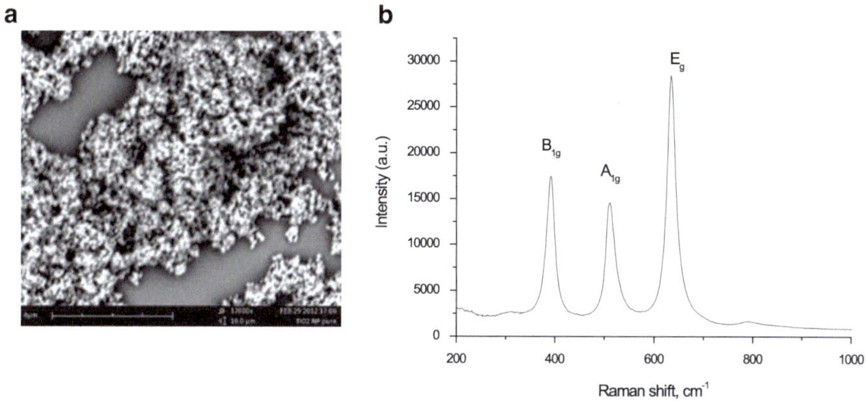

Fig. 22.1 (**a**) – SEM image of TiO_2 nanostructures deposited on glass; (**b**) – Raman spectrum of TiO_2 nanostructures

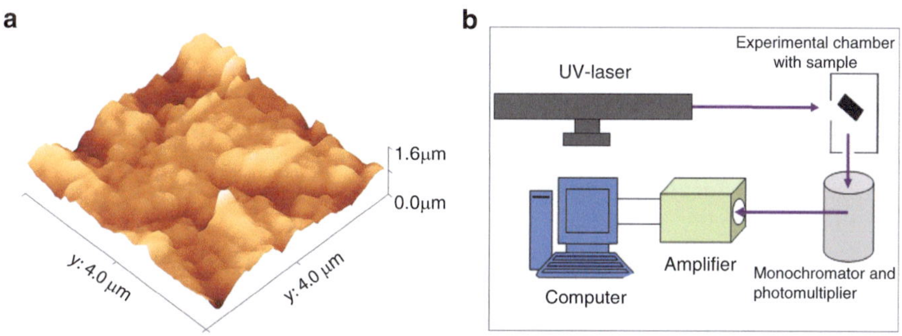

Fig. 22.2 (**a**) – AFM image of surface of TiO_2 nanostructures (Asylum Research MFP-3D); (**b**) – photoluminescence setup

deposited samples has been investigated with AFM (Fig. 22.2a). The obtained TiO_2 nanostructures had high active surface area. Mean square surface roughness (Rsq), measured with free software Gwiddion, was 140 nm for prepared TiO_2 nanostructures. Photoluminescence spectra were measured by setup shown on Fig. 22.2b. The photoluminescence was stimulated by UV laser LCS-DTL-374QT with excitation wavelenth $\lambda = 355$ nm. The emission spectra were amplified and recorded in the range of 370–800 nm.

22.3 Results and Discussion

To fabricate biosensitive layer, the antigens of BLV were immobilized on TiO_2 surface. TiO_2nanostructures were exposed to water solution of BLV antigens (Ag) for 10 min and then were washed two times in distilled water and dried in air at room temperature. The backside of TiO_2 sample was sealed to prevent immobilization of Ag on it.

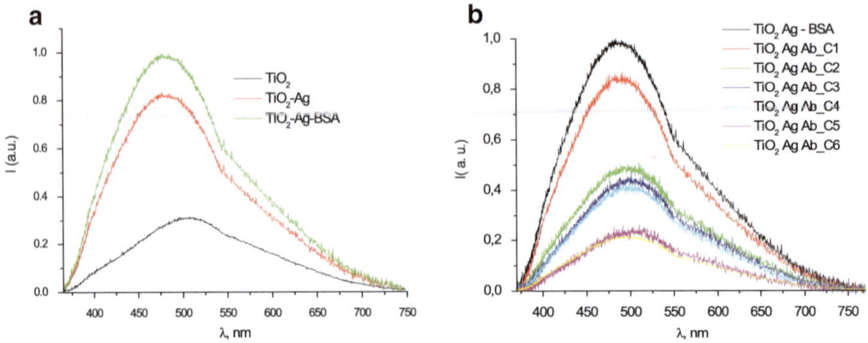

Fig. 22.3 (**a**) – PL spectra of TiO_2 nanoparticles before and after immobilization of BLV antigens, and after BSA adsorption; (**b**) – PL spectra of TiO_2-Ag-BSA layer under different concentrations of BLV Ab

The photoluminescence spectrum of pure TiO_2 samples is characterized by broad peak centered at 510 nm. A number of papers reported on room temperature PL in TiO_2 nanostructures [22, 26–28]. Usually TiO_2 nanostructures demonstrate emission in the range 430–560 nm. Two mechanisms of luminescence are proposed: self-trapped excitons (STE) (430–510 nm) and oxygen vacancies (530–560 nm).

Photoluminescence (PL) spectra of TiO_2 nanoparticles before and after immobilization of antigens are shown on Fig. 22.3a. As one can see, the immobilization of BLV antigens lead to the significant changes of PL spectra in intensity and peak position. It was found that after Ag immobilization PL spectra was shifted to shorter wavelengths, what can be a proof of formation links between TiO_2 and Ag. Increase of PL intensity could result from charge transfer between Ag molecules and conductance band of TiO_2. And the UV shift of PL maximum can be explained by additional dipole-dipole interaction, what can change energetic position of recombination centers into TiO_2. After immobilization of antigens, the bovine serum albumin (BSA) was deposited on the biosensitive layer as a blocking agent to prevent nonspecific protein adsorption. The intensity of PL after BSA adsorption has been increased.

After forming of biosensitive layer by immobilization BLV antigens, the BLV antibodies (which play role of analyte) were deposited on the functionalized surface from water solutions with different concentrations. PL spectra of TiO_2-Ag-BSA biosensor, measured under different Ab concentrations are shown in Fig. 22.3b. It was found that PL intensity decreased with the increase of analyte concentration. At the same time, peak position moved to higher wavelengths.

Thus, the biosensor response to leucosis Ab can be a function of two parameters: PL intensity and position of PL peak. To analyze the sensor response we calculated the changes biosensor signal S according the following equation

$$S = \frac{S_{\text{Ag-BSA}} - S_{\text{Ab}}}{S_{\text{Ag-BSA}}}, \tag{22.1}$$

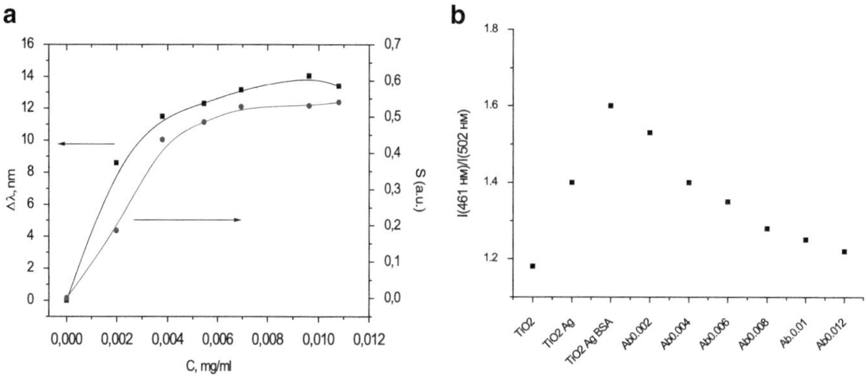

Fig. 22.4 (**a**) – Response of the biosensor signal S and peak shift to different concentrations of BLV antibodies; (**b**) Ratio of $I_{STE}/I_{V[O]}$ for TiO_2 NP before and after interaction with biomolecules

where $S_{\text{Ag-BSA}}$ and S_{Ab} are PL peak's intensities of TiO_2 nanostructures with immobilized BLV antigens before and after interaction with BLV antibodies, correspondently.

The changes of peak position after adsorption of Ab were calculated according following equation:

$$\Delta\lambda = \lambda_{\text{Ag-BSA}} - \lambda_{\text{Ab}}, \tag{22.2}$$

where $\lambda_{\text{Ag-BSA}}$ and λ_{Ab} are PL peak positions of TiO_2 nanostructures with immobilized BLV antigens before and after interaction with BLV antibodies, correspondently.

The results, obtained with the use of equations (22.1) and (22.2) are plotted in Fig. 22.4a. The analysis of the results showed that the changes of biosensor parameters had similar behavior. The obtained experimental curves increased at the range of Ab concentrations from 2–10 mkg/ml. The further increase of Ab concentration led to the saturation of signal changes.

It is known that photoluminescence spectrum of TiO_2 nanostructures can be split into two peaks, related to self-trapped excitons (STE) and oxygen vacancies $V_{[O]}$. We performed the fitting of the obtained PL spectra before and after interaction with biological molecules (the fitting is not shown here) and plotted the ratio of PL intensities related to STE and oxygen vacancies (Fig. 22.4b). It was found that the positions of both peaks have not been changed for all steps of the experiment (1–1.5 nm). At the same time the redistribution of the integrated intensity between peaks caused by STE at 461 nm and oxygen vacancies $V_{[O]}$ at 502 nm was observed.

The formation of biosensitive layer and BSA molecules adsorption was accompanied by an intensity increase of the peak at 461 nm and a decrease of the peak at 502 nm, leading to a shift in the overall spectrum to shorter wavelengths. This indicates that the biomolecules adsorption by the surface of TiO_2 reduces the rate of radiative recombination caused by oxygen vacancies and increases the photoluminescence intensity caused by STE. After immune reaction between antigens and antibodies, the ratio of PL intensities related to STE and oxygen vacancies decreased with increasing concentration of antibodies.

The formation of antibody-antigen complex can lead to the changes in molecular structure of antigens previously immobilized on TiO_2 surface which is accompanied by link changes between antigens and TiO_2 surface.

Successful results were obtained for Salmonella antigens detection using similar procedure and biosensor, based on photoluminescence from ZnO nanorods [29–31]. The studied ZnO nanorods can be used as transducers in optical biosensors for Salmonella detection. The optimal response of the fabricated biosensor is observed at concentrations 10^2–10^6 cells/ml.

22.4 Mechanism of Interaction Between TiO_2 Surface and Bio-molecules

Reaction of TiO_2 with proteins (Ab, Ag and BSA) is due to non-covalent binding. Van der Waals and hydrophobic bonds are suggested as the mechanism of interaction between TiO_2 surface and BLV molecules. An increase of PL emission after biomolecules immobilization is observed as a result of charge transfer from bio-molecules to TiO_2 surface. The proteins are bound to the surface by several functional groups affecting the surface band bending. The changes in UV PL emission after target molecules adsorption could occur due to the immune reaction between Ag-Ab couple. The decrease in intensity of the PL peak after target molecules adsorption is due to elimination and/or weakening of link between TiO_2 and bio-molecules, caused by structural modification of previously adsorbed Ab molecules as a result of antigen-antibody reaction. However, the mechanism of interaction cannot be determined from only measurements of PL. Additional methods, such as XPS confocal microscopy are needed to investigate it.

22.5 Conclusions

Specific selective interaction between immobilized antibodies and antigens couples ('key'-'lock' principle) can be monitored by PL of TiO_2 and ZnO. Photoluminescence from these nanomaterials is a suitable method for characterizing sample surface, surface defects and the changes of the surface properties as a result of biological impact. TiO_2 and ZnO nanostructures can be successfully used as a platform for the immobilization of biologically active substances on their surface which is confirmed by the changes in PL intensity and PL peak shift. Photoluminescence method of analyte detection is not the highest one comparably to SERS, SPR or fluorescence however it is easier to be applied, does not need label system and can become the next generation of sensing devices. Obtained results provide a basis for the prospective application of metal oxide based nanostrucutures in immune biosensors for rapid diagnosis of such viruses as Bovine leucosis, Salmonella and other dangerous biological species.

References

1. Ming-Chung Wu, Sápi A, Avila A, Szabó M, Hiltunen J, Huuhtanen M, Tóth G, Kukovecz Á, Kónya Z, Keiski R, Su W-F, Jantunen H, Kordás K (2011) Enhanced photocatalytic activity of TiO_2 nanofibers and their flexible composite films: decomposition of organic dyes and efficient H2 generation from ethanol–water mixtures. Nano Res 4(4):360–369
2. Comini E, Baratto C, Faglia G, Ferroni M, Vomiero A, Sberveglieri G (2009) Quasi-one dimensional metal oxide semiconductors: preparation, characterization and application as chemical sensors. Prog Mater Sci 54:1–67
3. Alla T, Mikhael B, Roman V, Volodymyr K, Valentyn S, Nikolay S, Rositza Y (2016) Optical biosensors based on ZnO nanostructures: advantages and perspectives. A review. Sensors and Actuators B 229:664–677
4. Qiu J, Zhang Sh, Zhao H (2011) Recent applications of TiO_2 nanomaterials in chemical sensing in aqueous media. Sens Actuators B 160:875–890
5. Plugaru R, Cremades A, Piqueras J (2004) The effect of annealing in different atmospheres on the luminescence of polycrystalline TiO_2. J Phys Condens Matter 16:S261–S268. PII: S0953-8984(04), 67008-1
6. Holford TRJ, Davis F, Higson SPJ (2012) Recent trends in antibody based sensors. Biosens Bioelectron 34:12–24
7. Viswanathan S, Rani Ch, Ja-an Annie Ho (2012) Electrochemical immunosensor for multiplexed detection of food-borne pathogens using nanocrystal bioconjugates and MWCNT screen-printed electrode. Talanta 94:315–319
8. Brandão D, Liébana S, Campoy S, Cortés P, Alegret S, Pividori MI (2013) Electrochemical magneto-immunosensing of Salmonella based on nano and micro-sized magnetic particles. J Phys Conf Ser 421:012020. doi:10.1088/1742-6596/421/1/012020
9. Stevanovic A, Büttner M, Zhang Zh, Yates JT Jr (2012) Photoluminescence of TiO_2: effect of UV light and adsorbed molecules on surface band structure. J Am Chem Soc 134: 324–332
10. Drbohlavova J, Chomoucka J, Hrdy R, Prasek J, Janu L, Ryvolova M, Adam V, Kizek R, Halasova T, Hubalek J (2012) Effect of nucleic acid and albumin on luminescence properties of deposited TiO_2 quantum dots. Int J Electrochem Sci 7:1424–1432
11. Si P, Ding S, Yuan J, Lou XW, Kim DH (2011) Hierarchically structured one-dimensional TiO_2 for protein immobilization, direct electrochemistry, and mediator-free glucose sensing. ACS Nano 5(9):7617–7626
12. Setaro A, Lettieri S, Diamare D, Maddalena P, Malagù C, Carotta MC, Martinelli G (2008) Nanograined anatase titania-based optochemical gas detection. New J Phys 10:053030
13. Mun K-Sh, Alvarez SD, Choi W-Yo, Sailor MJ (2010) A stable, label-free optical interferometric biosensor based on TiO_2. ACS NANO 4(4):2070–2076
14. Rodionov VE, Shnidko IN, Zolotovsky A, Kruchinin SP (2013) Electroluminescence of Y_2O_3:Eu and Y_2O_3:Sm films. Mater Sci 31:232–239
15. Balida V, Ferrer JF (1977) Expression of the bovine leukemia virus and its internal antigen in blood lymphocytes. Proc Soc Exp Biol Med 156:388–391
16. Matthews R (1979) Classification and nomenclature of viruses. Int Virol 12:128–296

17. Wu M-Ch, Sápi A, Avila A, Szabó M, Hiltunen J, Huuhtanen M, Tóth G, Kukovecz Á, Kónya Z, Keiski R, Su W-F, Jantunen H, Kordás K (2011) Enhanced photocatalytic activity of TiO_2 nanofibers and their flexible composite films: decomposition of organic dyes and efficient H2 generation from ethanol–water mixtures. Nano Res 4(4):360–369
18. Comini E, Baratto C, Faglia G, Ferroni M, Vomiero A, Sberveglieri G (2009) Quasi-one dimensional metal oxide semiconductors: preparation, characterization and application as chemical sensors. Prog Mater Sci 54:1–67
19. Chen X, Mao SS (2007) Titanium dioxide nanomaterials: synthesis, properties, modifications, and applications. Chem Rev 107(7):2891–2959
20. Qiu J, Zhang Sh, Zhao H (2011) Recent applications of TiO_2 nanomaterials in chemical sensing in aqueous media. Sens Actuators B 160:875–890
21. Plugaru R, Cremades A, Piqueras J (2004) The effect of annealing in different atmospheres on the luminescence of polycrystalline TiO_2. J Phys Condens Matter 16:S261–S268. PII: S0953-8984(04) 67008-1
22. Preclíková J, Galá P, Trojánek F, Daniš S, Rezek B, Gregora I, Němcová Y, Malý P (2010) Nanocrystalline titanium dioxide films: influence of ambient conditions on surface- and volume-related photoluminescence. J Appl Phys 108:113502
23. Stevanovic A, Büttner M, Zhang Zh, Yates JT Jr (2012) Photoluminescence of TiO_2: effect of UV light and adsorbed molecules on surface band structure. J Am Chem Soc 134:324–332
24. Drbohlavova J, Chomoucka J, Hrdy R, Prasek J, Janu L, Ryvolova M, Adam V, Kizek R, Halasova T, Hubalek J (2012) Effect of nucleic acid and albumin on luminescence properties of deposited TiO_2 quantum dots. Int J Electrochem Sci 7:1424–1432
25. Si P, Ding S, Yuan J, Lou XW, Kim DH (2011) Hierarchically structured one-dimensional TiO_2 for protein immobilization, direct electrochemistry, and mediator-free glucose sensing. ACS Nano 5(9):7617–7626
26. Mercado C, Seeley Z, Bandyopadhyay A, Bose S, McHale JL (2011) Photoluminescence of dense nanocrystalline titanium dioxide thin films: effect of doping and thickness and relation to gas sensing. ACS Appl Mater Interfaces 3:2281–2288
27. Nair PB, Justinvictor VB, Daniel GP, Joy K, Ramakrishnan V, Thomas PV (2011) Effect of RF power and sputtering pressure on the structural and optical properties of TiO_2 thin films prepared by RF magnetron sputtering. Appl Surf Sci 257:10869–10875
28. Li X, Gao C, Wang J, Lu B, Chen W, Song J, Zhang Sh, Zhang Zh, Pan X, Xie E (2012) TiO_2 films with rich bulk oxygen vacancies prepared by electrospinning for dye-sensitized solar cells. J Power Sources 214:244–250
29. Viter R et al (2012) Novel immune TiO_2 photoluminescence biosensors for leucosis detection. Procedia Eng 47:338–341
30. Smyntyna V et al (2012) ZnO nanorods room temperature photoluminescence biosensors for salmonella detection. In: Frontiers in optics, Rochester. Laser Science, vol XXVIII, 14–18 Oct 2012
31. Viter R et al (2014) Application of room temperature photoluminescence from ZnO nanorods for salmonella detection. IEEE Sens J 14(6):2028–2034

Chapter 23
Cerium Dioxide CeO_{2-x} and Orthovanadate ($Gd_{0.9}Eu_{0.1}VO_4$) Nanoparticles for Protection of Living Body from X-Ray Induced Damage

G. Grygorova, V. Klochkov, Ye. Mamotyuk, and Yu. Malyukin

Abstract The effectiveness of CeO_{2-x} and $GdEuVO_4$ nanoparticles in radioprotection in model animals during radiation exposure has been shown. Nanoparticles were injected per os by two different schemes – once and during 15 days. For the 15-days scheme of $GdEuVO_4$ nanoparticles injection during 30 days of observation the survival of animals irradiated with a dose of 6.0 Gy was 100 %; for ceria it was 90 % for ceria. It was revealed that 7.0 Gy radiation dose is lethal. These results can serve as a motivation to consider inorganic nanoparticles as the basis for new drugs possessing by radioprotective effect.

23.1 Introduction

During last years the radiobiology has achieved a number of great successes. Radioprotective effect was revealed for the range of substances with different chemical structure. To the most important ones from the viewpoint of the possible practical applications belong cystamine, serotonin, mexamine and some others organic compounds which action is based as on the chemical modification of biologically active target molecules, so on the inactivation of oxide radicals which form at the interaction of ionizing radiation with water and biomolecules. However, these radioprotectors had only limited success due to short half-lives (hours or even minutes), lack of penetration to the site of radical production, and daily dosing requirements [2]. Moreover, the majority of traditional radioprotectors with stable chemical structure are effective only at high concentrations and often possess a toxic side action.

G. Grygorova (✉) • V. Klochkov • Yu. Malyukin
Institute for Scintillation Materials of NAS of Ukraine, 60 Lenin ave., 61001 Kharkiv, Ukraine
e-mail: grigorova@isma.kharkov.ua

Ye. Mamotyuk
Grigoriev Institute for Medical Radiology-National Academy of Medical Science of Ukraine, 82 Pushkins'ka Str., 61024 Kharkiv, Ukraine

J. Bonča, S. Kruchinin (eds.), *Nanomaterials for Security*, NATO Science for Peace and Security Series A: Chemistry and Biology, DOI 10.1007/978-94-017-7593-9_23

In the last years new materials for radiation protection with high stability and ability to self-regeneration were proposed [9]. Such materials are based on the inorganic nanocrystals which are able to participate in redox reactions as mediators at electron transfer. At interaction with radicals possessing with non-paired electrons nanoparticles can as to take, so to give an electron turning thereby a radical into an ion or neutral molecule. Besides, near the surface of nanoparticle recombination of radicals accompanied by formation of neutral molecules can take place. It is remarkable that the chemical composition of nanoparticle remains unchangeable during interaction, and the number of acts of interaction is unlimited. So, contrary to organic antioxidants, for which neutralization of radicals is accompanied by inactivation of antioxidant molecule itself, nanoparticle can neutralize unlimited number of radicals. So, for an effective suppression of free radicals in an organism concentrations of nanoparticles can be several times less than for conveniently used organic antioxidants. Also, due to concentration effect activity of nanoparticles increases at increase of free radical concentration and vice versa, deceases at free radical concentration decease, so nanoparticles play a role of regulators of concentration of free radicals in an organism allowing to maintain the required level of reactive oxygen species.

The role of nanoparticles as radioprotectants is a cutting-edge development in decades of scientific interest regarding the protection of normal cells and tissues. In both in vitro and in vivo experiments it was shown that the cerium dioxide nanoparticles, fullerenes and their derivatives [1, 3, 5, 11] protect cells and tissues from radiation injury. The ability of cerium oxide nanoparticles to protect living cells from the influence of ionizing radiation was shown in the number of papers [10]. For instance, in [10] both human normal and tumor cells were treated with cerium oxide nanoparticles and irradiated, and cell survival was measured. It was shown that the presence of cerium oxide in normal cells provided almost 99 % protection from radiation induced cell death, while for tumor cells the same concentration of nanoceria have shown almost no protection.

Another and almost unexplored type of inorganic nanocrystals is rare-earth orthovanadates. Vanadium compounds recently have attracted a lot of attention as potential base for innovative medical agents, and our own preliminary investigations have shown that efficiency of radiation protection for $GdEuVO_4$ nanoparticles is comparable to efficiency of nanoceria and even exceeds it for some schemes of NPs injection [8].

23.2 Experimental

Colloidal solutions based on orthovanadates $Gd_{0.9}Eu_{0.1}VO_4$ and cerium dioxide CeO_{2-x} nanoparticles used in the work were obtained by methods previously described in [6, 7]. The colloidal solutions are stabilized by sodium citrate (NaCit) with molar ratio CeO_2/NaCit as 1:1. All ñolloidal solutions were transparent in transmitted light and opalescent in lateral light (Tyndall cone). Colloidal particles

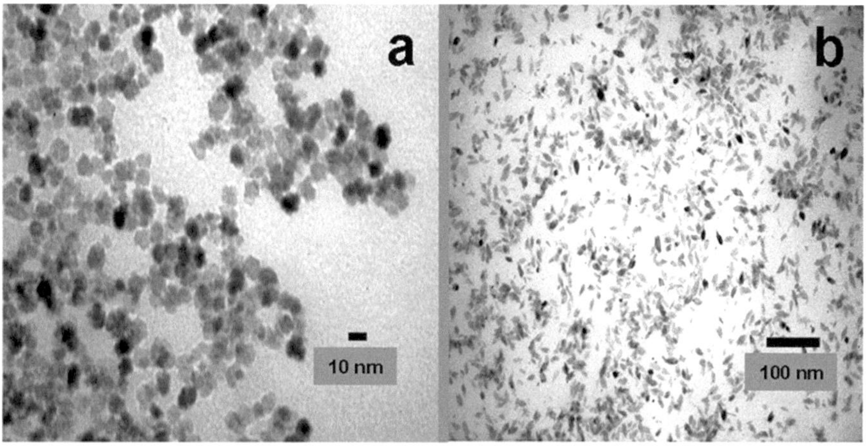

Fig. 23.1 TEM images of CeO_{2-x} (**a**) and $GdEuVO_4$ (**b**) nanoparticles

pass easily through nitrocellulose ultrafilters with a pore diameter of 100 nm. The process solutions have physiological value pH = 7.2–7.8 and, therefore, they can be used for biological testing. The solutions were stored in sealed ampules without changing their properties for more than 2 months at normal conditions.

In the experiments, colloidal solutions of nanoparticles with average sizes of 10 nm for CeO_{2-x} and 8×20 nm for $GdEuVO_4$ were used (Fig. 23.1).

The work was conducted in accordance with international principles of the "European Convention for the Protection of Vertebrate Animals used for Experimental and other Scientific Purposes" [4].

The investigation was performed on 90 white male rats weight in g 150–180 g. The X-ray irradiation of the animals was performed using a x-ray therapeutic unit model RUM 17 (USSR) in standard technical conditions. X-ray irradiation conditions: tube voltage 190 kV, current 10 mA, filters 0.5 mm Cu + 1.0 mm Al, field $20 \times 20\,cm^2$, the absorbed doses 6.0 and 7.0 Gy, the absorbed dose rate in air 0.57 Gy/min, $E_{ef} = 79.0$ keV.

Solutions of nanoparticles with concentration of 0.2 g/l were injected per os by 0.5 ml using the schemes:

Scheme 1 – injection of NPs during 5 days before irradiation, then irradiation with doses of 6.0 and 7.0 Gy followed by injection of solution for another 10 days (CeO_{2-x}5 + Irradiation by 6.0 Gy + CeO_{2-x} 10; $GdEuVO_4$ 5 + Irradiation by 6.0 Gy + $GdEuVO_4$ 10).

Scheme 2 – single injection of NPs for 40 min before irradiation with dose of 7.0 Gy (CeO_{2-x}5 + Irradiation by 7.0 Gy + CeO_{2-x} 10; $GdEuVO_4$ 5 + Irradiation by 7.0 Gy + $GdEuVO_4$ 10).

Control groups were irradiated without injection of nanoparticles.

The following parameters were analyzed: 30-day survival rate (% mortality); the average life duration; integral indicators (cumulative frequency).

By developed in the laboratory (in Grigoriev Institute for Medical Radiology, Ukraine) system of accounting for various signs of acute radiation sickness (ARS) one can quantitatively predict positive or negative radioprotective effect of the tested substance. These integral frequency index (IFI, %) of signs of acute radiation sickness was considered separately for survived and dead rats for 30 days after irradiation. In this case, we took into account cumulative survival rate and frequency of characteristic signs of radiation damage of body: a ruffed fur (a common indicator for the state of the body), bloating, diarrhea, swelling of mouth, peel eyes. All these symptoms may indicate the severity of radiation sickness

23.3 Results and Discussion

23.3.1 Radioprotective Activity of Nanoparticles in Experiments with Rats

Main experimental results are shown in Fig. 23.1. The data show that at a dose of 6.0 Gy injection of two types of nanoparticles by Scheme 1 reduces the 30-day mortality from 60 % to 0–10 %. Radiation with a dose of 7.0 Gy is lethal, so in control 2 almost all animals died. However, after injection of nanoparticles according to schemes 1 and 2 radioprotective effect of nanoparticles is observed, although it is quite negligible.

In addition to the analysis of the 30-day survivability, different signs of acute radiation sickness were monitored, such as a ruffled fur (a common indicator for the state of the body), bloating, diarrhea, swelling of mouth and eyes. Figures 23.1, 23.2, 23.3, and 23.4 show changes of frequencies of investigated signs of acute radiation sickness for nanoparticles injected to rats irradiated by 6.0 Gy. By all these characteristics a protective effect of the nanoparticles was observed in the case of per os injection of both types of nanoparticles and irradiation by 6.0 Gy.

Figures 23.5 and 23.6 show changes of frequencies of investigated signs of acute radiation sickness for nanoparticles injected to rats irradiated by 7.0 Gy. For an irradiation by 7.0 Gy the signs of acute radiation sickness did not differ from those of the controls (Fig. 23.2).

Thus, in in vivo experiments it was found that per os injection of nanoparticles based on orthovanadates of rare earths and cerium dioxide protects the organism of rats following exposure to a lethal dose of radiation.

In general, the radioprotective effect of considered nanoparticles corresponds to, and even possibly exceeds the level of protection provided by conventionally used organic radioprotectors. Probably, the radioprotective effect of nanoparticles is related to their antioxidant action. Moreover, unlike organic antioxidants, inorganic nanoparticles can provide the "traps" for free radicals. Also, participating in the redox processes following by neutralization of free radicals they can play the role of mediators of electron transfer without changing their physical and chemical

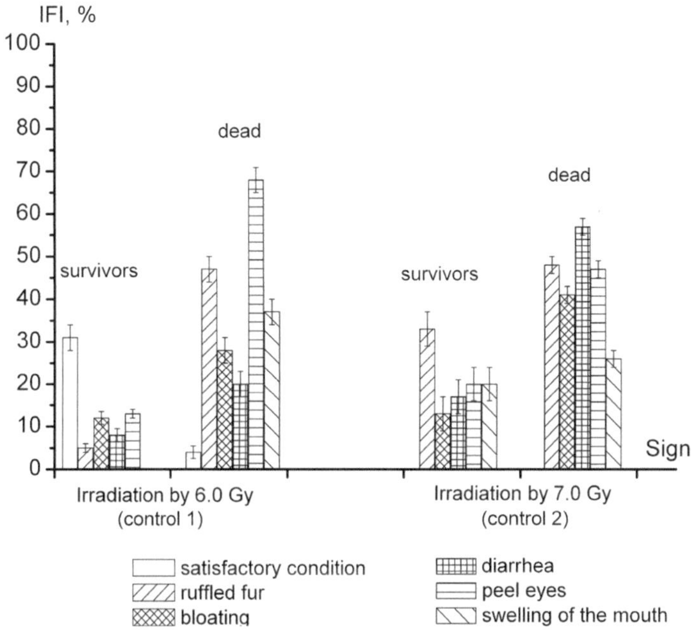

Fig. 23.2 Integral frequency index (IFI, %) of signs of acute radiation sickness for rats in the experiments with X-Ray irradiation by 6.0 Gy and injection of CeO_{2-x}

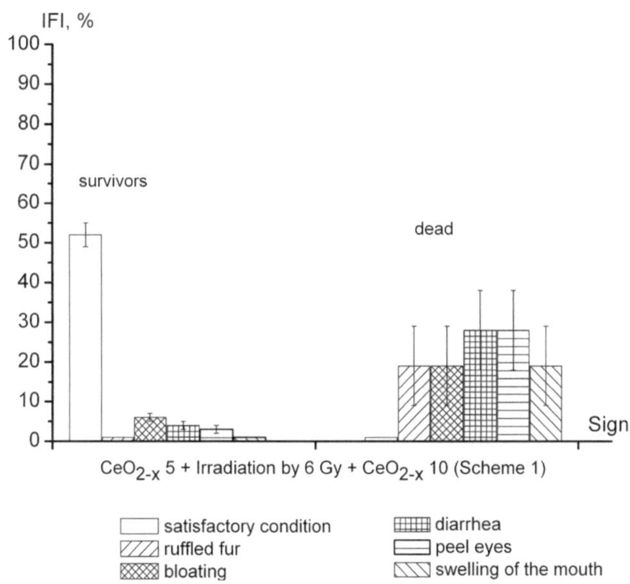

Fig. 23.3 Integral frequency index (IFI, %) of signs of acute radiation sickness for rats in the experiments with X-Ray irradiation by 6.0 and 7.0 Gy

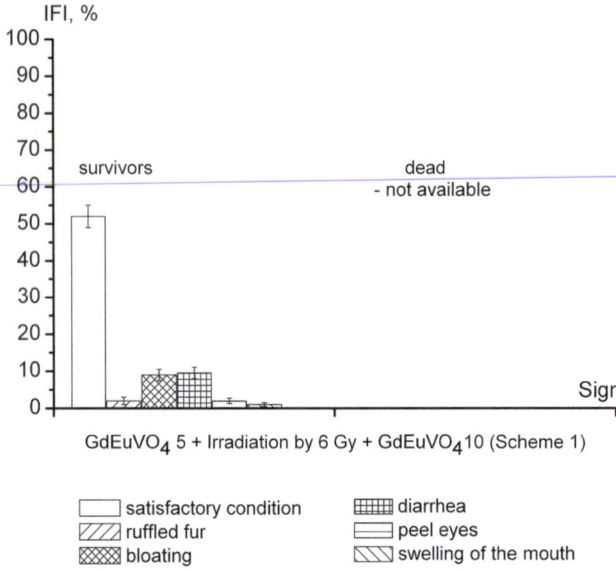

Fig. 23.4 Integral frequency index (IFI, %) of signs of acute radiation sickness for rats in the experiments with X-Ray irradiation by 6.0 Gy and injection of $GdEuVO_4$

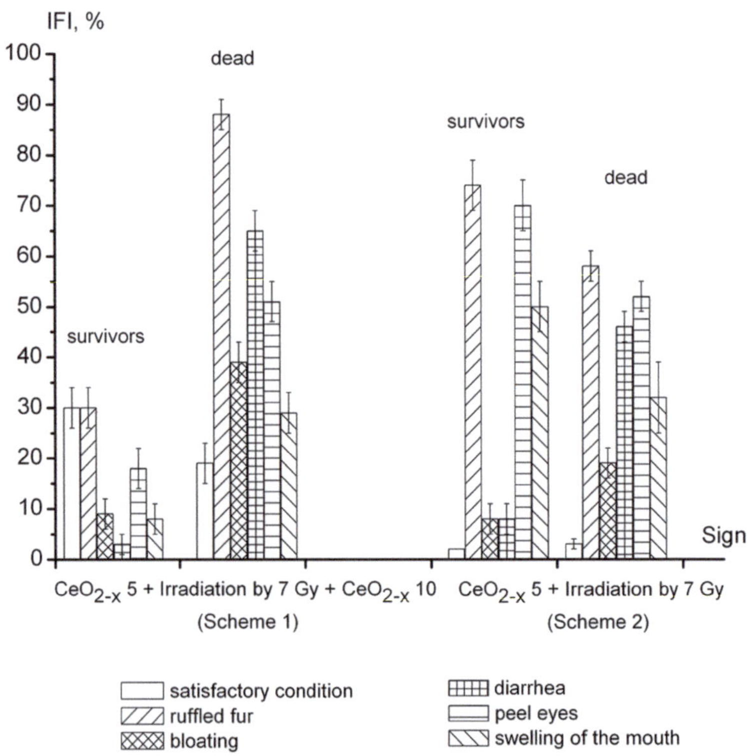

Fig. 23.5 Integral frequency index (IFI, %) of signs of acute radiation sickness for rats in the experiments with X-Ray irradiation by 7.0 Gy and injection of CeO_{2-x}

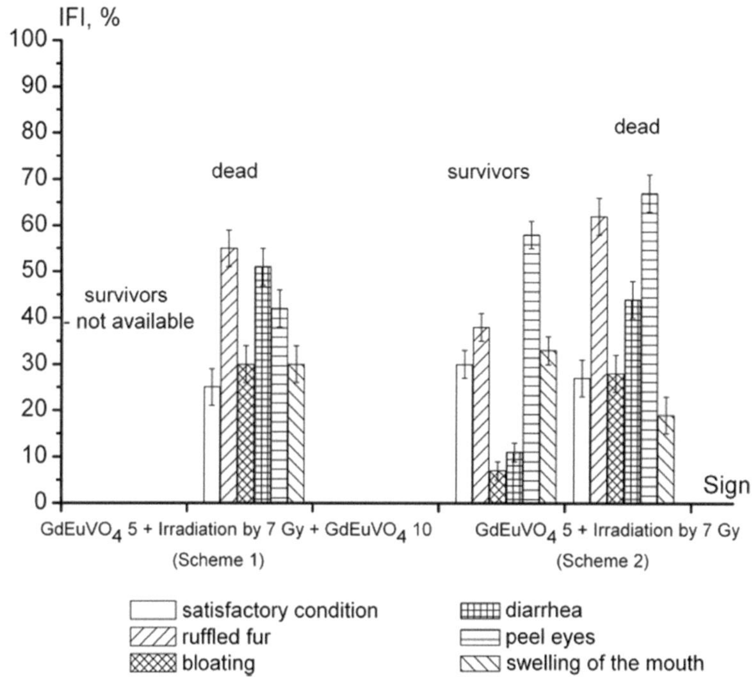

Fig. 23.6 Integral frequency index (IFI, %) of signs of acute radiation sickness for rats in the experiments with X-Ray irradiation by 7.0 Gy and injection of $GdEuVO_4$

properties. Thus, one inorganic nanoparticle is able to participate in the unlimited number of acts of capture and neutralization of reactive radicals.

But one can not exclude the possibility of induction of own protective functions of the organism at injection of nanoparticles. The mechanism of radioprotective action of considered types of nanoparticles requires further more in-depth research.

References

1. Cheryl HB (2013) Harnessing cerium oxide nanoparticles to protect normal tissue from radiation damage. Transl Cancer Res 2:343–358
2. Colon J, Herrera L, Smith J, Patil S, Komanski C, Kupelian P, Baker CH (2009) Protection from radiation-induced pneumonitis using cerium oxide nanoparticles. Nanomed Nanotechnol Biol Med 5:225–231
3. Daroczi B, Kari G, McAleer MF, Wolf JC, Rodeck U, Dicker AP (2006) In vivo radioprotection by the fullerene nanoparticle DF-1 as assessed in a zebrafish model. Clin Cancer Res 12:7086–7091
4. European convention for the protection of vertebrate animals used for experimental and other scientific purposes (1986) Council of Europe, Strasbourg, p 53

5. Karakoti A, Singh S, Dowding JM, Seal S, Self WT (2010) Redox-active radical scavenging nanomaterials. Chem Soc Rev 39:4422–4432
6. Klochkov VK, Malyshenko AI, Sedyh OO, Malyukin, Yu V (2011) Wet-chemical synthesis and characterization of luminescent colloidal nanoparticles: $ReVO_4:Eu^{3+}$ (Re = La, Gd, Y) with rod-like and spindle-like shape. Funct Mater 1:111–115
7. Klochkov VK, Grigorova AV, Sedyh OO, Malyukin YV (2012) The influence of agglomeration of nanoparticles on their superoxide dismutase-mimetic activity. Colloids Surf A Physicochem Eng Asp 409:176–182
8. Mamotyuk EM, Klochkov VK, Grygorova GV, Yefimova SL, Malyukin Yu V (2015) Radioprotective effect of CeO_2 and $GdEuVO_4$ nanoparticles in "in vivo" experiments. In: Nanoscience advances in CBRN agents detection, information and energy security. Springer, Dordrecht, pp 193–197
9. Menon A, Nair CKK, Dhanya KC (2012) Nanomedicine in radiation protection. Pushpagiri Med J 3:137–140
10. Tarnuzzer RW, Colon J, Patil S, Seal S (2005) Vacancy engineered ceria nanostructures for protection from radiation-induced cellular damage. Nano Lett 5:2573–2577
11. Vavrova J, Řezačova M, Pejchal J (2012) Fullerene nanoparticles and their anti-oxidative effects: a comparison to other radioprotective agents. J Appl Biomed 10:1–8

Chapter 24
Anthology and Genesis of Nanodimensional Objects and GM Food as the Threats for Human Security

O. Kharlamov, M. Bondarenko, O. Khyzhun, and G. Kharlamova

Abstract At the end of the last century two important scientific discovery in biochemistry and nanochemistry were made which essentially have changed the quality and quality of modern food. In 1981–1982 years were created the first genetically modified organism (GMO) and hardly later was discovered a new state of matter as nanodimensional objects (NDO) and nanomaterials. With 1996–2009 year the area of crops of GM cultures have increased with 2.8 up to 134.0 million ha. It corresponds almost to all 2/3 of cultivated fertile grounds. NDO are widely used as various external decoration and the additives in food and packings. Hence, a new epoch of preparation of artificial modified by GMO and NDO food have began, as from lack of organic food can and already millions inhabitants of a planet perish. However, at hit into an organism of people through skin, nose and mouth NDO also easy overcome practically all biological of barriers and can interact and destroy different human organs. The first results of study of an influence of NDO and GMO on the organism of people and plants are threatening. The objectives of this report is to demonstrate the experimental results of nanoecological nanothreats related with a scale usage of GMO and NDO as the additives in nano-food and in it packings.

24.1 Introduction

"Maintain humanity under 500,000,000 in perpetual balance with nature." It is one of ten conunandments engraved in eight modern and four ancient languages on granite plates of Georgia Guidestones. The monument was ordered for a civil

O. Kharlamov (✉) • M. Bondarenko • O. Khyzhun
Frantsevich Institute for Problems of Materials Science of NASU, Krzhyzhanovsky St. 3, 03142 Kiev, Ukraine
e-mail: akharlamov@ukr.net

G. Kharlamova
Taras Shevchenko National University of Kyiv, 64 Volodymyrsska Street, 03001 Kyiv, Ukraine

J. Bonča, S. Kruchinin (eds.), *Nanomaterials for Security*, NATO Science for Peace and Security Series A: Chemistry and Biology, DOI 10.1007/978-94-017-7593-9_24

engineering firm in 1979 ostensibly Robert C. Christian. Is remarkable, that in 1974 year the number of people on all planet has already overcome 4 billion. There is a natural question. What will with the people that will appear outside of balance with a nature? In present time (2015 year) such people already more than 6.5 billion. Consequently, it are required the civilizational qualitative changes of technologies, and first of all, technologies of the production and the preparation of food.

Just in this period of time the first genetically modified objects (GMO) and nanotechnologies are created. Moreover, new state of substance as the hollow spherical closed molecules (fullerenes, single-walled carbon nanotubes (SWCNT)), nanostructures and nanodimensional objects (NDO) of the simple and complex substances was discovered. These unique scientific achievements, first of all, were used not only in various systems of safety, but also for the decision of a problem of food maintenance of the people of all planet by accessible products of a feed. Nanodimensional food not only is much more useful, but also the illusion of its huge abundance creates. Probably, therefore the transition from laboratory researches GMO and NDO up to their industrial application has appeared extremely short. So, the first GM culture (GM tomato FlavrSavr) was created in 1993 [1], and during 1 year it was begun industrial cultivation by company Calgene [2].

Manufacture of GM agricultural cultures is represented by an effective way of the complete decision of a food problem of mankind. Indeed, by numerous experiments was convincingly shown, that GM agricultural cultures have the raised productivity because of stability to a drought, herbicides, insect and viruses. However the question on biological influence of GM products of a feed on living organism and, first of all, on the man is opened. The researches of a similar direction not so much complex, how many require rather long period of study of object: the results of study of influence of GMO (as well as NDO) on the organism of the man during life of one generation not always are rather correct. Therefore it is very important attentively to consider the results, in which the toxicity or the harmful influence of GMO and NDO is so definitely demonstrated already in a short interval of life of the living organism.

In the given paper the researches concerning spheres of an application of GMO and NDO in foodstuff and possible consequences of their uncontrollable use are systematized. Earlier [3–5] by us were classified NDO, the nanochemical and nanooxicological features and the possible routes of the penetration them in organism of the man are described. Is remarkable, that GMO can get in organism of people only with products of a feed, whereas NDO also through a skin, a nose and an ear. The special attention here is given to the description of harmful influence such NDO as carbon (nanotubes, fullerenes and quasi-fullerenes) and heterocarbon (carbon nitride (C_3N_4) and silicon carbide SiC) molecules and nanostructures. Just these NDO we synthesize [6–14] by the original methods and investigate their special, nanochemical properties.

24.2 Threats of Human Security from Genetically Modified Products

Paul Berg in 1974 year for the first time has been synthesized recombinant (cloned from DNA various organisms) DNA [15]. For this discovery to him was awarded of the Nobel premium. The opportunity of creation of the genetically modified organisms has appeared. Herbert Boyer and Stanley Cohen in 1972 year for the first time have entered an allogenic gene of salmonella [16] in a unicellular bacterium of Escherichia coli. It has become the beginning of the creation of gen engineering as biotechnology of a direct transfer of DNA from one organism to another.

The first alive the transgenic essences (mouse and fly Drosophila) were created in 1981–1982 years, and the first genetically modified agriculture (a GM tomato FlavrSavr) has appeared in 1993 year [1]. From the moment of a beginning of the cultivating by the company Calgene [2] of a GM tomato FlavrSavr the quantity of the names of GM agricultures (Fig. 24.1) and, as a consequence made foodstuff with GMO (Fig. 24.2), grows avalanchely.

The sowing areas of the engaged GM cultures have increased more than in 100 times for 14 years. Only for the period with 1996 for 2009 of the area of crops of GM cultures have increased with 2.8 up to 134.0 ml hectares, and in 2013, on the data Institute of science in society (University of Nottingham, UK) have reached already 175.2 ml hectares [17]. It corresponds almost to 12 % all cultivated fertile grounds. For today soybean is most widespread GM culture. On the data Institute of science in society on size of the areas of engaged under GM cultures, the leaders are 5 countries: USA, Brazil, Argentina, India and Canada (Fig. 24.3) [17]. So, in USA now 91 % soybean, 85 % of corn and 80 % of sugar beet are GMO (Fig. 24.4) [18]. The list of the countries, produced GM plants, continuously is extended. The question on an opportunity to sow the genetically modified plants [19] in territory of Russia is also decided (Resolution Russian Federation Government No. 839, September 23, 2013). (Russia else recently considered as the basic manufacturer of safe products of a feed.) The similar rates of increase of the areas of crops of GM plants, in particular soybean, can result in complete disappearance of traditional unmodified agricultures.

Fig. 24.1 Most widespread GM cultures

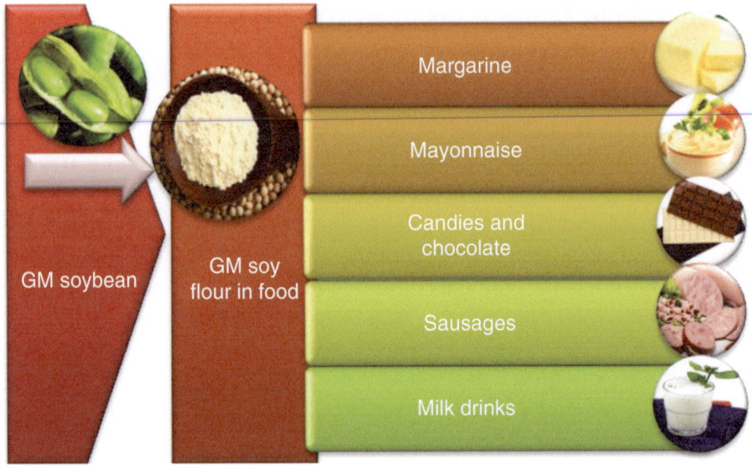

Fig. 24.2 Most widespread GM food

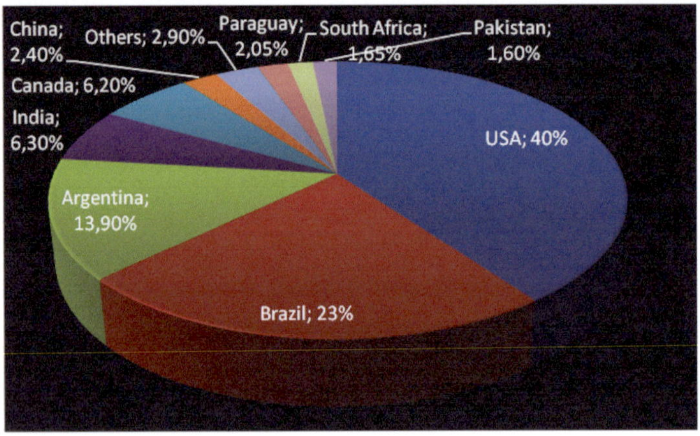

Fig. 24.3 Share of areas planted with GM crops (of the world), %

The uncontrollable growth of the consumption of a genetically modified food has caused an active discussion about safety of such products. Deeply it is symbolical, that for the first time the fears about possible toxicological influence of GMO [20] were stated already in 1974 year by the founder of recombinant DNA, which the opening and has allowed to create the first GMO.

Indeed, the risks for health of the man with an increase of the areas with GM plants can be caused by food allergies and horizontal gene transfer from GMO to the consumer (Fig. 24.5). The pollution of an environment by GMO can be caused by migration of genes owing to cross-pollination or horizontal gene transfer. At the same time, in the reports of the World organization of public health services (World Health Organization (WHO)) [21], General Management of the European

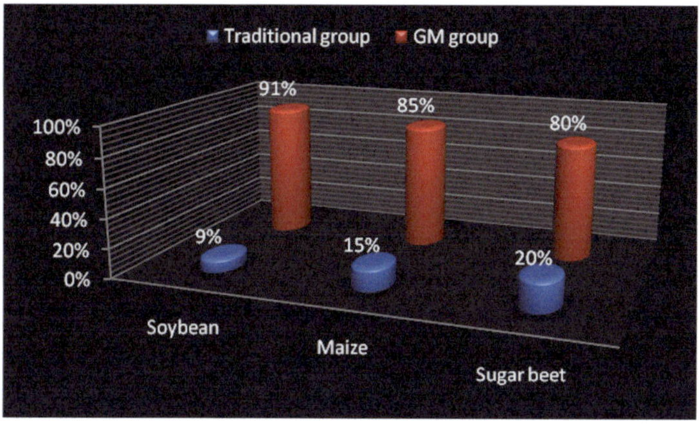

Fig. 24.4 The area of sowings in USA of traditional and GM soybean, maize, sugar beet

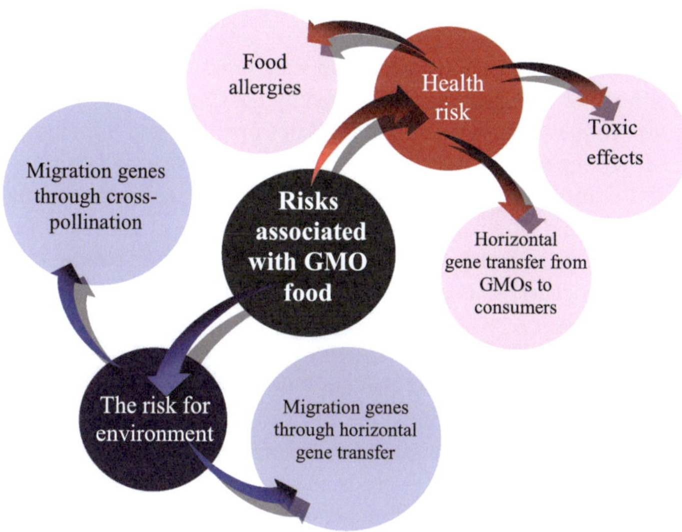

Fig. 24.5 The basic risks connected with cultivation of GM plants

commission on a science and information (European Commission Directorate-General for Research and Innovation) [22] the safety of the use of GM products is marked. In more than 300 scientific works the conclusion about harmlessness of GM products experimentally also proves to be true.

However there are publications of some scientists such as Arpad Pusztai (Britain) [23], Gilles-Eric Seralini (France) [24, 25], Irina Ermakova (Russia) [26] and William Engdahl (USA) [27], which testify about the unsafety of the use of GÌ products. So, in 1999 Arpad Pusztai [23] (Britain) experimentally has established, that the consumption of a transgenic potatoes with the built – in gene lectin (from snowdrop Galanthus nivalis) negatively influences on immune system of the rats: the abnormal changes bowel, illness of a liver, kidneys, head brain are observed. Is remarkable, that after this publication the company – developer of the transgenic potatoes with a gene of lectin immediately stopped it the cultivation: probably, the experimental results Arpad Pusztai were convincing.

The researches of Irina Ermakova [26] (Russia) demonstrated, that GM soybean (RR, the line 40.3.2) negatively influences on a health of the rats and their posterity. According to research by Ermakova[26] the consumption of GM soybean (RR, the line 40.3.2) causes the pathologies of sexual organs and reproductive functions of the animals, results in infringement of hormonal balance, infertility, the formation of tumours.

Scientific experiments of Seralini [24, 25] (France) have shown, that the some sorts of GM corn cause at experimental rats (Sprague-Dawley) cancerous tumours, defeats of pituitary, necrosis of a liver and nephropathy of kidneys. The death at the females of the rats in all processed groups was observed in 2–3 times more, than in control group.

The researches demonstrating the toxicity of GM food not only for laboratory animals, but also for the man also have appeared. So, the toxic influence of GM corn on organism of the man was established by the Canadian researchers Aris and Leblanc [28]. To the study were exposed 30 not pregnant and 30 pregnant women in Eastern Townships of Quebec (Canada). Was established [28], that a toxin Bt, developed by corn genetically modified pesticide, collects in organism both not pregnant women, and pregnant women. Moreover, a toxin Bt penetrates through placenta and collects in embryos. Thus, a health of the future people can be in the same danger that occurs, as is established by Ermakova [26], with the posterity of the experimental ruts.

As it is possible to see, it is not enough of researches on an establishment of a harmful influence of GM food. At the same time, these researches, probably, are so convincing that the interdiction on the cultivating of some sorts of soybean [29], corn and pasturable peas [30] nevertheless has appeared. However, the increase of new farmland under GM culture proceeds and, hence, the manufacture of GM food is promptly increased. Only we shall hope, that other sorts of GM plants (soybean, corn and peas) are not such harmful as directly for the consumer of GM food, and his posterity.

24.3 Threats of Human Security from Nanodimentional Objects

As against GMO the toxicological features of products with NDO are investigated very detailed. Application of NDO in comparison with GMO is more various: nanoparticles and nanomaterials on their basis are widely applied in electronics, chemical, textile and food-processing industry (Fig. 24.6), biomedicine and

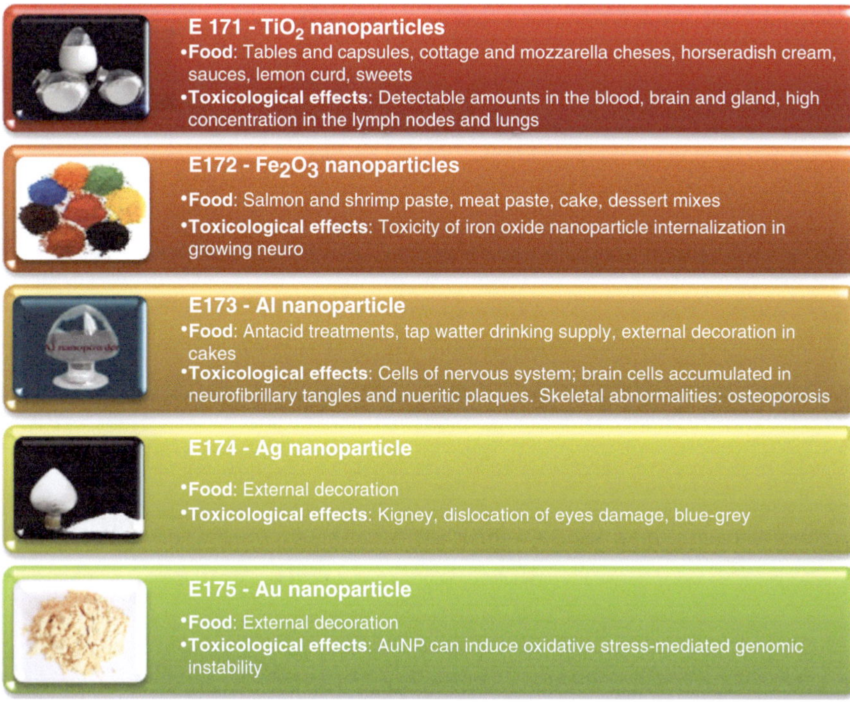

Fig. 24.6 The use of nanoparticles as food additive

cosmetology. Now already more than 1000 products in the market are made with the use of nanomaterials [31], the part from which can easy penetrate in the organism of the man through a digestive path, organs of breath, a skin and also through mucous environments of an eye and acoustical pass [3–5]. Therefore in the given review the special attention is given on the researches of toxicological properties as metallic (Fig. 24.7), and carbon [6–11, 13] (Fig. 24.8) nanoparticles, which intensively are used in a cosmetology, nanomedicins, a textile and food-processing industry. Besides the environment in an epoch of the nanotechnological enterprises also contains various NDO, which of air and waters penetrate in organism of all people. At the industrial enterprises up to 82 % of the workers receive the seen damages of an epidermal barrier [32], as well as the disease of organs of a breath [33] at an influence of metallic and nonmetallic nanoparticles.

In a food-processing industry the food additives on a basis of nanoparticles of oxides of iron (Fe_2O_3, FeO) and titanium (TiO_2) as well as such metals as aluminum (Al), silver (Ag) and gold (Au) are used in plenties. Toxicological effects of using in food the additives of nanoparticles of TiO_2, Al, Ag [34], Au [35] and iron oxides [36] is shown in Figs. 24.6, and 24.8. Addition of a dye on a basis of nanoparticles of Fe_2O_3, (the food additive E 172) in a forage farmed on farms, for example, shrimps and red fish provides an appetizing kind of a salmon and shrimp paste. Fractional

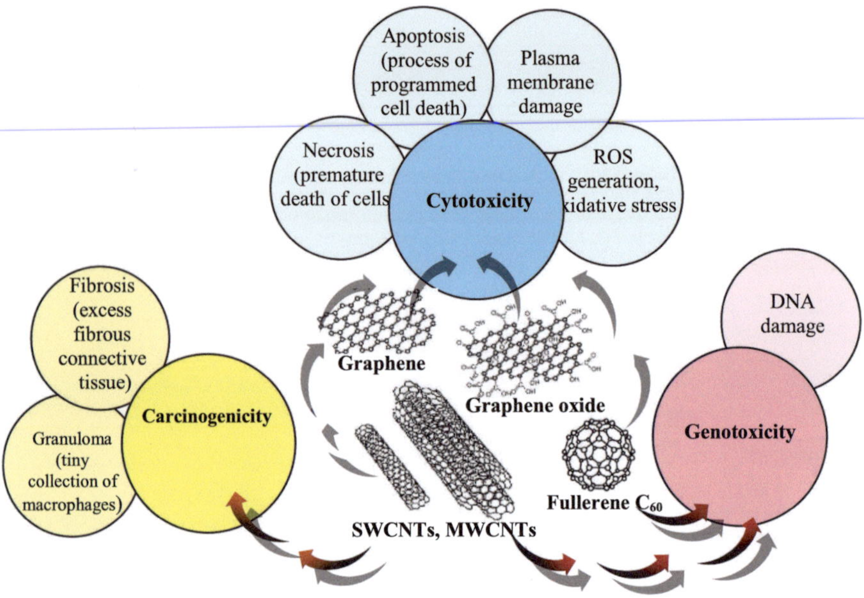

Fig. 24.7 Some unusual toxic peculiarities of carbon nanostructures

introduction of suspension of iron nanoparticles (dozes 1000, 2000 and 5000 mg/kg) results to the development of inflammatory processes on a mucous stomach and the intestines (of mice, birds and fish) as well as the change during a formation, a development and a maturing of blood cells [3–5]. Nanoparticles of Fe_2O_3 render toxic influence on cells of a bone marrow [37] both a mice, and people [38].

Nanoparticles of TiO_2 frequently are used in a food industry as the food additive E 171, which provides a visual whiteness or density to sauces, pastes, confectionery creams and cheeses. Besides, nanoparticles of TiO_2 are used in a cosmetology (creams) and pharmacology (tablet and capsules). However these nanoparticles penetrating in the organism through the pores of a skin collect in high concentration in the blood, brain, gland, lymph nodes and lungs and promote a photooxidation and cause oxidative stress [3–5]. Besides histopathological researches have shown [39] the serious damages of a skin to all groups of mice exposed to the influence of TiO_2 nanoparticles. Nanoribbons and nanoparticles of TiO_2 after an introduction in trachea of mice and rats cause neutrophilia, inflammation easy [40] and pulmonary fibrosis [41]. At the treatment by nanoparticles of ZnO of metastatic lymphatic tumours was established their unselective toxicity for healthy cells of organism [42].

Nanoparticles of silver are widely applied in electronics, medicine (antiseptic), food (additive E 174 and packing of products) and a textile industry. They are capable to penetrate in an organism of people and to cause the damage of cells (macrophages) [43] and to infringe of the functions of a liver and kidneys [3–5]. Ability of nanoparticles of silver to generate the active forms of oxygen leads to the

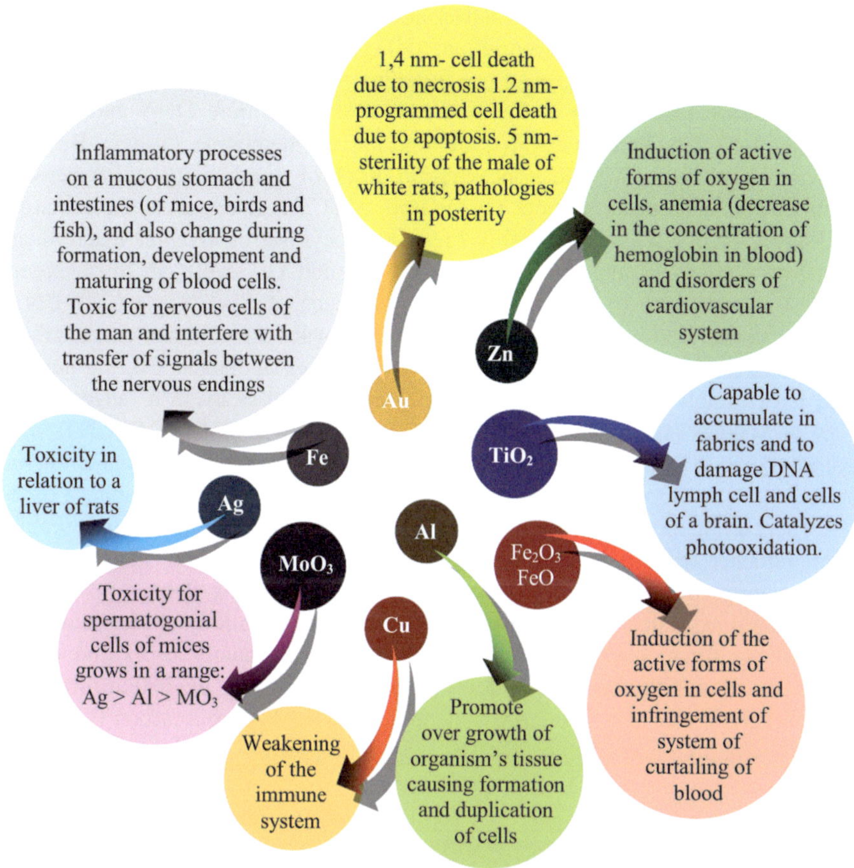

Fig. 24.8 Peculiarities of toxic impact of nanoparticles of various metals and oxides on live organism

damages of biomolecules such as fibers, ferments and DNA [44]. So, nanoparticles of silver (Ag NPs) cause an induction of apoptosis and an injure a mitochondrial membrane of cells HepG2 of a liver of people [45]. Ag nanoparticles can inhibit the viability of cells HepG2 and cause a stop of a cell cycle.

Nanoparticles of gold exert a toxic influence at them the collection in the vacuoles of fat cells [46], bone marrow, cells of blood and liver [47]. Thus [48] is established, that the degree of a toxicity depends on the size, surface chemistry, and shape of the gold nanoparticles.

Mezoporous nanoparticles of silicon oxide (SiO_2) even at their very low concentration are extremely toxic for tumoral cells of the man [49].

Molecules (single-walled nanotubes, fullerenes and its derivatives) and nanostructures (multi-walled nanotubes (MWCNT), graphene and graphene oxide) carbon generate the active forms of oxygen and induce an oxidizing damage of a membrane both nucleus of cells of a liver and blood of the man [3–5]. So, fullerene C_{60} is toxic at very low concentration (0.02 ppb). The intravenous injection of small dozes (on 15 and 25 mg/kg) of fullerene resulted within 5 min in death of two of twenty rats. Fullerene C_{60} and MWCNT also injure DNA and nucleus of cells [3–5]. Graphene and MWCNT injure a cell membrane stronger. Besides, as was established [50] the concentration limit of cytotoxicity of suspension from carbon nanotubes is about 0.01 mg/ml. SWCNT also causes the mutating of RNA [51]. SWNTs considerably break phagocytosis of alveolar macrophage at a low doze (0.38 mkg/sm^2), while MWCNT and C_{60} cause an infringement only at a high doze (3.06 mkg/sm^2). The influence of SWCNTs or MWCNTs at their concentration 3.06 mkg/sm^2 on macrophages has shown characteristic features necrosis and degeneration and also attributes apoptosis. So, the authors [52] have established the increase of a degree of the toxicity in a range: SWCNTs > MWCNT10 > quartz > C_{60}. Fullerenes, SWCNTs and MWCNTs also immunotoxic [53], cause a strong inflammation in alveolar and bronchial epithelial cells. In [54] in vitro and in vivo was investigated a cytotoxicity of graphene oxide (GO) depending on the size of nanoflakes (diameter 89 and 277 nm) and the duration of an influence for cells. Greater nanosized flakes of GO reduce the viability of cells faster. Nanoparticles of GO of any size destroy of a cell during 48 h.

It should be noted that the toxicity of the most potentially fit for numerous present and future applications (for catalysts, carriers, nanosensors and others nanodevices) nanomaterials based on hetero-carbon, such as hetero-fullerenes [9, 10], hetero-nanotubes, hetero-graphene, onion-like N-doped carbon (Fig. 24.9) [11], O-doped carbon nitride (O-C_3N [13]) still practically is not investigated.

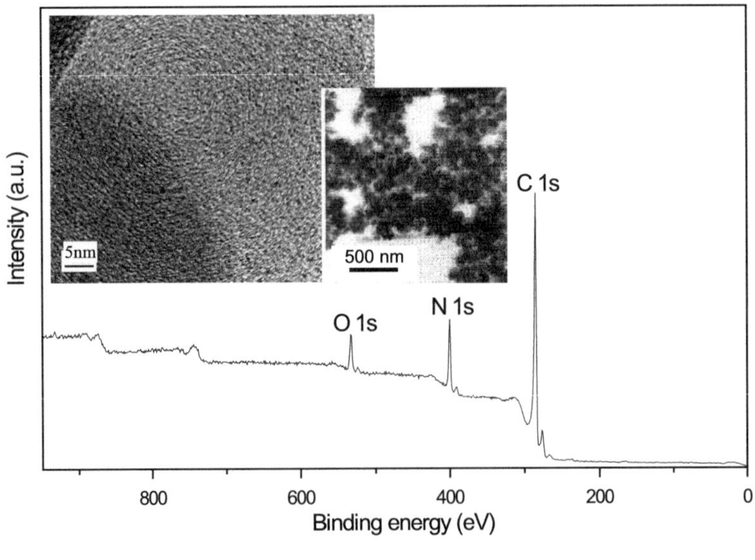

Fig. 24.9 XPS survey spectrum, TEM and HR TEM images (*insets*) of onion-like N-doped carbon

Heteroatomic (carbon-nitrogen-oxygen (Fig. 24.9)) derivatives because of brightly expressed asymmetry of the distribution of electronic density in a graphene layer have the unique electronic characteristics.

As it is possible to see, some results of researches of a degree of the toxicity of nanoparticles in vitro and in vivo are rather inconsistent. It is possible, that the toxicity of NDO essentially depends on the size, a surface state, a form and the method of preparation of a researched nanochemical object, as well as from a technique of study of the toxicity.

24.4 Conclusions

At the end of the last century first were created the generically changed organisms and was discovered a nanostate of substance: the epoch bio- and nanotechnologies began. The advanced countries alongside with the purposes of a military safety in epoch "nano" have begun actively to decide the questions of food safety. The basic manufacturer of GM plants is Northern America (USA and Canada), where the basic researches about a possible toxicity of GM products also intensively carry out. However, GM products use in many other countries. The products with the additives NDO are made, probably, in each country. Therefore to study and to supervise a harmful influence both GMO, and NDO it is extremely difficult as in technological and methodical relation. The results of researches as GMO, so NDO are not always unequivocal. However there are researches, which convincingly demonstrate a high toxicity of NDO and GM products. At the penetration in organism of people through a skin, the nose and mouth, NDO easily is overcome practically all biological barriers and can injure various human organs. Therefore, in a new epoch, the epoch of a nanotechnological and artificial food, probably, is possible to supplied with food even of those people, with which the destiny, "outside of balance with a nature". The question while remains open – that will be with their posterity?

References

1. Keith R, Hiatt B, Martineau B, Kramer M (1992) Safety assessment of genetically engineered fruits and vegetables: a Case study of the Flavr Savr tomato. CRC Press, Boca Raton
2. Weasel LH (2009) Food fray. Amacom Publishing, New York
3. Kharlamov A, Skripnichenko A, Gubareny N et al (2011) Toxicology of nano-objects: nanoparticles, nanostructures and nanophases. In: Mikhalovsky S, Khajibaev A (eds) Biodefence. NATO science for peace and security series A: chemistry and biology, Part 1. Springer, Dordrecht, pp 23–32
4. Kharlamov O, Bondarenko M, Kharlamova G et al (2015) Nanoecological security of foodstuffs and human. In: Bonča J, Kruchinin S (eds) Nanotechnology in the security systems. NATO science for peace and security series c: environmental security. Springer, Dordrecht, Chapter 19, pp 215–229

5. Kharlamov A, Bondarenko M, Skripnichenko A, Kharlamova G (2013) Nanothreats and nanotoxicological peculiarities of nanoobjects as one of the future trends of terrorist threat. In: Voica DR, Duyan A (eds) Trends and developments in contemporary terrorism. IOS Press, Amsterdam, pp 33–47
6. Kharlamov AI, Bondarenko ME, Kirillova NV (2012) New method for synthesis of fullerenes and fullerene hydrides from benzene. Russ J Appl Chem 85(2):233–238
7. Kharlamov AI, Kharlamova GA, Bondarenko ME (2013) New low-temperature method for joint synthesis of C_{60} fullerene and new carbon molecules in the form of C_3–C_{15} and quasi-fullerenes C_{48}, C_{42}, C_{40}. Russ J Appl Chem 86(8):1174–1183
8. Kharlamov O, Kharlamova G, Bondarenko M, Fomenko V (2013) Small carbon molecules and quasi-fullerenes as products of new method of hydrocarbons pyrolysis. In: Vaseashta A, Khudaverdyan S (eds) Advanced sensors for safety and security. NATO science for peace and security series B: physics and biophysics. Springer, Dordrecht, Chapter 30, pp 329–338.
9. Kharlamov AI, Kharlamova GA, Bondarenko ME (2013) New products of a new method for pyrolysis of pyridine. Russ J Appl Chem 86(2):167–175
10. Kharlamova G, Kharlamov O, Bondarenko M (2013) Hetero-carbon: heteroatomic molecules and nano-structures of carbon. In: Vaseashta A, Khudaverdyan S (eds) Advanced sensors for safety and security. NATO science for peace and security series B: physics and biophysics. Springer, Dordrecht, Chapter 31, pp 339–357
11. Kharlamov AI, Kharlamov GA, Bondarenko ME (2013) Preparation of onion-like carbon with high nitrogen content (∼15 %) from Pyridine. Russ J Appl Chem 86(10):1493–1503
12. Kharlamov O, Bondarenko M, Kharlamova G, Fomenko V (2015) Quasi-fulleranes and fulleranes as main products of fullerenizasion of molecules of Benzene, Toluene and Pyridine. In: Camesano TA (ed) Nanotechnology to aid chemical and biological defense. NATO science for peace and security series A: chemistry and biology. Springer, Dordrecht, Chapter 13, pp 191–205
13. Kharlamov O, Bondarenko M, Kharlamova G (2015) O-doped carbon nitride (O-g-C_3N) with high oxygen content (11.1 mass%) synthesized by pyrolysis of Pyridine. In: Camesano TA (ed) Nanotechnology to aid chemical and biological defense. NATO science for peace and security series A: chemistry and biology. Springer, Dordrecht, Chapter 9, pp 129–145
14. Ermakov V, Kruchinin S, Fujiwara A (2008) Electronic nanosensors based on nanotransistor with bistability behaviour. In: Bonca J, Kruchinin S (eds) Proceedings NATO ARW "Electron transport in nanosystems". Springer, Berlin, pp 341–349
15. Jackson DA, Symons RH, Berg P (1972) Biochemical method for inserting new genetic information into DNA of Simian virus 40: circular SV40 DNA molecules containing lambda phage genes and the galactose operon of Escherichia coli. Proc Natl Acad Sci 69:2904–2909
16. Cohen SN, Chang ACY, Boyer HW, Helling RB (1973) Construction of biologically functional bacterial plasmids in vitro. Proc Natl Acad Sci 70(11):3240–3244
17. Institute of Science in Society, University of Nottingham. http://www.isaaa.org.
18. Buravchikova D, Korobitsyna O, Melnikov A (2013) Genetically mined. Murderous law. Why officials and business allowed GMOs in Russia, Argumenty i Fakty 51
19. On state registration of genetically modified organisms intended for release into the environment, as well as products derived from the use of such organisms or containing such organisms, "Government of the Russian Federation, the decision", No. 839, September 23 (2013)
20. Berg P, Baltimore D, Boyer HW, Cohen SN (1974) Potential biohazards of recombinant DNA molecules. Science 185(4148):303
21. Modern food biotechnology, human health and development: a study based on the facts. WHO report (Russian) (2005)
22. European Commission Directorate-General for Research and Innovation, Directorate E – Biotechnologies, Agriculture, Food; Unit E2 – Biotechnologies (2010)
23. Ewen SW, Pusztai A (1999) Effect of diets containing genetically modified potatoes expressing Galanthus nivalis lectin on rat small intestine. Lancet 354(9187):1353–1354
24. Séralini GE, Clair E, Mesnage R (2012) Long term toxicity of a Roundup herbicide and a Roundup-tolerant genetically modified maize. Food Chem Toxicol 50:4221–4231

25. Séralini GE, Clair E, Mesnage R et al (2013) Answers to critics: why there is a long term toxicity due to a Roundup-tolerant genetically modified maize and to a Roundup herbicide. Food Chem Toxicol 53:476–483
26. Ermakova IV (2007) GM soybeans revisiting a controversial format. Nat Biotechnol 25(12):1351–1354
27. Engdahl W (2007) Seeds of destruction: the hidden agenda of GMO. Global Research, Montreal
28. Aris A, Leblanc S (2011) Maternal and fetal exposure to pesticides associated to genetically modified foods in Eastern townships of Quebec, Canada. Rep Toxicol 31(4):528–533
29. Nordlee JA, Taylor SL, Townsend JA et al (1996) Identification of a Brazil-nut allergen in transgenic soybeans. N Engl J Med 334(11):688–692
30. Prescott VE, Campbell PM, Moore A (2005) Expression of Bean-Amylase inhibitor in peas results in altered structure and immunogenicity. J Agric Food Chem 53(23):9023–9030
31. Lin M (2013) Toxic nanoparticles might be entering human food supply. University of Missouri-Columbia. Science Daily, 22 Aug 2013. www.sciencedaily.com/releases/2013/08/130822194530.htm
32. Korinth G, Weiss T, Penkert S et al (2007) Percutaneous absorption of aromatic amines in rubber industry workers: impact of impaired skin and skin barrier creams. Occup Environ Med 64:366–372
33. Yah CS, Simate GS, Iyuke SE (2012) Nanoparticles toxicity and their routes of exposures. Pak J Pharm Sci 25(2):477–491
34. Casals E, Vazquez-Campos S, Bastus NG, Punte V (2008) Distribution and potential toxicity of engineered inorganic nanoparticles and carbon nanostructures in biological systems. Trends Anal Chem 8(27):672–683
35. Li JJ, Gurung RL, Hartono D et al (2011) Genomic instability of gold nanoparticle treated human lung fibroblast cells. Biomaterials 23(32):5515–5523
36. Pisanic TR, Blackwell JD, Shubayev VI et al (2007) Nanotoxicity of iron oxide nanoparticle internalization in growing neuron. Biomaterials 28:2572–2581
37. Lak A, Dieckhoff J, Ludwig F et al (2013) Highly stable monodisperse PEGylated iron oxide nanoparticle aqueous suspensions: a nontoxic tracer for homogeneous magnetic bioassays. Nanoscale 5(23):11447–11455
38. Luengo Y, Nardecchia S, Puerto M et al (2013) Different cell responses induced by exposure to maghemite nanoparticles. Nanoscale 5(23):11428–11437
39. Korinth G and Drexler H (2013) Penetration of Nanoparticles through Intact and Compromised Skin. In: Nanomaterials. edited by Deutsche Forschungsgemeinschaft (DFG). Wiley-VCH Verlag GmbH & Co, KGaA, Weinheim, Chapter 1.4, pp 37–42
40. Bonner JC, Silva RM, Taylor AJ et al (2013) Interlaboratory evaluation of rodent pulmonary responses to engineered nanomaterials: the NIEHS nano GO consortium. Environ Health Perspect 121(6):676–682
41. Porter DW, Wu N, Hubbs AF et al (2013) Differential mouse pulmonary dose and time course responses to titanium dioxide nanospheres and nanobelts. Toxicol Sci 13:1179–193
42. Zeng K, Li J, Zhang Z et al (2015) Lipid-coated ZnO nanoparticles as lymphatic-targeted drug carriers: study on cell-specific toxicity in vitro and lymphatic targeting in vivo. J Mater Chem B 26(3):5249–5260
43. Pratsinis A, Hervella P, Leroux JC et al (2013) Toxicity of silver nanoparticles in macrophages. Small 9(15):2576–2584
44. Reidy B, Haase A, Luch A et al (2013) Mechanisms of silver nanoparticle release, transformation and toxicity: a critical review of current knowledge and recommendations for future studies and applications. Materials 6(6):2295–2350
45. Xue Y, Zhang T, Zhang B et al (2015) Cytotoxicity and apoptosis induced by silver nanoparticles in human liver HepG2 cells in different dispersion media. J Appl Toxicol 36(3):352–360
46. Mironava T, Hadjiargyrou M, Simon M, Rafailovich MH (2014) Gold nanoparticles cellular toxicity and recovery: adipose derived stromal cells. Nanotoxicology 8(2):189–201

47. Zhang XD, Wu D, Shen X et al (2011) Size-dependent in vivo toxicity of PEG-coated gold nanoparticles. Int J Nanomed 2011(6):2071–2081
48. Favi PM, Gao M, Arango LJS et al (2015) Shape and surface effects on the cytotoxicity of nanoparticles: gold nanospheres versus gold nanostars. J Biomed Mater Res A 103(11): 3449–3462
49. Martínez-Carmona M, Baeza A, Rodriguez-Milla MA et al (2015) Mesoporous silica nanoparticles grafted with a light-responsive protein shell for highly cytotoxic antitumoral therapy. J Mater Chem B 28(3):5746–5752
50. Kartel MT, Ivanov LV, Kovalenko SN, Tereschenko VP (2011) Carbon nanotubes: biorisks and biodefence. In: Mikhalovsky S, Khajibaev A (eds) Biodefence. NATO science for peace and security series A: chemistry and biology. Springer, Dordrecht, pp 11–22
51. Zhao Y, Wu Q, Li Y et al (2014) In vivo translocation and toxicity of multi-walled carbon nanotubes are regulated by microRNAs. Nanoscale 6:4275–4284
52. Jia G, Wang HF, Yan L et al (2005) Cytotoxicity of carbon nanomaterials: single-wall nanotube, multi-wall nanotube and fullerene. Env Sci Technol 39:1378–1383
53. Turabekova M, Rasulev B, Theodore M et al (2014) Immunotoxicity of nanoparticles: a computational study suggests that CNTs and C_60 fullerenes might be recognized as pathogens by Toll-like receptors. Nanoscale 6:3488–3495
54. Mendes RG, Koch B, Bachmatiuk A et al (2015) A size dependent evaluation of the cytotoxicity and uptake of nanographene oxide. J Mater Chem B 12(3):2522–2529

Chapter 25
Transmission of Three Resistance Sensor Signals Over Four Wire Line with Losses

A. Penin and A. Sidorenko

Abstract The invariant relationship between the sets of load conductivity values at the output of the balanced four wire communication line with losses and the corresponding current values at the input of this line is shown. This relationship does not depend on accuracy of measuring devices. It allows transmitting three signals over this communication line.

25.1 Introduction

Different sensors of physical values are used for monitoring technical or natural objects. For these usually remote devices, it is necessary to provide of transmission of measuring signals [1], for example, by multi-wire line [2].

At present time, methods for transmission of discrete electrical signals in binary code are used. Also, researches and elaborations of systems for transmitting discrete electrical signals in the analog form are important [3].

The theoretical researches show that invariant relationship takes place between the sets of load conductivity (resistance) values at the output of communication line with losses and the corresponding conductivity or current values at the input of this line [4]. The cross ratio of four points, known in projective geometry, is this invariant relationship. Therefore, the value of cross ratio (is the ratio of two proportions) does not depend on parameters of two wire line and accuracy of measuring devices. These invariants are used for transmission or, more precisely, for restoration of measuring signals over the two and three wire lines [5, 6].

It is naturally to use four wire lines for transmission of three signals. But, these lines must be as the balanced networks for the output terminals. Therefore, we may introduce projective coordinates and use the above invariant relationships. There are two variants of the balanced networks [7, 8]. In this regard, it is possible to consider these balanced networks in the general way.

A. Penin (✉) • A. Sidorenko
"D. Ghitu" Institute of Electronic Engineering and Nanotechnologies, Academy of sciences of Moldova, Chisinau, Moldova
e-mail: aapenin@mail.ru; anatoli.sidorenko@kit.edu

J. Bonča, S. Kruchinin (eds.), *Nanomaterials for Security*, NATO Science for Peace and Security Series A: Chemistry and Biology, DOI 10.1007/978-94-017-7593-9_25

25.2 First Balanced Network

Let us consider the equivalent circuit or model of a four wire line with losses in Fig. 25.1. Resistance sensors are the loads with conductivities Y_{L1}, Y_{L2}, and Y_{L3}, which connect to the output of this line. Each voltage source V_4, V_5, V_6 is the source for the corresponding sensor. A mutual influence of the loads on the input I_4, I_5, I_6 and output I_1, I_2, I_3 currents takes place because of loss resistances. For known values of the loss resistances or transmission parameters, we may calculate the load values by the measured input currents.

To determine the transmission parameters, we must test this line by complex manipulations at the input and output that is often inconvenience. Further, we will show how to calculate the load values using the measured input currents by manipulations at the output terminals only.

To do this, we consider our line as an active multiport relatively to load conductivities. The circuit is described by the following system of Y parameter equation

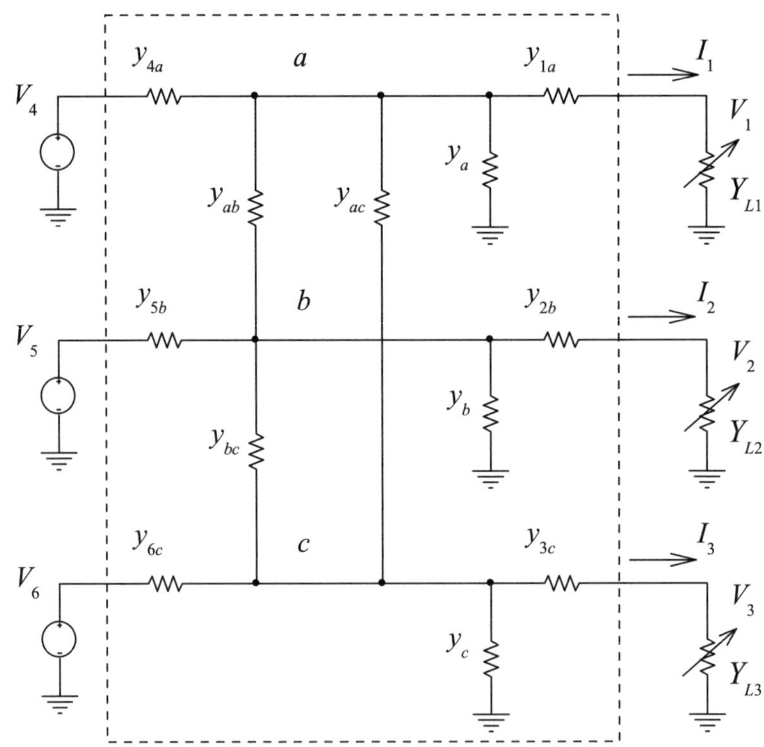

Fig. 25.1 Four wire line with longitudinal and lateral loss resistances

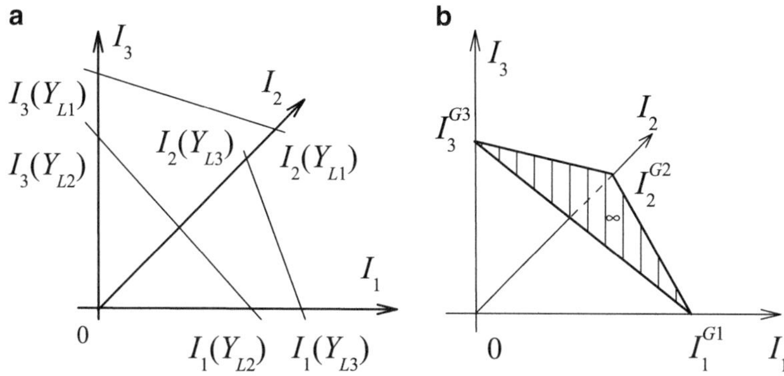

Fig. 25.2 (**a**) Points of intersection for each axis do not coincide among themselves. (**b**) Points of intersection coincide among themselves and form the plane of infinity ∞

$$\begin{pmatrix} I_1 \\ I_2 \\ I_3 \end{pmatrix} = \begin{pmatrix} -Y_{11} & Y_{12} & Y_{13} \\ Y_{12} & -Y_{22} & Y_{23} \\ Y_{13} & Y_{23} & -Y_{33} \end{pmatrix} \cdot \begin{pmatrix} V_1 \\ V_2 \\ V_3 \end{pmatrix} + \begin{pmatrix} I_1^{SC} \\ I_2^{SC} \\ I_3^{SC} \end{pmatrix}, \tag{25.1}$$

where I_1^{SC}, I_2^{SC}, and $I_3^{SC}$are the short circuitSC currents. Taking into account the voltages $V_1 = I_1/Y_{L1}$, $V_2 = I_2/Y_{L2}$, and $V_3 = I_2/Y_{L3}$, the equations of three bunches of planes are obtained in the form $(I_1, I_2, I_3, Y_{L1}) = 0$, $(I_1, I_2, I_3, Y_{L2}) = 0$, $(I_1, I_2, I_3, Y_{L3}) = 0$.

Crossing of the planes of one bunch among themselves defines a bunch axis. The equation of the axis of the bunch Y_{L1} corresponds to the condition $I_1 = 0$, $V_1 = 0$ and to equation $(I_2, I_3) = 0$. Therefore, this axis is located in the plane I_2, I_3 in Fig. 25.2a.

The points $I_2(Y_{L1})$, $I_3(Y_{L1})$ are the points of intersection with corresponding axis. Similarly, we obtain the points $I_1(Y_{L2})$, $I_3(Y_{L2})$ of intersection of the bunch axis Y_{L2} and the points $I_1(Y_{L3})$, $I_2(Y_{L3})$ of intersection of the bunch axis Y_{L3}.

On the other hand, the projective system of coordinates has to represent a tetrahedron $0\,I_1^{G1}I_2^{G2}I_3^{G3}$ in Fig. 25.2b. In this case, we accept the plane, which passes through three base points I_1^{G1}, I_2^{G2}, and I_3^{G3}on the axes of coordinates, as the plane of infinity ∞.

Therefore, the next conditions have to be satisfied

$$I_1(Y_{L2}) = I_1(Y_{L3}) = I_1^{G1}, \quad I_2(Y_{L1}) = I_2(Y_{L3}) = I_2^{G2},$$

$$I_3(Y_{L1}) = I_3(Y_{L2}) = I_3^{G3}.$$

We must determine requirements for Y parameters.

Let us consider the base point or base values

$$I_1 = I_1^{G1}, \quad V_1 = V_1^{G1}.$$

Then, $I_2 = 0$, $V_2 = 0$, $I_3 = 0$, $V_3 = 0$. Using (25.1), we get

$$-V_1^{G1} = \frac{I_2^{SC}}{Y_{12}} = \frac{I_3^{SC}}{Y_{13}}.$$

Similarly, we consider

$$I_2 = I_2^{G2}, \quad V_2 = V_2^{G2}, \quad -V_2^{G2} = \frac{I_1^{SC}}{Y_{12}} = \frac{I_3^{SC}}{Y_{23}};$$

$$I_3 = I_3^{G3}, \quad V_3 = V_3^{G3}, \quad -V_3^{G3} = \frac{I_1^{SC}}{Y_{13}} = \frac{I_2^{SC}}{Y_{23}}.$$

From here, we must obtain or complete the given line conductivities with the aid of Fig. 25.3.

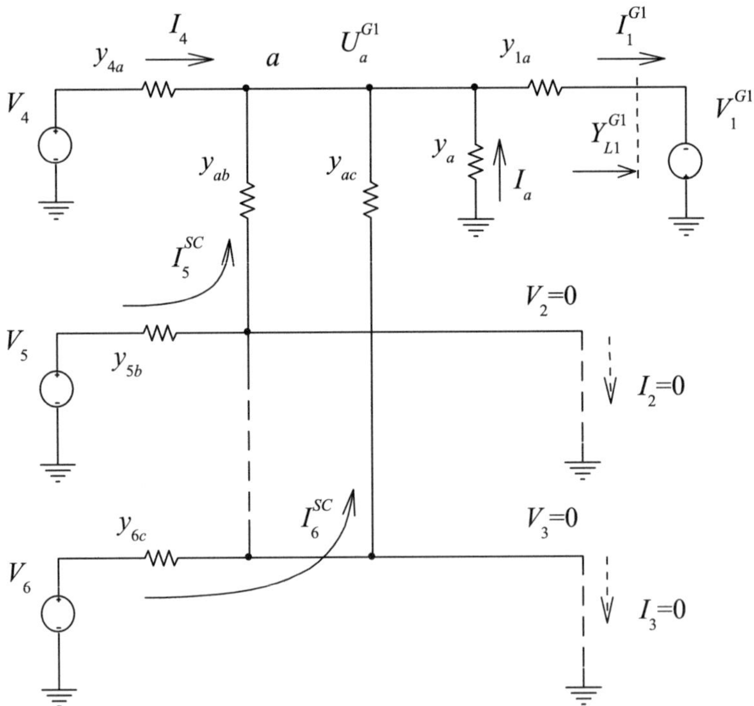

Fig. 25.3 Equivalent circuit of the line for calculation the base values I_1^{G1}, V_1^{G1}

Let the first load voltage $V_1 = -V_1^{G1}$. For the others, we consider

$$V_2 = 0,\ I_2 = 0; \quad V_3 = 0,\ I_3 = 0.$$

This condition gives the following relationship

$$U_a^{G1} = \frac{I_5^{SC}}{y_{ab}} = \frac{I_6^{SC}}{y_{ac}}.$$

In turn, the currents

$$I_5^{SC} = y_{5b}V_5, \quad I_6^{SC} = y_{6c}V_6.$$

Therefore, we get

$$\frac{y_{5b}}{y_{ab}}V_5 = \frac{y_{6c}}{y_{ac}}V_6. \tag{25.2}$$

We determine the base values

$$V_1^{G1} = U_a^{G1} + \frac{I_1^{G1}}{y_{1a}}, \quad I_1^{G1} = I_4 + I_5^{SC} + I_6^{SC} + I_a, \quad Y_{L1}^{G1} = \frac{I_1^{G1}}{V_1^{G1}},$$

where

$$I_4 = (V_4 + U_a^{G1})y_{4a}, \quad I_a = y_a U_a^{G1}.$$

Let the second load voltage $V_2 = -V_2^{G2}$. For the others, we consider

$$V_1 = 0,\ I_1 = 0; \quad V_3 = 0,\ I_3 = 0.$$

This condition gives the following relationship

$$U_b^{G2} = \frac{I_4^{SC}}{y_{ab}} = \frac{I_6^{SC}}{y_{bc}},$$

In turn, the current

$$I_4^{SC} = y_{4a}V_4.$$

Therefore, we get

$$\frac{y_{4a}}{y_{ab}}V_4 = \frac{y_{6c}}{y_{bc}}V_6. \tag{25.3}$$

From here, we determine the base values

$$V_2^{G2} = U_b^{G2} + \frac{I_2^{G2}}{y_{2b}}, \quad I_2^{G2} = I_4^{SC} + I_5 + I_6^{SC} + I_b, \quad Y_{L2}^{G2} = \frac{I_2^{G2}}{V_2^{G2}},$$

where

$$I_5 = (V_5 + U_b^{G2})y_{5b}, \quad I_b = y_b U_b^{G2}.$$

Let the third load voltage $V_3 = -V_3^{G3}$ and

$$V_1 = 0, \ I_1 = 0; \quad V_2 = 0, \ I_2 = 0.$$

Therefore,

$$U_c^{G3} = \frac{I_4^{SC}}{y_{ac}} = \frac{I_5^{SC}}{y_{bc}}.$$

From here

$$\frac{y_{4a}}{y_{ac}} V_4 = \frac{y_{5b}}{y_{bc}} V_5. \tag{25.4}$$

The base values

$$V_3^{G3} = U_c^{G3} + \frac{I_3^{G3}}{y_{3c}}, \quad I_3^{G3} = I_4^{SC} + I_5^{SC} + I_6 + I_c, \quad Y_{L3}^{G3} = \frac{I_3^{G3}}{V_3^{G3}},$$

where

$$I_6 = (V_6 + U_c^{G3})y_{6c}, \quad I_c = y_c U_c^{G3}.$$

We may consider that the voltage sources have equal voltage values and different output powers. Let the voltage source V_5 and load Y_{L2} be the most powerful elements. Therefore, the values of the conductivities y_{5b}, y_{bc}, y_{ab}, y_{4a} can be given independently. Then, relatively to these conductivities, using (25.2), (25.3) and (25.4), we get the values of the rest conductivities

$$y_{6c} = \frac{y_{bc}}{y_{ab}} y_{4a}, \quad y_{ac} = \frac{y_{bc}}{y_{5b}} y_{4a} = \frac{y_{ab}}{y_{5b}} y_{6c}. \tag{25.5}$$

The obtained conductivity values do not limit especially possibilities of this circuit, but allow simplifying essentially calculation of loads. We must note that access to line elements is necessary.

25.3 Second Balanced Network

If there is no access to the line, we may introduce two additional bias current sources ΔI_1, ΔI_3 at the output of this line as it is shown in Fig. 25.4.

Our line is an active multi-port A for the output terminals and is also described by the following system of the equations

$$\begin{pmatrix} I_1 \\ I_2 \\ I_3 \end{pmatrix} = \begin{pmatrix} -Y_{11} & Y_{12} & Y_{13} \\ Y_{12} & -Y_{22} & Y_{23} \\ Y_{13} & Y_{23} & -Y_{33} \end{pmatrix} \cdot \begin{pmatrix} V_1 \\ V_2 \\ V_3 \end{pmatrix} + \begin{pmatrix} J_1^{\mathrm{SC,SC,SC}} - \Delta I_1 \\ I_2^{\mathrm{SC,SC,SC}} \\ J_3^{\mathrm{SC,SC,SC}} - \Delta I_3 \end{pmatrix}, \tag{25.6}$$

where $J_1^{\mathrm{SC,SC,SC}}$, $J_2^{\mathrm{SC,SC,SC}}$, $J_3^{\mathrm{SC,SC,SC}}$ are the short circuit SC currents.

We take into account that

$$J_1 = \Delta I_1 + I_1, \quad J_2 = I_2, \quad J_3 = \Delta I_3 + I_3.$$

We determine the values ΔI_1, ΔI_3 and the base value Y_{L1}^{G1}, Y_{L2}^{G2}, Y_{L3}^{G3}.

Let $Y_{L2} = Y_{L2}^{G2}$. Then,

$$I_2 = I_2^{G2}, \quad V_2 = V_2^{G2},$$
$$I_1 = 0,\ V_1 = 0, \quad I_3 = 0,\ V_3 = 0.$$

These requirements determine the circuit in Fig. 25.5.

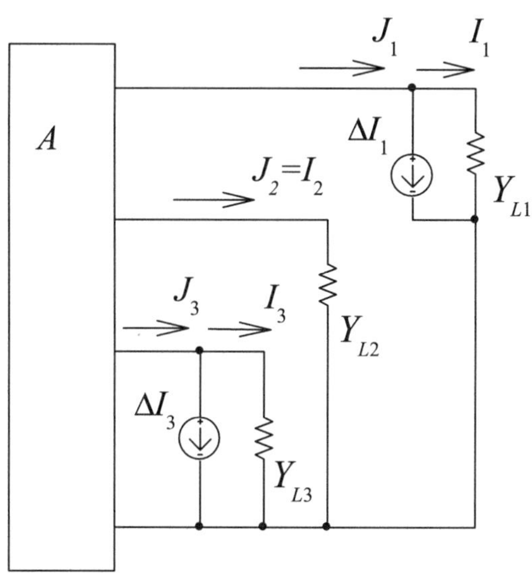

Fig. 25.4 Equivalent circuit of the line with bias current sources ΔI_1, ΔI_3

Fig. 25.5 Equivalent circuit of the line with the base value of the second load

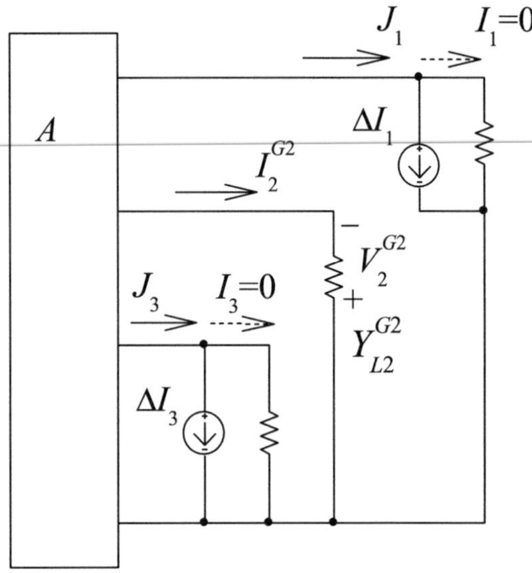

Using (25.6), we get

$$\begin{cases} 0 = Y_{12}V_2^{G2} + J_1^{\text{SC,SC,SC}} - \Delta I_1 \\ I_2^{G2} = -Y_{22}V_2^{G2} + I_2^{\text{SC,SC,SC}} \\ 0 = Y_{23}V_2^{G2} + J_3^{\text{SC,SC,SC}} - \Delta I_3. \end{cases}$$

Similarly, we consider in series $Y_{L1} = Y_{L1}^{G1}$ and $Y_{L3} = Y_{L3}^{G3}$.

Finally, we get

$$-V_1^{G1} = \frac{I_2^{\text{SC,SC,SC}}}{Y_{12}}, \quad I_1^{G1} = \left(\frac{Y_{11}}{Y_{12}} + \frac{Y_{13}}{Y_{23}}\right) I_2^{\text{SC,SC,SC}},$$

$$-Y_{L1}^{G1} = \frac{I_1^{G1}}{-V_1^{G1}} = Y_{11} + \frac{Y_{13}Y_{12}}{Y_{23}}, \tag{25.7}$$

$$-\Delta I_1 = \frac{Y_{13}}{Y_{23}} I_2^{\text{SC,SC,SC}} - J_1^{\text{SC,SC,SC}}.$$

$$-V_2^{G2} = \frac{Y_{13}}{Y_{12}Y_{23}} I_2^{\text{SC,SC,SC}}, \quad I_2^{G2} = \left(\frac{Y_{22}Y_{13}}{Y_{12}Y_{23}} + 1\right) I_2^{\text{SC,SC,SC}},$$

$$-Y_{L2}^{G2} = \frac{I_2^{G2}}{-V_2^{G2}} = Y_{22} + \frac{Y_{12}Y_{23}}{Y_{13}}. \tag{25.8}$$

$$-V_3^{G3} = \frac{I_2^{SC,SC,SC}}{Y_{23}}, \quad I_3^{G3} = \left(\frac{Y_{33}}{Y_{23}} + \frac{Y_{13}}{Y_{12}}\right) I_2^{SC,SC,SC},$$

$$-Y_{L3}^{G3} = \frac{I_3^{G3}}{-V_3^{G3}} = Y_{33} + \frac{Y_{13}Y_{23}}{Y_{12}}, \tag{25.9}$$

$$-\Delta I_3 = \frac{Y_{13}}{Y_{12}} I_2^{SC,SC,SC} - J_3^{SC,SC,SC}.$$

25.4 Projective Coordinates of the Output

We accept that coordinate axes I_1, I_2, I_3 determine the Cartesian coordinate system (I_1, I_2, I_3) in Fig. 25.6.

Let a running regime correspond to a point M^1, which is set by load conductivities $Y_{L1}^1, Y_{L2}^1, Y_{L3}^1$ and currents I_1^1, I_2^1, I_3^1. The short circuit regime of all the loads is presented by the point SC. As it was showed above, the plane $G_1G_2G_3$ passes through the base currents I_1^{G1}, I_2^{G2}, I_3^{G3}. Then, the running value Y_{L3}^1 corresponds to the plane which passes through the point M^1 and the straight line G_1G_2. This line corresponds to the intersection of the planes $G_1G_2G_3$ and $0G_1G_2$. Similarly, Y_{L2}^1 agrees to the plane, which passes through the point M^1 and the straight line G_1G_3. Also, Y_{L1}^1 matches to the point M^1 and the line G_2G_3. We remind that this line is the axis of the bunch planes Y_{L1}. In turn, the values $Y_{L1} = 0$, $Y_{L1} = \infty$, Y_{L1}^{G1} are the characteristic values of Y_{L1}. Therefore, the running value Y_{L1}^1 corresponds to the non-uniform coordinate m_1^1 in the form of the cross ratio of the four points

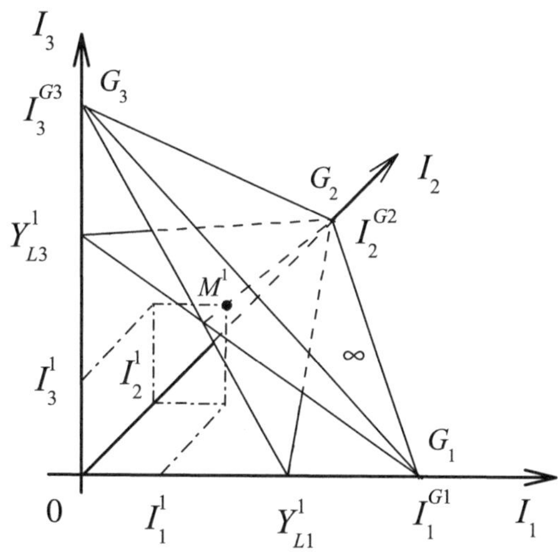

Fig. 25.6 Cartesian coordinate system (I_1, I_2, I_3) and projective coordinate $0G_1G_2G_3$

$$m_1^1 = (0\, Y_{L1}^1 \infty\, Y_{L1}^{G1}) = \frac{Y_{L1}^1}{Y_{L1}^1 - Y_{L1}^{G1}}. \tag{25.10}$$

There, $Y_{L1} = 0$, $Y_{L1} = Y_{L1}^{G1}$ are the base points and $Y_{L1} = \infty$ is a unit point. The cross ratio for m_2^1, m_3^1 is expressed similarly

$$m_2^1 = \frac{Y_{L2}^1}{Y_{L2}^1 - Y_{L2}^{G2}}, \quad m_3^1 = \frac{Y_{L3}^1}{Y_{L3}^1 - Y_{L3}^{G3}}. \tag{25.11}$$

Besides the non-uniform coordinates m_1^1, m_2^1, m_3^1 of the point M^1, there are homogeneous projective coordinates ξ_1^1, ξ_2^1, ξ_3^1, ξ_4^1, which are set by a coordinate tetrahedron and a unit point SC. These homogeneous coordinates are defined as the ratio of the distances δ_1^1, δ_2^1, δ_3^1, δ_4 for the point M^1 and distances δ_1^{SC}, δ_2^{SC}, δ_3^{SC}, δ_4^{SC} for a unit point SC to the planes of the coordinate tetrahedron $0G_1G_2G_3$. Then, the distances δ_1^1, δ_1^{SC} correspond to the plane $0G_2G_3$; therefore, $\delta_1^1 = I_1^1$, $\delta_1^{SC} = I_1^{SC}$. Similarly, δ_2^1, δ_2^{SC} match to the plane $0G_1G_3$ and $\delta_2^1 = I_2^1$, $\delta_2^{SC} = I_2^{SC}$. Also, δ_3^1, δ_3^{SC} correspond to $0G_1G_2$ and $\delta_3^1 = I_3^1$, $\delta_3^{SC} = I_3^{SC}$; δ_4^1, δ_4^{SC} agree with the plane $G_1G_2G_3$. Therefore, we obtain

$$\begin{gathered} \rho\,\xi_1^1 = \frac{\delta_1^1}{\delta_1^{SC}} = \frac{I_1^1}{I_1^{SC}}, \quad \rho\,\xi_2^1 = \frac{\delta_2^1}{\delta_2^{SC}} = \frac{I_2^1}{I_2^{SC}}, \\ \rho\,\xi_3^1 = \frac{\delta_3^1}{\delta_3^{SC}} = \frac{I_3^1}{I_3^{SC}}, \quad \rho\,\xi_4^1 = \frac{\delta_4^1}{\delta_4^{SC}}, \end{gathered} \tag{25.12}$$

where ρ is a coefficient of proportionality.

In turn, the homogeneous projective coordinates set the non-uniform coordinates as follows

$$m_1^1 = \frac{\rho\xi_1^1}{\rho\xi_4^1}, \quad m_2^1 = \frac{\rho\xi_2^1}{\rho\xi_4^1}, \quad m_3^1 = \frac{\rho\xi_3^1}{\rho\xi_4^1}. \tag{25.13}$$

25.5 Projective Coordinates of the Input

Let us superpose the coordinate system (I_4, I_5, I_6) with (I_1, I_2, I_3) in Fig. 25.7.

Then, any point (I_1, I_2, I_3) corresponds to a point (I_4, I_5, I_6). In particular, the running regime point M^1, a unit point SC and the coordinate tetrahedron $0G_1G_2G_3$ correspond to the points $\bar{M}^1$, $\overline{SC}$ and $\bar{0}\bar{G}_1\bar{G}_2\bar{G}_3$. For convenience, the conformity of the output and input characteristic regime points is shown in Table 25.1.

We consider that $I_1^{SC} = I_1^{SC,SC,SC}$, $I_2^{SC} = I_2^{SC,SC,SC}$, $I_3^{SC} = I_3^{SC,SC,SC}$.

It is known that projective coordinates of points M^1 and $\bar{M}^1$, concerning the own coordinate systems, are equal among themselves. Therefore, we must introduce the known coordinates ξ_1^1, ξ_2^1, ξ_3^1, ξ_4^1 of the point $\bar{M}^1$.

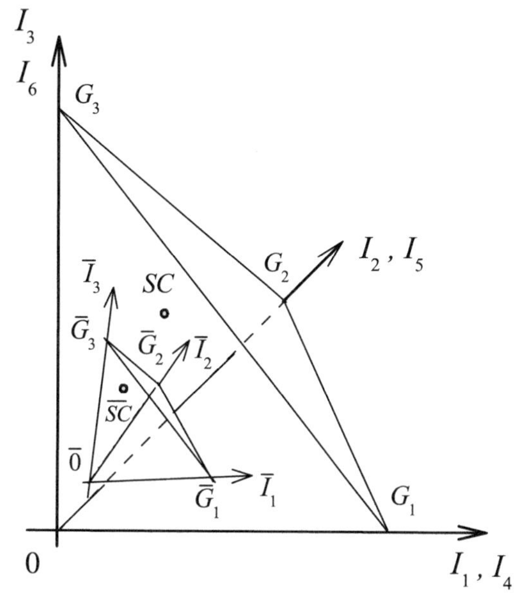

Fig. 25.7 Superposition of coordinate systems of output (I_1, I_2, I_3) and input (I_4, I_5, I_6) currents

Table 25.1 Output–input conformity of the line

	Output currents				Input currents		
Output points	I_1	I_2	I_3	Input points	I_4	I_5	I_6
0	0	0	0	$\bar{0}$	I_4^{OC}	I_5^{OC}	I_6^{OC}
G_1	I_1^{G1}	0	0	$\bar{G}_1$	I_4^{G1}	I_5^{G1}	I_6^{G1}
G_2	0	I_2^{G2}	0	$\bar{G}_2$	I_4^{G2}	I_5^{G2}	I_6^{G2}
G_3	0	0	I_3^{G3}	$\bar{G}_3$	I_4^{G3}	I_5^{G3}	I_6^{G3}
SC	I_1^{SC}	I_2^{SC}	I_3^{SC}	$\overline{SC}$	I_4^{SC}	I_5^{SC}	I_6^{SC}

These homogeneous coordinates are defined as the ratio of the distances $\bar{\delta}_1^1$, $\bar{\delta}_2^1$, $\bar{\delta}_3^1$, $\bar{\delta}_4$ for the point $\bar{M}^1$ and the distances $\bar{\delta}_1^{SC}$, $\bar{\delta}_2^{SC}$, $\bar{\delta}_3^{SC}$, $\bar{\delta}_4^{SC}$ for a unit point $\overline{SC}$ to the planes of the coordinate tetrahedron $\bar{0}\bar{G}_1\bar{G}_2\bar{G}_3$

$$\begin{aligned}
\rho\,\xi_1 &= \frac{\bar{\delta}_1}{\bar{\delta}_1^{SC}} = \frac{A_{23}I_4 - B_{23}I_5 - C_{23}I_6 - D_{23}}{A_{23}I_4^{SC} - B_{23}I_5^{SC} - C_{23}I_6^{SC} - D_{23}}, \\
\rho\,\xi_2 &= \frac{\bar{\delta}_2}{\bar{\delta}_2^{SC}} = \frac{A_{13}I_4 - B_{13}I_5 + C_{13}I_6 + D_{13}}{A_{13}I_4^{SC} - B_{13}I_5^{SC} + C_{13}I_6^{SC} + D_{13}}, \\
\rho\,\xi_3 &= \frac{\bar{\delta}_3}{\bar{\delta}_3^{SC}} = \frac{-A_{12}I_4 - B_{12}I_5 + C_{12}I_6 - D_{12}}{-A_{12}I_4^{SC} - B_{12}I_5^{SC} + C_{12}I_6^{SC} - D_{12}}, \\
\rho\,\xi_4 &= \frac{\bar{\delta}_4}{\bar{\delta}_4^{SC}} = \frac{A_\infty I_4 + B_\infty I_5 + C_\infty I_6 + D_\infty}{A_\infty I_4^{SC} + B_\infty I_5^{SC} + C_\infty I_6^{SC} + D_\infty},
\end{aligned} \tag{25.14}$$

where matrix determinants have the view

$$A_{23} = \begin{vmatrix} I_5^{OC} & I_6^{OC} & 1 \\ I_5^{G2} & I_6^{G2} & 1 \\ I_5^{G3} & I_6^{G3} & 1 \end{vmatrix}, \quad B_{23} = \begin{vmatrix} I_6^{OC} & I_4^{OC} & 1 \\ I_6^{G2} & I_4^{G2} & 1 \\ I_6^{G3} & I_4^{G3} & 1 \end{vmatrix},$$

$$C_{23} = \begin{vmatrix} I_4^{OC} & I_5^{OC} & 1 \\ I_4^{G2} & I_5^{G2} & 1 \\ I_4^{G3} & I_5^{G3} & 1 \end{vmatrix}, \quad D_{23} = \begin{vmatrix} I_4^{OC} & I_5^{OC} & I_6^{OC} \\ I_4^{G2} & I_5^{G2} & I_6^{G2} \\ I_4^{G3} & I_5^{G3} & I_6^{G3} \end{vmatrix};$$

$$A_{13} = \begin{vmatrix} I_5^{OC} & I_6^{OC} & 1 \\ I_5^{G1} & I_6^{G1} & 1 \\ I_5^{G3} & I_6^{G3} & 1 \end{vmatrix}, \quad B_{13} = \begin{vmatrix} I_6^{OC} & I_4^{OC} & 1 \\ I_6^{G1} & I_4^{G1} & 1 \\ I_6^{G3} & I_4^{G3} & 1 \end{vmatrix},$$

$$C_{13} = \begin{vmatrix} I_4^{OC} & I_5^{OC} & 1 \\ I_4^{G1} & I_5^{G1} & 1 \\ I_4^{G3} & I_5^{G3} & 1 \end{vmatrix}, \quad D_{13} = \begin{vmatrix} I_4^{OC} & I_5^{OC} & I_6^{OC} \\ I_4^{G1} & I_5^{G1} & I_6^{G1} \\ I_4^{G3} & I_5^{G3} & I_6^{G3} \end{vmatrix};$$

$$A_{12} = \begin{vmatrix} I_5^{OC} & I_6^{OC} & 1 \\ I_5^{G1} & I_6^{G1} & 1 \\ I_5^{G2} & I_6^{G2} & 1 \end{vmatrix}, \quad B_{12} = \begin{vmatrix} I_6^{OC} & I_4^{OC} & 1 \\ I_6^{G1} & I_4^{G1} & 1 \\ I_6^{G2} & I_4^{G2} & 1 \end{vmatrix},$$

$$C_{12} = \begin{vmatrix} I_4^{OC} & I_5^{OC} & 1 \\ I_4^{G1} & I_5^{G1} & 1 \\ I_4^{G2} & I_5^{G2} & 1 \end{vmatrix}, \quad D_{12} = \begin{vmatrix} I_4^{OC} & I_5^{OC} & I_6^{OC} \\ I_4^{G1} & I_5^{G1} & I_6^{G1} \\ I_4^{G2} & I_5^{G2} & I_6^{G2} \end{vmatrix};$$

$$A_{\infty} = \begin{vmatrix} I_5^{G1} & I_6^{OG1} & 1 \\ I_5^{G2} & I_6^{G2} & 1 \\ I_5^{G3} & I_6^{G3} & 1 \end{vmatrix}, \quad B_{\infty} = \begin{vmatrix} I_6^{G1} & I_4^{G1} & 1 \\ I_6^{G2} & I_4^{G2} & 1 \\ I_6^{G3} & I_4^{G3} & 1 \end{vmatrix},$$

$$C_{\infty} = \begin{vmatrix} I_4^{G1} & I_5^{G1} & 1 \\ I_4^{G2} & I_5^{G2} & 1 \\ I_4^{G3} & I_5^{G3} & 1 \end{vmatrix}, \quad D_{\infty} = \begin{vmatrix} I_4^{G1} & I_5^{G1} & I_6^{G1} \\ I_4^{G2} & I_5^{G2} & I_6^{G2} \\ I_4^{G3} & I_5^{G3} & I_6^{G3} \end{vmatrix}.$$

In turn, in accordance to (25.12), (25.13) and (25.14), we obtain the non-uniform coordinates m_1^1, m_2^1, m_3^1. Finally, the load conductivities are restored by (25.10) and (25.11)

$$Y_{L1}^1 = \frac{Y_{L1}^{G1} m_1^1}{m_1^1 - 1}, \quad Y_{L2}^1 = \frac{Y_{L2}^{G2} m_2^1}{m_2^1 - 1}, \quad Y_{L3}^1 = \frac{Y_{L3}^{G3} m_3^1}{m_3^1 - 1}. \tag{25.15}$$

25.6 Appliance for Transmission of Signals

Figure 25.8 schematically shows a structure of appliance designed to transmit three signals by the second balanced network.

The transmission system contains:

a receiver 1 of three transmitted signals U_{S1}, U_{S2}, U_{S3} at the input of a four wire line 2 with resistance losses; a transmitter 3 of these signals at the line output.

In turn, the receiver 1 contains:

a voltage source 4; current sensors 5 of the input currents I_4, I_5, I_6;
a calculator 20 of the transmitted signalsU_{S1}, U_{S2}, U_{S3}.

The transmitter 3 includes:

current sensors 6 of the output currents I_1, I_2, I_3;
a calculator 7 of the bias currents $\Delta I_1, \Delta I_3$ and the base values $Y_{L1}^{G1}, Y_{L2}^{G2}, Y_{L3}^{G3}$;
formers 8-1, 8-2 of the bias currents $\Delta I_1, \Delta I_3$;
formers 9-1, 9-2, 9-3 of the base values $Y_{L1}^{G1}, Y_{L2}^{G2}, Y_{L3}^{G3}$;
a multi-channel generator 10 of control pulses; analog switching units 11, 12, 13;
formers 14, 15, 16 of the information conductivities Y_{L1}, Y_{L2}, Y_{L3};
sources 17, 18, 19 of the transmitted signalsU_{S1}, U_{S2}, U_{S3}.

Operation of the appliance consists of two stages.

At the first stage, the line 2 parameters are defined and the bias currents $\Delta I_1, \Delta I_3$ and the base values $Y_{L1}^{G1}, Y_{L2}^{G2}, Y_{L3}^{G3}$ are calculated by (25.7), (25.8) and (25.9).

At the second stage, the transmission of the signals U_{S1}, U_{S2}, U_{S3} is directly completed. The voltage source 4 through the current sensors 5 provides power supply of the line 2. The group of three of the break-make contacts of the units 11, 12, 13 connects one after another the base and the information conductivities to the line 2 wires.

Therefore, the six sets of three conductivities at the line output are obtained. The corresponding sets of the input and the output currents of the line are shown in Table 25.2, which is similar to Table 25.1.

We accept cross ratios (25.10), (25.11) are equal to the transmitted signals; that is,

$$m_1 = U_{S1}, \quad m_2 = U_{S2}, \quad m_3 = U_{S3}.$$

Then, the information conductivities (25.15) are expressed by the formers 14, 15, 16 and the sources 17, 18, 19 of the transmitted signals

$$Y_{L1} = \frac{Y_{L1}^{G1} U_{S1}}{U_{S1} - 1}, \quad Y_{L2} = \frac{Y_{L2}^{G2} U_{S2}}{U_{S2} - 1}, \quad Y_{L3} = \frac{Y_{L3}^{G3} U_{S3}}{U_{S3} - 1}.$$

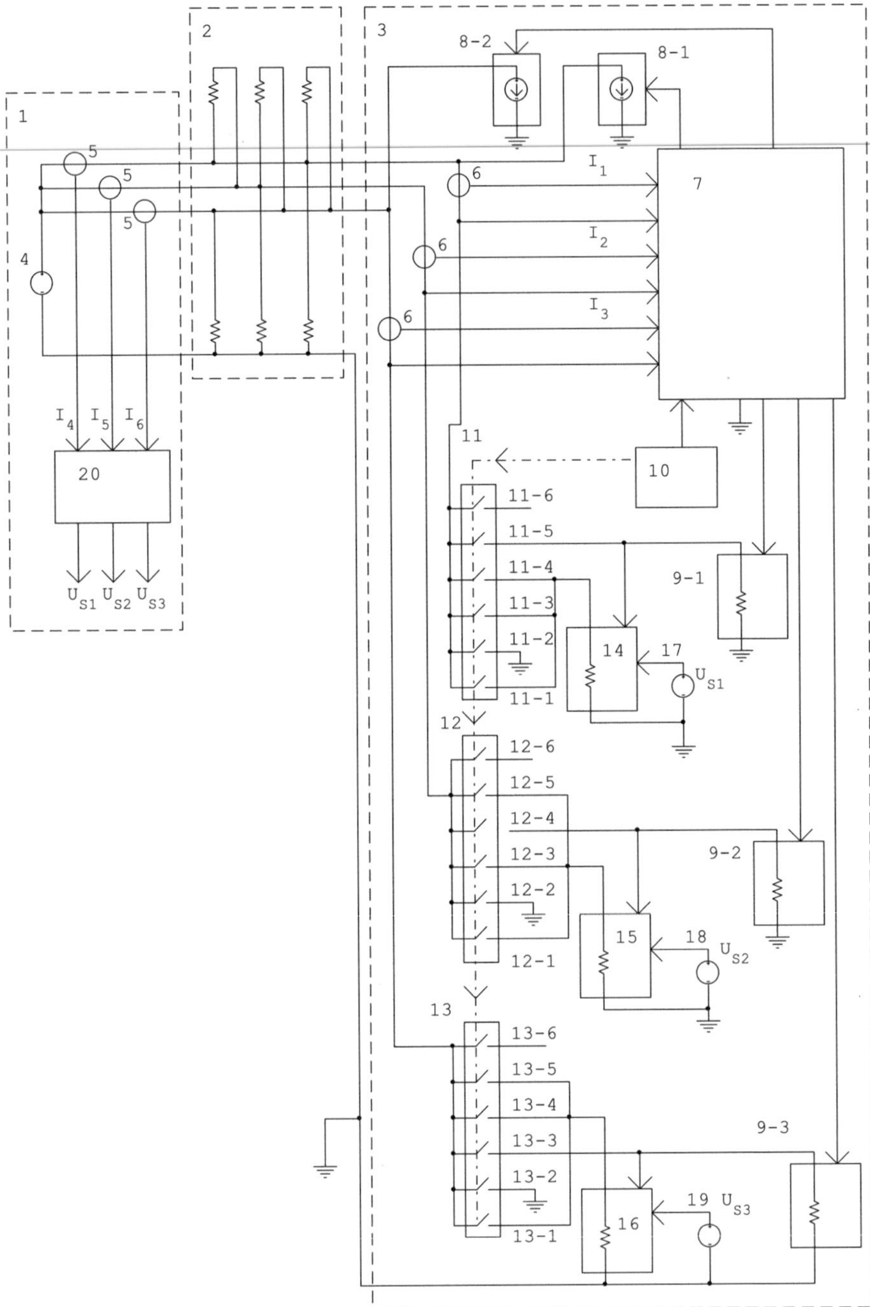

Fig. 25.8 Transmission system of signals U_{S1}, U_{S2}, U_{S3} over the four wire line

Table 25.2 Conductivity-currents conformity of the line

Set	Conductivities			Output currents			Input currents		
1	Y_{L1}	Y_{L2}	Y_{L3}	I_1	I_2	I_3	I_4	I_5	I_6
2	Y_{L1}^{SC}	Y_{L2}^{SC}	Y_{L3}^{SC}	I_1^{SC}	I_2^{SC}	I_3^{SC}	I_4^{SC}	I_5^{SC}	I_6^{SC}
3	Y_{L1}	Y_{L2}	Y_{L3}^{G3}	0	0	I_3^{G3}	I_4^{G3}	I_5^{G3}	I_6^{G3}
4	Y_{L1}	Y_{L2}^{G2}	Y_{L3}	0	I_2^{G2}	0	I_4^{G2}	I_5^{G2}	I_6^{G2}
5	Y_{L1}^{G1}	Y_{L2}	Y_{L3}	I_1^{G1}	0	0	I_4^{G1}	I_5^{G1}	I_6^{G1}
6	Y_{L1}^{OC}	Y_{L2}^{OC}	Y_{L3}^{OC}	0	0	0	I_4^{OC}	I_5^{OC}	I_6^{OC}

Next, the calculator 20 determines the homogeneous coordinates (25.14) and the transmitted signals in the following view

$$U_{S1} = \rho\xi_1 \div \rho\xi_4, \quad U_{S2} = \rho\,\xi_2 \div \rho\,\xi_4, \quad U_{S3} = \rho\,\xi_3 \div \rho\,\xi_4. \tag{25.16}$$

So, it is possible to separate (or restore) three signals by only input currents of a line.

We note also that multiplicative and additive errors of measurement of input currents are mutually reduced in (25.16).

25.7 Conclusions

The offered choice of line parameters permits to introduce projective coordinates. Input projective coordinates allow finding load conductivities only by the measured input currents without running determination of the transmission parameters of this line. It allows transmitting signals over a communication line. The obtained results can be a basis for carrying out applied researches and developments; in particular, results can be generalized for *AC* lines.

References

1. Walker K, Ramplin M (2002) US Patent 6, 459, 363B1, 1 Oct 2002
2. Alderegula A, Richter G, Williams J (2009) US Patent 7, 502, 991 B2, 10 Mar 2009
3. Ovchinnikov V (2013) US Patent 8, 446, 977B2, 21 May 2013
4. Penin A (2015) Analysis of electrical circuits with variable load regime parameters: projective geometry method. Springer, Cham
5. Penin A, Sidorenko A (2015) Transmission of measuring signals and power supply of remote sensor. In: Bonca J, Kruchinin S (eds) Nanotechnology in the security systems. NATO science for peace and security series C: environment security. Springer, Dordrecht, pp 267–281
6. Penin A (2012) Invariant properties of cascaded six-pole networks. Int J Circuits Syst Signal Process **6**(5), 305–312
7. Penin A, Sidorenko A (2016) Transmission of resistance sensor signals over multi-wire line with losses. In: 3rd international conference on nanotechnologies and biomedical engineering. Springer, Singapore, pp 332–335
8. Penin A, Sidorenko A, Donu S (2015) MD patent application S20150082

Author Index

A
Adiguzel O., 141

B
Bespalova I., 149
Bondarenko M., 239, 297
Borschak V., 227
Botton G.A., 45
Bourgeois O., 35
Brytavskyi Ie., 227

E
Eckmann A., 19

F
Filevska L.M., 259
Fiorillo A.S., 201
Frydman A., 35

G
Garden J.-L., 35
Gauquelin N., 45
Geveliuk S.A., 281
Grigaliūnas V., 159
Grigorieva I.V., 19
Grinevych V.S., 259
Grushevskaya H.V., 97
Grygorova G., 289
Guskos N., 109

H
Harea D.V., 127
Harea E.E., 127
Hawthorn D.G., 45

I
Iaseniuc O.V., 127
Iovu M.S., 127

J
Janušas G., 159

K
Kharlamov O., 239, 297
Kharlamova G., 239, 297
Khyzhun O., 239, 297
Klochkov V., 149, 289
Kolomzarov Yu.V., 267
Kompitsas M., 187
Konup I.P., 281
Koralli P., 187
Kruchinin S.P., 67
Krylova N.G., 97
Kusiak-Nejman E., 109

L
Laarraj M., 35
Lipnevich I.V., 97

J. Bonča, S. Kruchinin (eds.), *Nanomaterials for Security*, NATO Science for Peace and Security Series A: Chemistry and Biology, DOI 10.1007/978-94-017-7593-9

M
Maksimchuk P., 149
Maksimenko L.S., 259, 267
Malyukin Yu., 149, 289
Mamotyuk Ye., 289
Masalov A., 149
Matyash I.E., 267
Mcmahon C., 45
Melnik R., 25
Mischuk O.M., 267
Mishchenko A., 19
Moldovan D., 3
Morawski A.W., 109
Mousdis G.A., 187
Mozia S., 109

N
Nawrocki W., 215
Nguyen-Duc T., 35
Novoselov K.S., 19

O
Orekhovskaja T.I., 97

P
Palevičius A., 159
Palevičius P., 159
Peeters F.M., 3
Penin A., 311
Petropoulou G., 187
Prabhakar S., 25
Pullano S.A., 201

R
Repetsky , 67
Richard J., 35
Rudenko S.P., 259, 267

S
Šakalys R., 159
Salomon P., 175
Seminko V., 149
Serdega B.K., 259, 267
Sherman A., 57
Shulitski B.G., 97
Sidorenko A., 311
Smyntyna V.A., 227, 259, 281
Stamataki M., 187
Starodub M.F., 281
Stetsenko M.O., 259, 267

T
Tavakoli A., 35
Tereshchenko A.V., 281
Tiriolo R., 201
Tsamakis D., 187
Typek J., 109

V
van Heeren H., 175
Viagin O., 149
Vinko J.D., 201
Volodina G.F., 127
Vyshyvana I.G., 67

W
Wei J.Y.T., 45

Z
Zhang H., 45
Zolnierkiewicz G., 109

Subject Index

Symbol
3D microstructure, 159

A
acetone sensor, 187
amorphous films, 127
artificial nuclei, 3
Au, 187
Au-PTFE, 267
azagraphene, 239

B
Bain distortion, 141
biomems, 175
biosensing, 175
biosensors, 175, 281
Bovine leucosis, 281

C
carbon nanotubes, 68
carbon nitride oxide, 239
ceria, 149
charged vacancy, 3
communication line, 311
computer generated holography, 159
conductive AFM, 19
coupled mathematical models, 25
CO sensor, 187
Cu-enrichment, 46
Cu_xO, 187

D
density of states, 57
detector, 215
diluted magnetic semiconductors, 109

E
electron beam lithography, 159
electron-phonon interaction, 25

F
first sharp diffraction peak, 127
fluorinated graphene, 19
four wire communication line, 311

G
gene modified organism, 297
glass transition, 35
glasses, 127
graphene, 3, 239

H
heterojunction, 227
high sensitivity, 35

J. Bonča, S. Kruchinin (eds.), *Nanomaterials for Security*, NATO Science for Peace and Security Series A: Chemistry and Biology, DOI 10.1007/978-94-017-7593-9

Hubbard model, 57
hydrogen peroxide, 149
hydrogen sensor, 187

I
in vivo, 289
infrared spectra, 201
irradiation, 289

L
Langmuir–Blodgett technique, 97
lattice invariant shear, 141
layered structures, 141
luminescence, 149

M
magnetic interference, 215
magnetic properties, 109
martensitic transformation, 141
meta-generalized gradient approximation, 68
methane sensor, 187
microfluidic, 175
micromorphology, 227
modulation polarization spectroscopy, 259, 267
multi-walled carbon nanotubes, 97

N
nanocalorimetry, 35
nanocomposite, 267
nanofood, 297
nanoJoule, 35
nanometer scale, 35
nanoobjects, 297
nanoparticles, 187, 289
nanoporous materials, 201
nanostructured sensor, 227
nanotechnology, 175
nanothreat, 297
noive level, 215

O
O-doped carbon nitride, 239

P
phase conversion, 46
photoluminescence, 281
plasma frequency, 267
Point of Care, 175
polarization effect, 97

Q
quasi-bound states, 3

R
radioprotective effect, 289
rats, 289
reactive oxygen species, 149
roadmap, 175

S
Salmonella, 281
security, 297
semiconductor quantum dots, 25
sensor application, 267
sensors, 201
shape memory effect, 141
small mass detection, 35
spectral functions, 57
spin-dependent transport, 68
spin-flip rates, 25
spin-orbit coupling, 25
spin relaxation mechanisms, 25
SQUID, 215
strong coupling diagram technique, 57
supercritical charge, 3
surface plasmon resonances, 259
surface plasmons, 267

T
thermal memory, 141
thin films, 35, 187
tin dioxide cluster films, 259
TiO_2, 281
titanium oxide nanocomposite, 109

W
waveguides, 175

X
X-ray diffraction, 127

Y
Y-B-Cu-O thin film, 46

Z
Zeolite, 201
ZnO, 187, 281

MIX
Papier aus verantwortungsvollen Quellen
Paper from responsible sources
FSC® C105338

If you have any concerns about our products,
you can contact us on
ProductSafety@springernature.com

In case Publisher is established outside the EU,
the EU authorized representative is:
Springer Nature Customer Service Center GmbH
Europaplatz 3, 69115 Heidelberg, Germany

Printed by Libri Plureos GmbH
in Hamburg, Germany